Ultrastructure of the digestive tract

ELECTRON MICROSCOPY IN BIOLOGY AND MEDICINE

Current Topics in Ultrastructural Research

SERIES EDITOR: P.M. MOTTA

Already published in this series

Vol. 1 Motta, P.M. (ed.): Ultrastructure of Endocrine Cells and Tissues.
ISBN: 0-89838-568-7.

Vol. 2 Van Blerkom, J. and Motta, P.M. (eds.): Ultrastructure of Reproduction: Gametogenesis, Fertilization, and Embryogenesis. ISBN: 0-89838-572-5.

Vol. 3 Ruggeri, A. and Motta, P.M. (eds.): Ultrastructure of the Connective Tissue Matrix.
ISBN: 0-89838-600-4

Ultrastructure of the Digestive Tract

edited by

P.M. Motta, M.D., Ph.D.
Department of Anatomy, Faculty of Medicine
University of Rome, Rome, Italy

and

H. Fujita, M.D.
Department of Anatomy, Osaka University
Medical School, Osaka, Japan

assistant editor

S. Correr, Ph.D.
Department of Anatomy, University
Medical School, Rome, Italy

1988 **MARTINUS NIJHOFF PUBLISHERS**
a member of the KLUWER ACADEMIC PUBLISHERS GROUP
BOSTON / DORDRECHT / LANCASTER

Distributors

for the United States and Canada: Kluwer Academic Publishers, 101 Philip Drive, Assinippi Park, Norwell, MA 02061
for the UK and Ireland: Kluwer Academic Publishers. MTP Press Limited, Falcon House, Queen Square, Lancaster LA1 1RN, UK
for all other countries: Kluwer Academic Publishers Group, Distribution Centre, P.O. Box 322, 3300 AH Dordrecht, The Netherlands

Library of Congress Cataloging-in-Publication Data

Ultrastructure of the digestive tract.

(Electron microscopy in biology and medicine)
Includes bibliographies and index.
1. Alimentary canal — Cytology. 2. Ultrastructure (Biology) 3. Electron microscopy. I. Motta, Pietro M. 1942– . II. Fujita, Hisao III. Series. [DNLM: 1. Digestive System — ultrastructure. 2. Microscopy, Electron. WI 101 U47]
QM301.U47 1987 611'.01893 87-12292

ISBN 0-89838-893-7

Preface

When established four years ago, the scope of this international series in electron microscopy essentially was to provide an opportunity for the publication of selected review contributions by specialists in ultrastructural research.

Previous volumes presented over the last three years have focused on special topics of present interest in contemporary biomedicine such as endocrine cells, reproduction, and connective tissues. In these fields, in fact, integrated methods of electron microscopy have contributed much to generate new ideas and concepts of general value in both basic and clinical applications.

The Ultrastructure of the Digestive Tract basically follows the same guidelines and style of the other books in the series and is an invited collection of selected contributions of authors from various laboratories active in the field of electron microscopy.

Therefore, although the various chapters consist of individual topics, they nevertheless should be considered as interrelated contributions of specific subjects in the field. The idea was to have critical reviews of aspects previously published elsewhere by experts in the field who, as a rule, include other relevant information in their articles in order to update and enrich the subject.

This book contains fifteen chapters by renowned electron microscopists. Each chapter, according to the policy of the editors, reviews a particular topic in great detail, providing updated information, study methods and results, authors' ideas on future investigative approaches, and possible guidelines for forthcoming work.

We hope that this book will be useful to cell biologists, morphologists, physiologists, and pathologists. Furthermore, it should prove to be of value to medical, graduate, and postgraduate students, as well as medical practitioners interested in clinical investigation and the basics of gastroenterology.

We would like to thank all those who have contributed to this volume and also those colleagues of the advisory board who have always responded to our requests with patience and sympathy during the preparation of the volume. Special thanks are due to Sharon Greene and David Finn, both of whom have had the patience to read and reread many chapters with editorial skill.

The fine cooperation of Mr. J.K. Smith and the production staff of Martinus Nijhoff is gratefully appreciated.

Contents

Contributing authors

F. Amenta, Department of Public Health and Cell Biology. 2nd University of Rome, Tor Vergata, 00173 Rome, Italy

F. Barberini, Department of Anatomy, Laboratory of Electron Microscopy, Faculty of Medicine, University of Rome "La Sapienza", Viale Regina Elena, 289, 00161 Rome, Italy

C. Cavallotti, Department of Neurological Sciences, Via Alfonso Borelli, 50, 00161 Rome, Italy

L. Chowdhury, Department of Pathology, Michael Reese Hospital and Medical Center, Division of Gastroenterology, Lake Shore Drive at 31st Street, Chicago, Illinois 60616, U.S.A.

S. Correr, Department of Anatomy, Faculty of Medicine, University of Rome "La Sapienza", Viale Regina Elena, 289, 00161 Rome, Italy

V. Desmet, Department of Medical Research, Laboratory of Histochemistry and Cytochemistry, U.Z. Sint Rafael, K.U. Leuven, Minderbroedersstraat 12, B-3000 Leuven, Belgium

C.M. Fenoglio-Preiser, University of New Mexico, School of Medicine, Veterans Administration Medical Center, Albuquerque, New Mexico 87131, U.S.A.

H. Fujita, Department of Anatomy, Osaka University Medical School, 4-3-57 Nakanoshima, Kita-Ku, Osaka 530, Japan

T. Fujita, Department of Anatomy, Niigata University, School of Medicine, Asahi-machi 951, Niigata, Japan

K. Geboes, Department of Medical Research, Laboratory of Histochemistry and Cytochemistry, U.Z. Sint Rafael, K.U. Leuven, Minderbroedersstraat 12, B-3000 Leuven, Belgium

H.F. Helander, Hassle Gastrointestinal Research Laboratories, Department of Biology, S-431 83 Mölndal, Sweden

H. Kavin, Department of Pathology, Michael Reese Hospital and Medical Center, Division of Gastroenterology, Lake Shore Drive at 31 st Street, Chicago, Illinois 60616, U.S.A.

W.J. Krause, Department of Anatomy, School of Medicine, University of Missouri, Columbia, Missouri 65212, U.S.A.

S.S. Lim, Department of Anatomy, Louisiana State University, Medical Center, 1901 Perdido Street, New Orleans, Lousiana, 70112, U.S.A.

M.B. Listrom, University of New Mexico, School of Medicine, Veterans Administration Medical Center, Albuquerque, New Mexico 87131, U.S.A.

F.N. Low, Department of Anatomy, Louisiana State University, Medical Center, 1901 Perdido Street New Orleans, Louisiana 70112, U.S.A.

B.R. MacPherson, Department of Anatomy, Faculty of Medicine, University of Alberta, 509 Medical Science Building, Edmonton, Alberta, Canada T6G 2H7

S.G. McClugage, Department of Anatomy, Louisiana State University, Medical Center, 1901 Perdido Street, New Orleans, Louisiana 70112, U.S.A.

J. Mebis, Department of Medical Research, Laboratory of Histochemistry and Cytochemistry, U.Z. Sint Rafael, K.U. Leuven, Minderbroedersstraat 12, B-3000 Leuven, Belgium

P.M. Motta, Department of Anatomy, University Medical School "La Sapienza" Viale Regina Elena, 289, 00161 Rome, Italy

T. Murakami, Department of Anatomy, Okayama University Medical School, 2-5-1 Shikata-cho, Okayama 700, Japan

O. Ohtani, Department of Anatomy, Okayama University Medical School, 2-5-1 Shikata-cho, Okayama 700, Japan

A. Ohtsuka, Department of Anatomy, Okayama University Medical School, 2-5-1 Shikata-cho, Okayama 700, Japan

R.L. Owen, Cell Biology and Aging Section (151 E), Veterans Administration Medical Center, 4150 Clement Street, San Francisco, California 94121, U.S.A.

J. Pappo, Cell Biology and Aging Section (151E), Veterans Administration Medical Center, 4150 Clement Street, San Francisco, California 94121, U.S.A.

C.J. Pfeiffer, Division of Veterinary Biology and Clinical Studies, Virginia-Maryland Regional College of Veterinary Medicine, Virginia Tech, Blacksburg, Virginia 24061, U.S.A.

H. Tatsumi, Department of Anatomy, Osaka University Medical School, 4-3-57 Nakanoshima, Kita-Ku, Osaka 530, Japan

T. Yamamoto, Department of Anatomy, Faculty of Medicine, Kyushu University, Fukuoka 812, Japan

Ultrastructure of the digestive tract

CHAPTER 1

Morphological changes in the developing alimentary canal: A review by scanning electron microscopy

SOO-SIANG LIM & FRANK N. LOW

1. Introduction

1.1. Light microscopy

The complex series of developmental events leading to the mature alimentary mucosa in higher mammals has interested investigators for over a century (1). Villi in the small intestine were discovered as early as 1861 by von Kölliker and somewhat later in 1868 in the developing large intestine by Barth. But systematic studies of the developmental process did not get underway until the twentieth century when Hilton (2) described the ontogeny of a large number of species in his doctoral thesis. Attention was turned somewhat later to a series of human embryos by Johnson (3, 4).

These early light microscopists were in general agreement regarding the changes in surface contours. These progressed from the smooth round surface of the gut lumen of early embryos to the complex pattern that characterized the adult. All agreed that the formation of longitudinal folds occurred in all portions of the gut as the earliest indication of developmental differentiation. These were followed by a complex series of surface changes in certain portions of the gut such as those involved in the formation of villi. Each level of the gut, however simple or complex, followed its own developmental pattern. These are described in some detail in a following section. However, even the early investigators (5) noted that all the progressive developmental stages found in the human repeated the sequence of developmental stages observed from the standpoint of comparative anatomy.

1.2. Transmission electron microscopy

The development of transmission electron microscopy focused attention on the internal cytology of the alimentary mucosa. The development of the esophageal epithelium was described by Kober and Herbst (6) and its cytology in the newborn and adult mouse by Parakkal (7). The fine structure of the gastric mucosa in the bat was described by Ito and Winchester (8). The small intestine received somewhat closer inspection with special attention paid to the morphogenesis of intestinal villi (9). Holman (10) turned his attention to the development of microvilli in the chick. The internal fine structure of the villus was described by Palay and Karlin (11). The classic study of Vollrath (12) summarized the morphological and histochemical development of the small intestine. Concerning the large intestine, the development of the colonic mucosa in the rat was described by Helander (13). The literature in transmission electron microscopy pertaining to both the developmental and adult fine structure in the alimentary canal is so voluminous that its most reliable source is now to be found in special textbooks on the subject.

1.3. Scanning electron microscopy

Renewed interest in the luminal contours of the alimentary canal came with application of the scanning electron microscope to biological research. Observations of regional luminal contours of the mature digestive tract largely confirmed the findings of light microscopists. This is well illustrated by the work of Burke and Holland (14) who examined the gastrointestinal tract of six adult monkeys. The surface topology of the stomach was described by Pfeiffer (15). The absorptive area of the intestine occupied the chief interest of early SEM investigators (16–19). Several investigations concentrated on villi (20–22). The adult cecum, colon, and rectum were also represented in this body of literature (23–28). Specialized areas such as intestinal microvilli (29, 30) and the epithelium covering lymphoid follicles (31) were also described.

Developmental studies tend to outnumber those of the adult alimentary mucosa with marked emphasis on the small intestine. The extent of the chicken gut from esophagus to rectum is described in its development by Lim and Low (32). Esophageal development is also recorded in the rat (6) and its postnatal growth in the opossum (33). The stomach of the hamster (34) and the opossum (35) also received attention. The developing small intestine is the subject of numerous publications, about half of which are devoted to the development of villi. Details of the somewhat complex changes that occur during development in the small intestine are offered in the following section on regional characteristics. The cecum and colon are also well represented in the developmental literature and are likewise cited below.

2. Regional characteristics (see accompanying drawings)

2.1. Esophagus

Developmental changes in the esophagus are very simple and have received very little investigative attention. In the very early stages of development, the proliferation of the esophageal lining results in an occluded lumen. In the chick embryo, reestablishment of a patent esophagus soon occurs and the mucosal epithelium is then thrown into distinct longitudinal folds. Shortly afterwards, profuse ciliation of the mucosal cells is observed (Fig. 1-1). Cilia and other cell surface elabora-

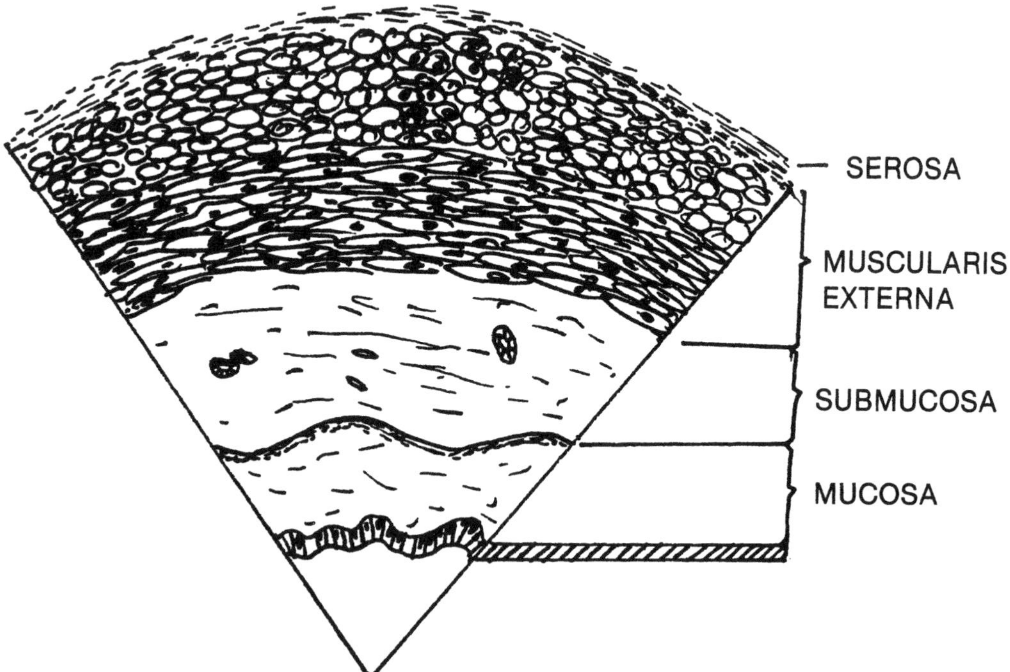

Diagram 1. Basic histology of the gastrointestinal tract.

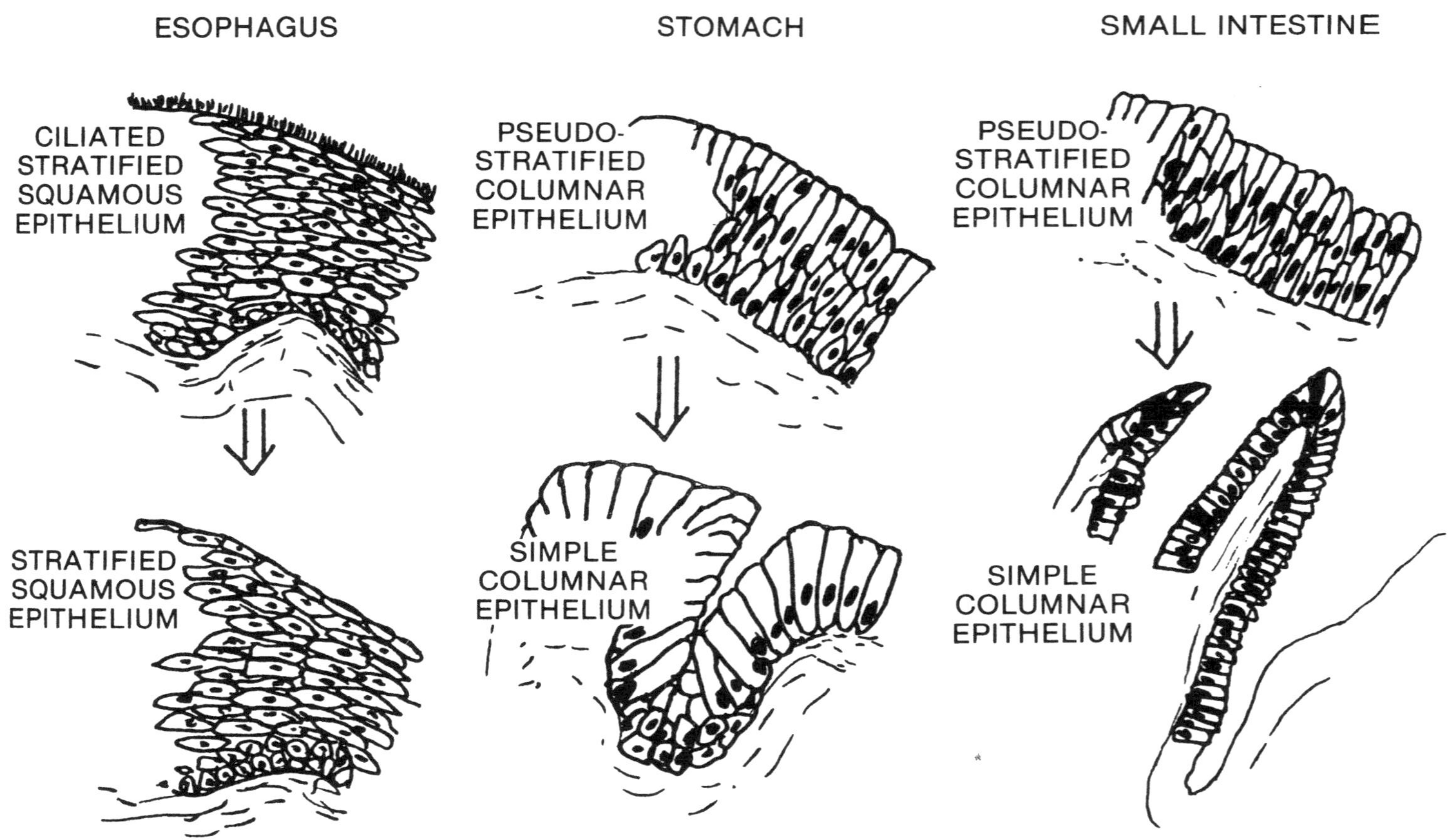

Diagram 2. Changes in the mucosa surface during development.

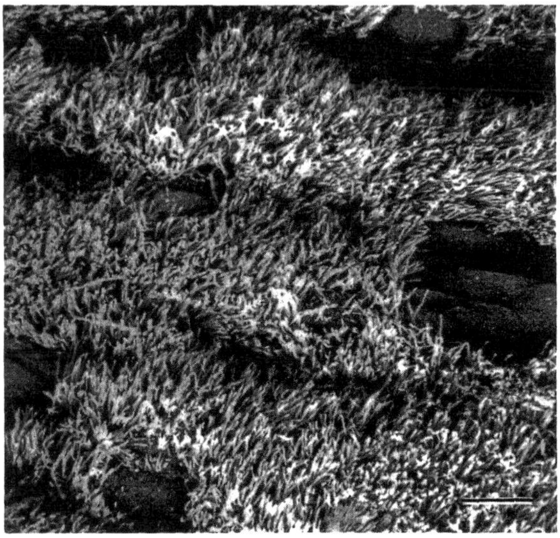

Figure 1-1. Luminal surface of esophagus (chick embryo, 13 days). Large tufts of cilia cover entire cell surfaces — only small "islands" of nonciliated cells are seen. Bar = 18 μm.

tions (microvilli and microplicae), as seen in Fig. 1-2, decrease in density at later stages of development and completely disappear by hatching time. Similar observations of the developing esophageal mucosa were made in rats (6). Ciliated cells appear about the same time as similar cells in the trachea. Cells with single cilia persist until 17 days' gestation in the rat. These cells form columnar epithelium which becomes squamous within 48 hours at about term. In general, keratinization and exfoliation of these cells occur after birth, as described by Krause et al. (33) in the postnatal development of the esophagus in the opossum.

2.2. Stomach

Development of surface contours in the stomach has also received very little attention in the literature. A description of the developing avian stomach has been documented (32). The transient yet profuse covering of "macrovillus-like" structures (Fig. 1-3), as noted in the early proventriculus (glandular stomach), disappears by 17 days of incubation. The only surface appendages of the rounded apices of the cells at this stage are blebs and microvilli (Fig. 1-4). Within the next two days, numerous papillae become apparent, each containing a proventriculus gland surrounded by smooth surfaced plical structures. The regions between adjacent glands acquire a unique conformation from the anastomoses of these concentric plical formations (Fig. 1-5). In contrast, the luminal surface of the developing gizzard or muscular stomach reveals little of topographical

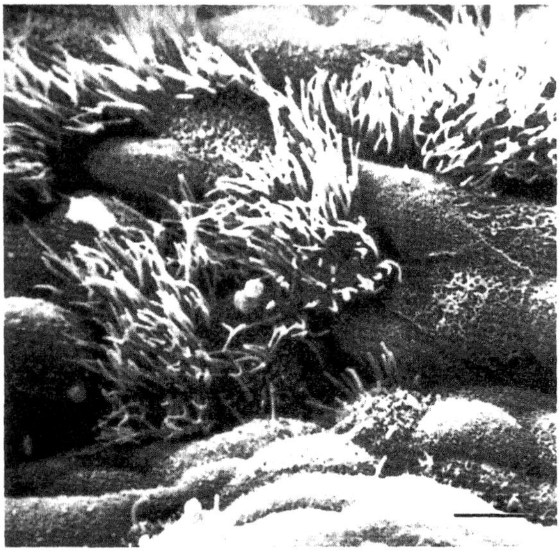

Figure 1-2. Luminal surface of esophagus (chick embryo, 17 days). In addition to the cilia, patches of microvilli and microplicae are evident on cell surfaces. Bar = 5.5 μm.

Figure 1-3. Luminal surface of proventriculus (chick embryo, 12 days). The mucosal surface is covered with thick, club-shaped processes. Bar = 5 μm.

Figure 1-4. Luminal surface of proventriculus (chick embryo, 17 days). Macrovilluslike processes have disappeared, revealing apical cell surfaces covered with numerous microvilli and blebs. Bar = 5 μm.

interest. The surface of the mucous membrane remains relatively flat throughout the entire embryonic period (32).

Aside from an account of the developing stomach in the human embryo (3), no information is available on the developing mucosal surfaces of other vertebrate stomachs. To fill this gap, we examined the developing stomach of the golden hamster (*Cricetus auratus*) which has a gestation period of 16 days. The chronological table of normal hamster development as described by Boyer (36) was utilized in the staging of the embryos.

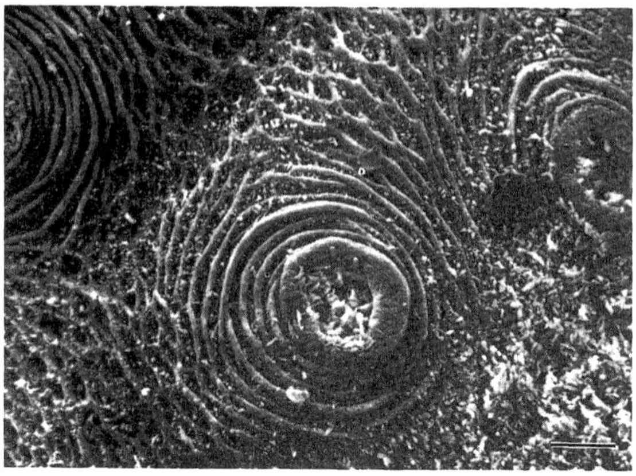

Figure 1-5. Luminal surface of proventriculus (chick embryo, 18 days). Numerous glands are apparent, characterized by the concentric arrangement of plicae around the openings of the glands. Anastomoses of the plicae occur in regions between adjacent glands. Bar = 17 μm.

Until 11 days of gestation, the entire surface of the gastric mucosa appears relatively smooth but broken by occasional small folds (Fig. 1-6). Higher magnifications reveal a single cilium in the center of many cells, with cellular boundaries well defined by numerous microvilli (Fig. 1-7). In the 13-day embryo, the surface mucosa of the stomach presents two morphologically distinguishable regions. A smooth, flattened mucosal surface is apparent in the forestomach (Fig. 1-8), with individual cellular surfaces covered with numerous stubby microvilli (Fig. 1-9). In the glandular stomach, tortuous convolutions of epithelial folds are prominent (Fig. 1-10). The apical surfaces of these cells vary in appearance but all reveal microvilli of differing height and sizes (Fig. 1-11). As the embryo approaches term, the differences in the two regions become less prominent but nonetheless obvious. The forestomach has now acquired distinct longitudinal folds (Fig. 1-12). These are also convoluted and are very similar in appearance to those observed in the glandular stomach of the previous stage (Fig. 1-10). Regional differences in these areas are

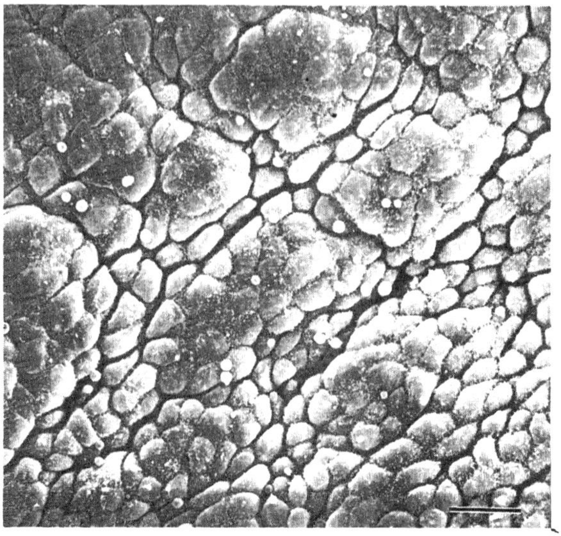

Figure 1-6. Luminal surface of stomach (hamster embryo, 13 days). The mucosal epithelium of the forestomach is relatively flat except for occasional folds. Bar = 21 μm.

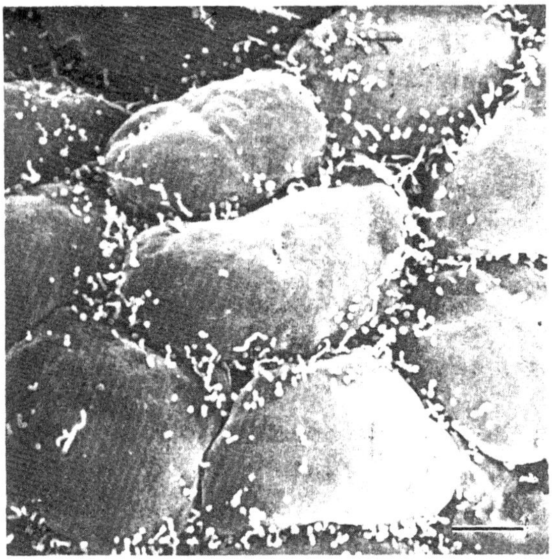

Figure 1-7. Luminal surface of stomach (hamster embryo, 11 days). Higher magnification of the epithelium reveals many cells with a single, centrally located cilium. Cell boundaries are well defined by microvilli. Bar = 3 μm.

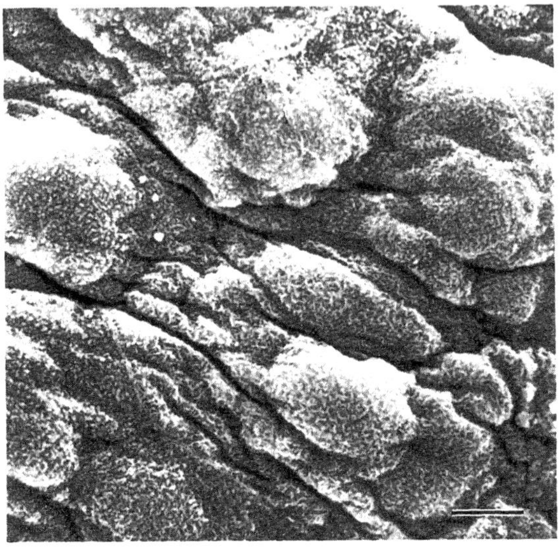

Figure 1-9. Luminal surface of stomach (hamster embryo, 13 days). Higher magnification reveals profuse microvilli covering the individual cell surfaces. Bar = 5 μm.

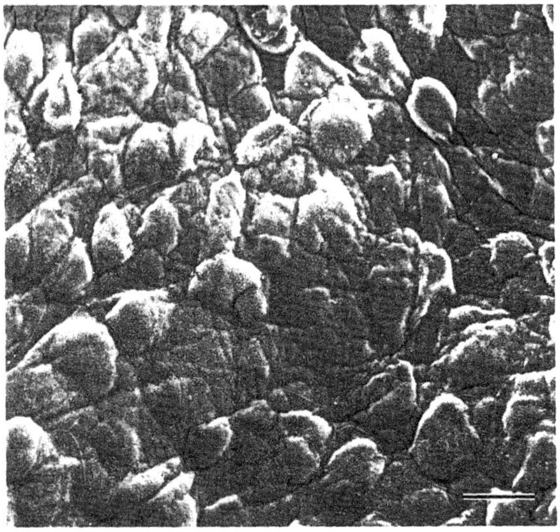

Figure 1-8. Luminal surface of stomach (hamster embryo, 13 days). Forestomach mucosa is still relatively flat. Bar = 17 μm.

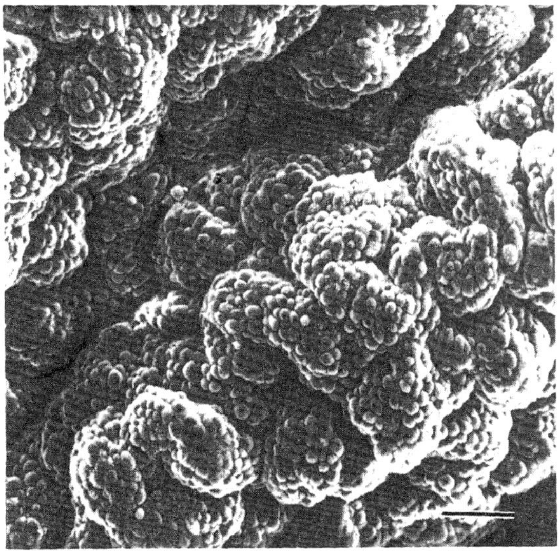

Figure 1-10. Luminal surface of stomach (hamster embryo, 13 days). Mucosa of glandular stomach is thrown into convoluted folds. Bar = 38 μm.

still apparent in newborn (8 hours postpartum) hamsters. The cells in the longitudinal folds of the forestomach are now keratinized and superficial cells can be observed in different states of desquamation (Figs. 1-13, 1-14). In the region of the glandular stomach, the rounded apical surfaces of the cells contribute a cobblestone look to the convoluted epithelial folds. No sloughing off of cells is observed (Fig. 1-15). Occasional slit-like openings of gastric glands can be seen (Fig.

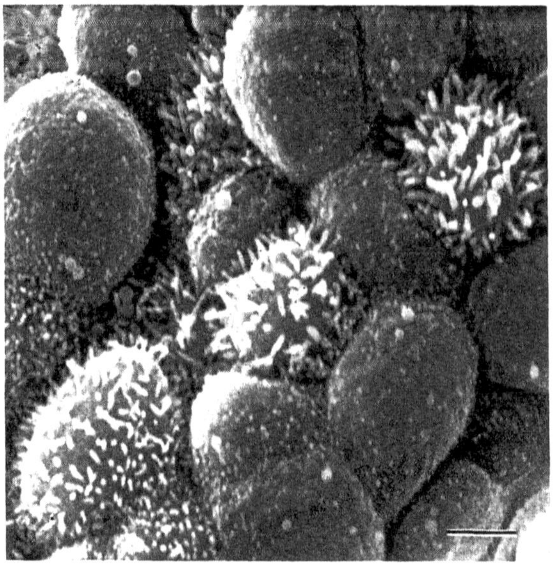

Figure 1-11. Luminal surface of stomach (hamster embryo, 13 days). The rounded apices of the cells of the glandular stomach show numerous surface appendages, blebs, and microvilli of differing height. Bar = 3 μm.

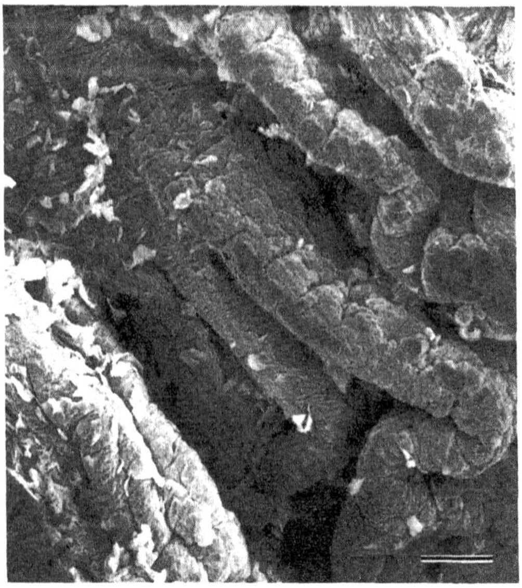

Figure 1-13. Luminal surface of stomach (newborn hamster). Keratinization of mucosal epithelium in forestomach. Bar = 170 μm.

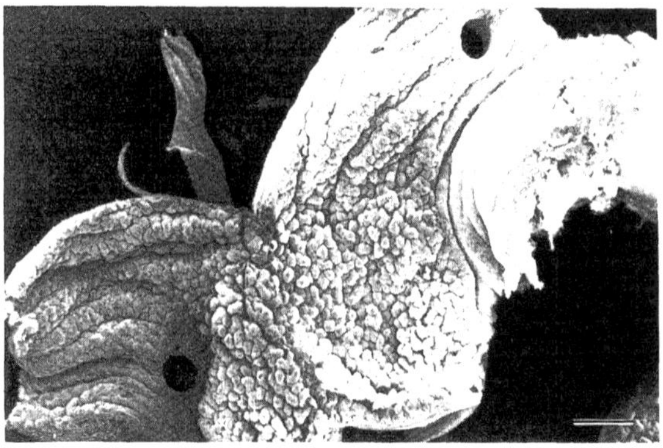

Figure 1-12. Luminal surface of stomach (hamster embryo, 14 days). Stomach was dissected in half, following lines of curvature and lain flat (2 pinholes are evident). The forestomach (top) has acquired folds at this time while the glandular stomach (bottom) reveals compacted protrusions. Bar = 330 μm.

Figure 1-14. Luminal surface of stomach (newborn hamster). Higher magnification reveals cells and sheets of cells in different stages of desquamation. Bar = 31 μm.

1-16). At this time they are few in number with ribbons of mucin around these glands.

In the human embryo up to a crown-rump length of 55 mm, the luminal surface of the stomach shows only a few gastric pits (3). Afterwards, longitudinal furrows begin to appear. These become larger until, at birth, the surface contours resemble those of the adult. Hattori and Fujita (34) describe the formation of gastric glands (and pits) in the postnatal hamster and

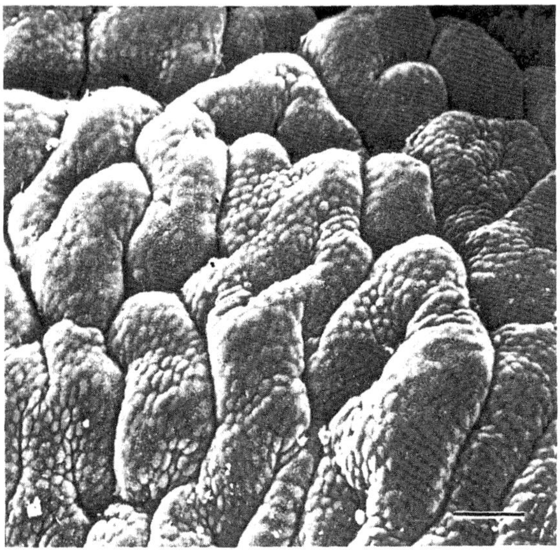

Figure 1-15. Luminal surface of stomach (newborn hamster). Glandular portion of stomach. The rounded convoluted folds of the earlier stages seem to be compacted. Bar = 28 μm.

Figure 1-17. Luminal surface of ileum (chick embryo, 12 days). Individual previllus ridges have acquired zigzag conformation. Bar = 25 μm.

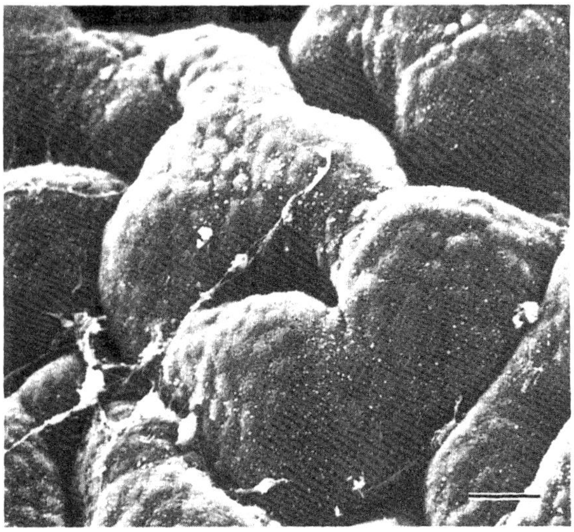

Figure 1-16. Luminal surface of stomach (newborn hamster). Slit-like opening of gland surrounded by strands of mucus. Bar = 13 μm.

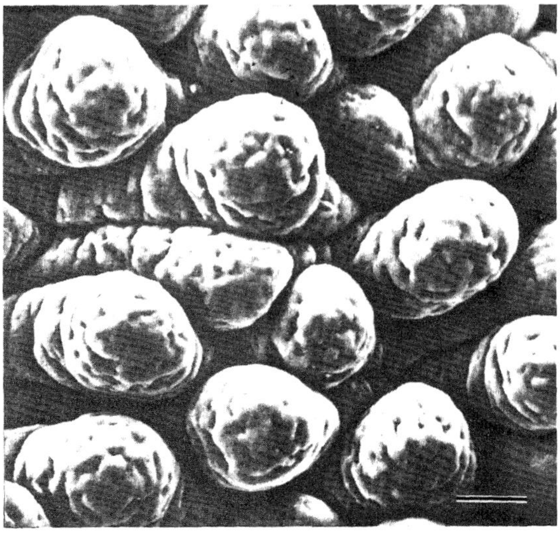

Figure 1-18. Luminal surface of ileum (chick embryo, 19 days). Each turn of the zigzag has given rise to a bulge from an emerging villus. Bar = 25 μm.

Krause et al. (35) describe postnatal development in the opossum.

2.3. *Small intestine (duodenum, jejunum, ileum)*

As the absorptive portions of the alimentary canal are approached, the developmental characteristics of the surface contours become quite complex. The formation of villi and their influence on luminal surface contours is essentially the same for all levels of the small intestine, but a fairly uniform progression of development courses from cephalic (duodenum) to caudal (ileocaecal junction) (37). The process begins with longitudinal

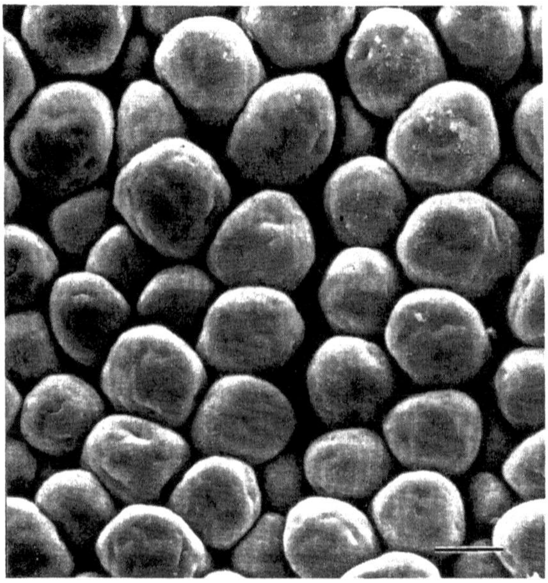

Figure 1-19. Luminal surface of ileum (chick embryo, 20 days). Two rows of villi have arisen from each zigzag ridge. Bar = 24 μm.

folds that become increasingly higher (38). Then, a zigzag pattern develops (Fig. 1-17), usually at about 11 or 12 days in the chick (32, 37, 39, 40). Individual villi appear by separation at angles of the zigzag pattern (Fig. 1-17) and thereafter grow into the various forms (leaflike, fingerlike, etc.) characteristic of the adult in that particular species (Figs. 1-18, 1-19).

Substantially, this pattern was observed even by early light microscopists who recognized the repetition in human development of all the stages found in comparative anatomy (5). Scanning electron microscopy added to this the zigzag pattern noted early in the development of villi (32), which seems to be generally characteristic. A number of investigators have devoted their studies to the duodenum (40–44), while others have concentrated on the ileum (37, 45). Most workers, however, have not restricted their observations to any particular part of the small intestine; Doran and Heydon in rat (46), Gheri-Bryk et al. in chick (47), Krause et al. in opossum (33, 35), Pacini and Gheri-Bryk in chick (38), Pacini et al. in chick (48), Belak et al. in quail (49, 50), Hoskins et al. in dog (51), and Kurohmaru et al. in vole (52). Lacroix et al. (53) studied the enzymology of the brush border in the developing human. All of these studies seem to be in general accord with the early light microscopic observations of Hilton (2) in numerous vertebrates and of Johnson (3) in the human gut.

2.4 Cecum and colon

Villous development in the prenatal human colon follows closely the pattern well recognized in the small intestine, even to the longitudinal folds in the mucosa and the zigzag pattern that precedes the definitive villi (54). In the cow, the mucosa of the cecum develops villi which decrease in length from about the twentieth week of pregnancy (27, 55). Unlike the small intestine, the large intestine in mammals loses its villi during the latter part of gestation or shortly after birth (26). In the rat, villi are present at birth but disappear entirely by 10 days of age (56). A similar observation is made in the large intestine of the newborn golden hamster (Figs. 1-20, 1-21). The leaflike villi that persist are very similar to those described by Kurohmaru et al. (43) in the adult hamster small intestine. Loss of villi is initiated by a strong constriction at the basal region, followed by villous separation and disappearance. These observations are completely in agreement with the early light microscopic observations of colonic villi by Johnson (4) in developing human large intestine. He recognized their essential stages of development and decline, declaring them to be entirely missing at birth.

3. Species differences

This subject can best be understood by reverting to the principle of recapitulation of phylogeny in ontogeny as expressed in the development of the human gut. It appears that nothing new or startling occurs in either the ontogeny or the persisting adult form of the surface contours of the gut as one progresses from lower to higher forms. In birds colonic villi persist in the adult (32); in rats they are present at birth but disappear in the first ten days (56); and in humans they are present only during fetal life, disappearing by birth (27). Other similar examples could be cited, but it should be remembered that all stages of development found in lower forms appear in human development. Moreover, some

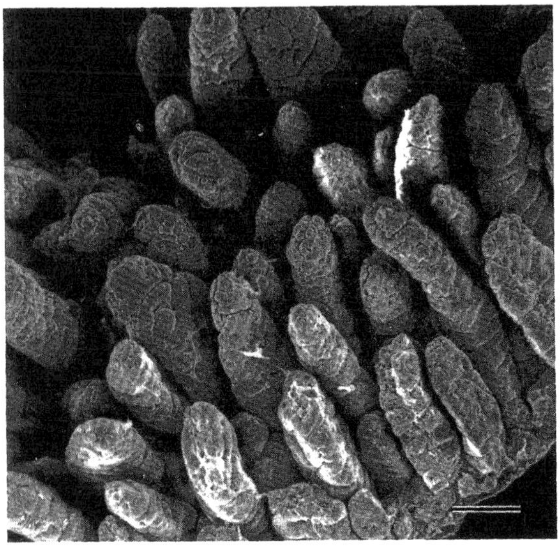

Figure 1-20. Luminal surface of ileum (newborn hamster). Leaflike villi have persisted after birth. Bar = 91 μm.

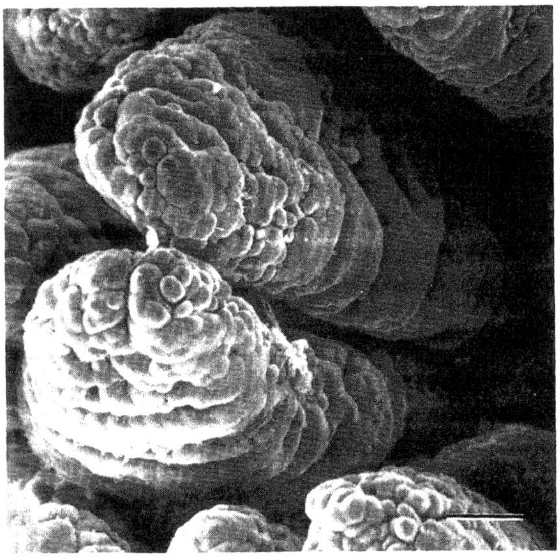

Figure 1-21. Luminal surface of large intestine (newborn hamster). Apical surfaces of cells at villous tips are more rounded than those at the base. Bar = 28 μm.

of the stages that are transitory in humans are permanent in many lower forms (5).

4. Experimental studies

A variety of experimental procedures have been undertaken using the alimentary mucosa. These reflect isolated subjects of interest and, unfortunately, do not form a well-organized body of work. However, we will cite those related to morphological changes that occur during the development of the digestive system.

Cellular proliferation, turnover, and migration in the fetal gastrointestinal tract are intimately associated with morphogenetic events that result in the organogenesis of specialized structures. Hence, factors including nutritional, hormonal, neural, and luminal influences that serve a regulatory role in the control of cellular kinetics have been of interest to investigators. The question is asked whether the continuously renewing epithelial mucosa is capable of alteration in response to these factors and other conditions such as irradiation, ischemia, surgical bypass, or partial resection.

It has been thought that one of the possible stimuli responsible for the complex events of villous development in the small intestine might be the swallowing of amniotic fluid. This, in the guinea pig, occurs between the sixth and seventh week (57). Amniotic fluid contains hormones and growth factors, including cortisol, endodermal growth factor (EGF), and other substances such as glucose, potassium, and somatomedin. The results obtained thus far have been inconsistent. Devletova (58) shows that in dog and rabbit fetuses, which have atresia of the upper parts of the digestive tract, the macroscopic and microscopic structure of the intestine do not differ from those of normally developing controls. On the other hand, recent studies by Calvert et al. (59) indicate that rat amniotic fluid contains a trophic factor which promotes formation of villi and induces the precocious differentiation of duodenal crypts and Paneth cells. This occurs in 15-day fetal mouse duodenal mucosa grown in vitro.

Cytochalasin B produces elongated microvilli spatially distributed among those of normal length (60). The influence of thyroxine and hydrocortisone on embryonic chick intestine does not seem to alter the surface morphology of explants (61). However, in chick embryos an effect of hydrocortisone (62), cortisone (63, 64), the thyroxine (65) on the differentiation of the duodenal enterocyte is demonstrable. Hydrocortisone produces increases in the length of microvilli along with greater cell height and enzymatic activity of the absorptive epithelial cells (66). A feasible interpretation is that this substance does exert a major force on growth and maturation. Application of cortisone affects the growth in height of the microvilli of the duodenum and jejunum in mice but has no effect on the width of the microvilli (67). Cortisone applied to newborn rats induces the morphology of the cytoplasm of the intestinal epithelial cells to resemble that in adult mammals (68).

Dietetic effects are also reported. Glucose is observed to be absorbed measurably by the fetal rat colon (69). Babak (70,71) showed that the length of the intestines in tadpoles varies with the type of protein consumed. Those consuming proteins of animal origin have a shorter gut than those feeding on plant proteins. "Germ-free" animals have a very different structure in the large intestine. Wostmann and Bruckner-Kardose (72) showed that the weight of the large intestine is considerably increased in "germ-free" animals while that of the ileum is not affected. Microvilli in the cecum of germ-free rats are longer (73), while short-term fasting causes sloughing of cells in the small intestine (74). The ischemic mucosa of mice showed jejunal villi with a bulged appearance (75).

Stimulation with microbial flora affects the microstructure of the small intestine of suckling piglets (76). Villi of the small intestine atrophy in sheep injected with *Trichostrongylus rugatus* (77). However, surface contours of the cecal mucosa of the rhesus monkey remain normal after injection with spirochaetes (78). Irradiation of 1,000 rads with cobalt 60 causes the mucosa of rats to show degeneration and fusion of atrophic villi in 3–4 days (79). Recovery can be detected at 5–6 days and appears complete at 8–12 days. Thymectomy in newborn hamsters disturbs the crypts of the small intestine in weaned animals (80), while hypophysectomy affects differentiation of the duodenal enterocyte (81).

5. Discussion

The various parts of the developing digestive tract undergo marked changes in geometric form during development. The use of the scanning electron microscope (SEM) in the examination of the vertebrate alimentary tract has revealed a plethora of changing surface patterns and contours. Combined with the information obtained from light microscopic and transmission electron microscopic observations of the internal histological components, a more comprehensive picture of the complex morphogenetic events leading to the molding of specialized structures is possible. The higher magnification, increased depth of field, and overall view of all exposed surfaces of cells as afforded by SEM provide information on cell number, shape, and surface structures that may play important roles in morphogenesis. The three-dimensional relationships of the direction and extent of cell movement and proliferation in the final structuring of a particular organ are easily visualized.

The formation and prenatal maturation of the vertebrate digestive tract can be conveniently divided into these general phases.

(a) After formation of the endodermal tube, a generalized growth in length, width, and mass occurs. Histologically, this ofter translates into a stratification of the existent epithelium.
(b) Longitudinal folds which protrude into the internal luminal surfaces of each segment of the digestive tube are formed. These folds persist in some form or other even after postnatal development.
(c) Concomitantly or shortly afterwards, a complex series of surface changes follows, e.g., ciliation of the esophageal epithelia or reshaping of the previllous ridges in the intestine to form villi and crypts. This phase is also coincident with a shift from a stratified epithelium to a simple epithelium in the stomach and intestines.
(d) Differentiation and proliferation of specific cell types within a particular part of the digestive system occurs. It must be emphasized that each level of the gut, however simple or complex, follows its own temporal sequence.

In the esophagus, the cellular proliferation which results in the increase in length and diameter also occludes the lumen of this structure. Reestablishment of the patent esophagus occurs at about 8–9 days in the chick (82) and at about 10 days in the golden hamster (36). The two-layered cuboidal epithelium is thrown into longitudinal folds at about 11 days. By about 17–18 days, however, stratified squamous epithelium prevails (83–85). In the interim, the profuse ciliation of the epithelium has come and gone, so that by the time of hatching at 21 days the epithelium has become devoid of cilia. The significance of this transient appearance of cilia is not known.

The avian stomach is divided into two morphologically and physiologically distinct portions, the glandular stomach or proventiculus and the muscular stomach or gizzard. Separation of the proventriculus from the gizzard occurs at about the fifth day of incubation in the chick (82). The stratified columnar epithelium of the mucosa of the 11-day embryo differentiates to that of a simple columnar epithelium between the seventeenth and twentieth day of incubation. The superficial luminal glands of the proventriculus which first appear in the lower part of the glandular stomach at 11 days become apparent in the upper part of the proventriculus at about the fifteenth day (82, 85). According to Romanoff (82), the very active proliferation of the epithelial cells during this period results in the characteristic folds between the openings of the glands (Fig. 1-5). In contrast, the mucosal surface of the gizzard appears flat throughout development, with a few small folds and no obvious glandular openings (32).

In the golden hamster, as in certain other mammals, the stomach is comprised of a single structure. However, morphologically distinct regions within the stomach are obvious throughout development as well as in the adult (86). The forestomach during the late stages of development is characterized by a keratinized mucosal epithelium (87, 88). These cells in various stages of desquamation (Figs. 1-13, 1-14) can be easily, distinguished from those of the heavily convoluted mucosa of the glandular stomach (Figs. 1-15, 1-16). Similar keratinization of the gastric mucosa is described in the Japanese field vole (86), a herbivorous rodent. Three distinct regions are present in its stomach, the esophageal sac, fundic stomach, and pyloric stomach. In comparing the gross characteristics of the stomach between the field vole and golden hamster, the major difference seems to be the presence of fimbria in the former (86). These fimbria distinguish the esophageal sac from the fundic stomach and probably serve to prevent a backflow of its contents. Hence, the esophageal sac may be a more efficient fermental vat than that of the forestomach of the golden hamster. More so than at any other level of the gut, variations in the gross characteristics of the stomach are apparent among the different vertebrates, perhaps a direct consequence of the animals' adaptations to different dietary habits.

The intestines exhibit dramatic changes in surface morphology during their development, resulting in the geometrically precise pattern of formation of the intestinal villi. In addition to the general proliferation of cells to increase the length and width of this portion of the digestive system, three major changes occur: (a) formation of the longitudinal previllous ridges, which increase in number by geometric progression (39), (b) the assumption of a precise zigzag shape of these ridges, and (c) the emergence of a double row of villi in an orderly relation to the zigzag pattern (Figs. 1-17, 1-18). This sequence of events has been particularly well investigated in the small intestine of various vertebrates, and no major deviation from the general pattern has been noted. Regional differences have been reported in the morphology of the small intestine within small animals — mostly in the shape and size of the previllous ridges and villi (30, 42, 89). For example, compared to the duodenum, the previllous ridges of the ileum are thicker and fuller and are noticeably constricted at each turn of the zigzag formation. The villi do not seem to be as tall as those in the duodenum, but the cell surface is likewise thickly covered with microvilli. Genesis of villi in the large intestine seems to follow the same pattern as that in the small intestine but on a less prominent scale.

In the chick, the geometry of formation of villi is less obvious, and there are no villi in the more distal portions of the colon. Postnatal development of the small intestine includes an increase in the size or variance in shape of the villi. In the large intestine, however, several studies report loss of colonic villi at birth in the human embryo (4) or several days after birth, as in the rat (56). This may perhaps be a reflection of the decrease in absorptive capacities in this part of the intestine.

A number of developmental mechanisms are proposed to account for the morphological rearrangements during formation of villi. Coulombre and Coulombre (39) suggest that mechanical pressure resulting from active proliferation of cells within the confines of the less active mesenchyme would result in the folds in the early stages of intestinal development. Others suggest changes in the shape and adhesiveness of individual cells (90) or the contractions of microfilaments within the epithelial cells (9). Tsai and Overton (91) note that spatial organization of the epithelium is related to the disposition of collagen in the extracellular matrix. It is suggested that this could account for the reorientation of cells during formation of villi.

The mechanisms and significance underlying the various morphological changes during development are not clear. However, the combination of the different microscopic techniques, especially that of the SEM, has greatly clarified the physical relationships of these events. This will pave the way for further studies directed at the clarification of cell differentiation and organ development.

6. Concluding remarks

All the physical and psychic attributes of living organisms are adapted to help them secure food. As an energy source, food must be ingested and then prepared for use by the process of digestion. In its simplest form, as seen in protozoans, digestion can be wholly intracellular; the food passes directly into the cell cytoplasm by phagocytosis and is attacked in food vacuoles by digestive enzymes. More commonly, however, extracellular digestion occurs, as seen in metazoans, where there is a permanent gut. Digestion becomes increasingly the work of enzymes secreted into the gut lumen.

In vertebrates, the most complex of metazoans, the permanence of the gut is established and the whole process of digestion has now become extracellular. The different facets of the digestive process have become localized in different parts of the gut and uniquely specialized cells are grouped together. This leads to compartmentalization of the gastrointestinal tract. Food is procured through the mouth and transported via the esophagus to the stomach. Here it is gathered and major digestion is initiated. In the intestine, intestinal juices are secreted, and, with the aid of secretions from liver and pancreas, chemical digestion of food is accomplished. The digested food and water are concurrently absorbed selectively throughout the entire intestine. The indigestible wastes are passed on and are finally evacuated through the rectum.

The basic histology of the alimentary tract is remarkably similar, characterized by the covering of the mesothelium (or adventitia) on the outside. Progressing internally, there is a connective tissue layer, the longitudinally and circularly oriented smooth muscle layers, nerve cell aggregates (myenteric plexus of Auerbach), the submucosa with its plexus of Meissner, and finally the mucosa. This consists of the muscularis mucosae, the tunica propria, and epithelium. This epithelium constitutes the luminal lining of the alimentary tract.

This luminal surface of the alimentary canal during development has been of interest to us. Here, the innovations of different vertebrates, those which improve or adapt the absorptive surfaces of the digestive system, can be seen clearly by SEM. Such surface changes include: (a) an increase in length, (b) an increase in diameter, (c) transient ciliation of esophageal epithelium, (d) formation of longitudinal folds in stomach and subsequent keratinization of epithelia, and (e) development of villi in the intestines.

The evolution of the gastrointestinal tract from lower to higher forms of vertebrates is reflected in the progressive developmental geometry of surface changes. These present a familiar and highly significant pattern. One cannot fail to recognize in this pattern the time-honored interpretation of holistic thinking, namely that phylogeny is recapitulated in ontogeny (92). It is, indeed, interesting to see the way that the end products of the developmental morphology of lower forms appear midway in prenatal life in the development of higher ones. Such structural expressions are subsequently discarded, only to be replaced by more advanced organizations that are better suited to the needs of more recent and complex organisms. This same phenomenon has been observed repeatedly over more than a century and a half in all species that have been closely analyzed with respect to both ontogeny and phylogeny. Perhaps even more fascinating (and rewarding) is the very evident expression of this same basic recapitulation in the micrographs of scanning electron microscopy. This complex electronic instrument, available to us for less than two decades, thus reconfirms certain basic principles of evolution and development. These have interested biologists since the beginning of biological research. It is therefore reasonable to anticipate that, whatever methods of observation and analysis may be developed in the future, the recapitulation of phylogeny in ontogeny will appear conspicuously in any overview of the final results.

Acknowledgments

The authors would like to thank Dr. Margaret Orsini, Department of Anatomy, University of Wisconsin for her assistance with hamster embryology and use of her hamster colony. We are also grateful to Dr. Edward Bersu, Department of Anatomy, University of Wisconsin for his helpful suggestions. We thank Pat Hanson, Sandy Damrauer, and Angela Pepin for typing the manuscript.

References

1. Barth: Beitrage zur entwicklung der darmwand. *Sitzungsb d Wein Akad d Wissensch* 58: 128–136, 1868.
2. Hilton WA: The morphology and development of intestinal folds and villi in vertebrates. *Am J Anat* 1: 459–505, 1902.
3. Johnson FP: The development of the mucous membrane of the esophagus, stomach and small intestine in the human embryo. *Am J Anat* 10: 521–561, 1910.
4. Johnson FP: The development of the mucous membrane of the large intestine and the vermiform process in the human embryo. *Am J Anat* 14: 187–233, 1913.
5. Berry, JM: On the development of the villi of the human intestine. *Anat Anz* 17: 242–249, 1900.
6. Kober HJ, Herbst R: Die lumenseitige oberflache des rattenesophagus wahrend der ontogenese. *Z Mikroskanat Forsch* 89: 702–714, 1975.
7. Parakkal PF: An electron microscopic study of esophageal epithelium in the newborn and adult mouse. *Am J Anat* 121: 175–195, 1967.
8. Ito S, Winchester RJ: The fine structure of the gastric mucosa in the bat. *J Cell Biol* 16: 541–577, 1963.
9. Burgess DR: Morphogenesis of intestinal villi. II. Mechanism of formation of previllous ridges. *J Embryol Exp Morph* 34: 723–740, 1975.
10. Holman J: Fine structure and dimension of intestinal microvilli during the early postincubation period in chicken. *Z Mikrosk Anat Forsch* 79: 14–26, 1968.
11. Palay SL, Karlin LJ: An electron microscope study of the intestinal villus. *J Biophys Biochem Cytol* 5: 363–372, 1959.
12. Vollrath L: Uber die entwicklungen des dünndarms der ratte. Morphologische, histochemische und experimentelle untersuchungen. *Ergebn Anat Entwickl Gesch* 41: 4–70, 1969.
13. Helander HF: Morphological studies on the development of the rat colonic mucosa. *Acta Anat* 85: 153–176, 1973.
14. Burke JA, Holland P: The epithelial surface of the monkey gastrointestinal tract. A scanning electron microscopic study. *Digestion* 14: 68–76, 1976.
15. Pfeiffer CJ: Surface topology of the stomach in man and the laboratory ferret. *J Ultrastruct Res* 33: 252–262, 1970.
16. Marsh MN, Swift JA, Williams ED: Studies of small intestinal mucosa with the scanning electron microscope. *Br Med J* 4: 95–96, 1968.
17. Marsh MN, Swift JA: A study of small intestinal mucosa using the scanning electron microscope. *Gut* 10: 940–949, 1969.
18. Balcerzak SP, Lane WC, Bullard JW: Surface structure of intestinal epithelium. *Gastroenterology* 58: 49–55, 1970.
19. Toner PG, Carr KE, Ferguson A, Makay C: Scanning and transmission electron microscopic studies of human intestinal mucosa. *Gut* 11: 471–481, 1970.
20. Carr KE, Toner PG: Scanning electron microscopy of rat intestinal villi. *Lancet* 3: 570–571, 1968.
21. Toner PG, Carr KE: The use of scanning electron microscopy in the study of intestinal villi. *J Path* 97: 611–617, 1969.
22. Anderson JH, Taylor AB: Observation of mammalian intestinal villi with the scanning electron microscope. *Physiologist* 13: 136, 1970.
23. Kavin H, Hamilton DG, Greasley RE, Eckert JD, Zuidema G: Scanning electron microscopy. A new method in the study of the rectal mucosa. *Gastroenterology* 59: 426–432, 1970.
24. Kavin H, Eckert JD, Zuidema G, Greasley R, Hamilton DG: Studies of human rectal mucosa by scanning electron microscopy. *Proc 3rd Ann Scanning Electron Microscopy Symp* (Chicago) p 233–240, 1970.
25. Crucioli V, Torsoli A, Pallone F, Paoluzi P, Bartocci G: Scanning electron microscopy of the normal human colon. *Rendic R Gastroenterol* 2: 184–188, 1970.
26. Wille KH: Scanning electron microscopic observations on the intestinal mucosa of fetal and adult cattle (*Bos primagenius taurus*). *Anat Anz* 138: 341–347, 1975.
27. Wille KH: Raster-electronenmikroskopischer beobachtungen an der blindarm-schleimhaut des rinderfetus. *Verh Anat Ges* 69: 891–898, 1975.
28. Wille KH: Uber die schleimhautoberfläsche des blinddarmes einiger haussäuger. Eine raster-elektronenmikroskopische untersuchung. *Zbl Vet Med* 4: 265–273, 1975.
29. Millington PF, Critchley DR, Tovell PWA, Pearson R: Scanning electron microscopy of intestinal microvilli. *J Microsc* (London) 89: 339–344, 1969.
30. Anderson JH, Taylor AB: Scanning and electron micro-

scopic studies of jejunal microvilli of the rat, hamster and dog. *J Morphol* 141: 281–291, 1973.
31. Torreo-Medina A: Morphologic characteristics of the epithelial surface of aggregated lymph follicles (Peyer's patches) in the small intestine of gnotobiotic calves and pigs. *Am J Vet Res* 42: 232–236, 1981.
32. Lim S-S, Low FN: Scanning electron microscopy of the developing alimentary canal in the chick. *Am J Anat* 150: 149–173, 1977.
33. Krause WJ, Cutts JH, Leeson CR: The postnatal development of the alimentary canal in the opossum. I. Oesophagus. *J Anat* 122: 293–314, 1976.
34. Hattori T, Fujita S: Fractographic study on the growth and multiplication of the gastric gland of the hamster. The gland division cycle. *Cell Tiss Res* 153: 145–149, 1974.
35. Krause WJ, Cutts JH, Leeson CR: The postnatal development of the alimentary canal in the opossum. II. Stomach. *J Anat* 122: 499–519, 1976.
36. Boyer CC: Chronology of development for the golden hamster. *J Morphol* 92: 1–37, 1953.
37. Desmeth M: Developmental variations of the villi along the ileum in newly hatched and adult pigeons. A scanning electron microscopic study. *Z Mikrosk Anat Forsch* 90: 489–498, 1976.
38. Pacini P, Gheri-Bryk S: The inner surface of chick small intestine by scanning electron microscope. *Bull Assoc Anat* (Nancy) 63: 317–322, 1979.
39. Coulombre AJ, Coulombre JL: Intestinal Development. I. Morphogenesis of the villi and musculature. *J Embryol Exp Morphol* 6: 403–411, 1958.
40. Grey RD: Morphogenesis of intestinal villi. I. Scanning electron microscopy of the duodenal epithelium of the developing chick embryo. *J Morphol* 137: 193–213, 1972.
41. Mathan M, Macey PC, Trier JS: Morphogenesis of fetal rat duodenal villi. *Am J Anat* 146: 73–92, 1976.
42. Magudda L, Micali A, Anastasi G, Mondello MR, DeLeo F: Morphogenesis of chick embryo duodenal villi studied by scanning electron microscope. *Arch Ital Anat Embryol* 84: 55–70, 1979.
43. Kurohmaru M, Nishida T, Mochizuki K: Morphological changes of developing duodenal villi in golden hamsters. *Jpn J Vet Sci* 41: 151–165, 1979.
44. Nakamura K, Komuro T: A three-dimensional study of the embryonic development and postnatal maturation of rat duodenal villi. *J Electron Miscrosc* (Tokyo) 32: 338–347, 1983.
45. Ono K, Takashio M: Scanning electron microscopic studies of ileal epithelial cells in suckling rats. *Anat Embyrol* 153: 1–8, 1978.
46. Doran GA, Heydon R: Scanning electron microscopy of developing villi in rat small intestine. *J Anat* 110: 507, 1971.
47. Gheri-Bryk S, Pacini P, Marchesi L: Scanning electron microscopy of the inner surface of small intestine in chick embryos. *Boll Soc Ital Biol Sper* 55: 236–240, 1979.
48. Pacini P, Gheri-Bryk S, Marchesi L: Lo sviluppo dell'intestino tenue mesenteriale di embrione di pollo al microscopio elletronico a scansione. *Arch Ital Anat e Embriol* 48: 285–295, 1979.
49. Belak M, Kocisova J, Boda K, Kosuth P, Longauer S, Maretta M: Examination of the development of the luminal relief of the small intestine in Japanese quail under the scanning electron microscope. *Arch Exp Veterinarmed* 35: 573–577, 1981.
50. Belak M, Kocisova J, Boda K, Kosuth P, Longauer S, Maretta M: Effect of hypergravitation upon the developing relief of the small intestine in Japanese quail. Studies with scanning electron microscope. *Arch Exp Veterinarmed* 35: 579–583, 1981.
51. Hoskins JD, Henk WG, Abdelboki YZ: Scanning electron microscopic study of the small intestine of dogs from birth to 337 days of age. *Am J Vet Res* 43: 1715–1720, 1982.
52. Kurohmaru M, Hayakawa T, Seki M, Zyo K, Nishida T, Mochizuki K: Postnatal development of intestinal villi in the field vole (*Microtus montebelli*). A scanning electron microscopic study. *Okajima's Folia Anat Jpn* 60: 73–85, 1983.
53. Lacroix P, Kedinger M, Simon-Ossman P, Haffen K: Early organogenesis of human small intestine: Scanning electron microscopy and brush border enzymology. *Gut* 25: 925–930, 1984.
54. Bell L, Williams L: A scanning and transmission electron microscopical study of the morphogenesis of human colonic villi. *Anat embryol* (Berlin) 165: 437–455, 1982.
55. Wille KH: The development of the bovine cecal mucosa. *Zntrlbl f Veterinarmed* 5: 103, 1976.
56. Ono K: Changes in the cecal villi during postnatal development in rats. *Cell Tiss Res* 208: 253–259, 1980.
57. Becker RE, Windle WF, Barth EE, Schulz MD: Fetal swallowing, gastrointestinal activity and defecation in amnio. *Surg Gynec Obstet* 70: 607–611, 1940.
58. Davletova LV: Concerning the role of digestive organs of mammals in the processes of intrauterine nutrition. *Zh Obshch Biol USSR* 22: 201–209, 1961.
59. Calvert R, Lehoux JG, Arsenault P, Menard D: Extracts of rat amniotic fluid contain a potent inducer of intestinal crypt formation. *Anat Rec* 205: 27–37, 1983.
60. Burgess DR, Grey RD: Alterations in morphology of developing microvilli elicited by cytochalasin B. Studies of embryonic chick intestine in organ culture. *J Cell Biol* 62: 566–574, 1974.
61. Black BL: Morphological development of the epithelium of the embryonic chick intestine in culture: Influence of thyroxine and hydrocortisone. *Am J Anat* 153: 573–600, 1978.
62. Moog F, Thomas ER: The influence of various adrenal and gonadal steriods on the accumulation of alkaline phosphatase in the duodenum of suckling mouse. *Endocrinology* 56: 187–197, 1955.
63. Hayes RL: The induction of cortisone-treated embryonic duodenum in vitro. I. The villus. *J Embryol Exp Morph* 14: 161–168, 1965.
64. Hayes RL: The induction of cortisone-treated embryonic duodenum in vitro. II. The striated border. *J Embryol Exp Morphol* 14: 169–175, 1965.
65. Moog F: Regional differences in the alkaline phosphatases of the mouse from birth to one year. *Dev Biol* 3: 153–174, 1961.
66. Noda S, Yoshizawa S, Moriuchi S, Kubota K: Effect of

hydrocortisone on the development of embryonic chick duodenum. Scanning electron microscopic and enzymatic studies. *Okajima's Folia Anat Jpn* 56: 67–80, 1979.
67. Overton J: Fine structure of free cell surface in developing mouse intestinal mucosa. *J Exp Zool* 159: 195–201, 1965.
68. Clark SL: the ingestion of proteins and colloidal materials by columnar absorptive cells of the small intestine in suckling rats and mice. *J Biophys Biochem Cytol* 5: 41–50, 1959.
69. Potter GD, Schmidt KL, Lester R: Glucose absorption by in vitro perfused colon of the fetal rat. *Am J Physiol* 245: G424–G430, 1983.
70. Babak E: Eber den einfulss der nahrung auf die lange des darmkanals. *Biol Zbl* 23: 477–486, 1903.
71. Babak E: Uber die morphogenetische reaktion des darmkanals der froschlarve auf muskelproteine vershiedener tierklassen. *Hoffmeister's Beitr zür Chem Physiol umd Pathol* 7: 323–360, 1905.
72. Wostmann B, Bruckner-Kardose E: Development of cecal distension in germ-free baby rats. *Am J Physiol* 197: 1345–1346, 1959.
73. Gustafsson BE, Maunsbach AB: Ultrastructure of the enlarged cecum in germ-free rats. *Z Zellforsch* 120: 555–578, 1971.
74. Bayer PC, Rittenburg JH, Bird FH, Chawan CB, Allan M: Influence of short-term fasting on chicken alimentary canal mucosa. *Poult Sci* 60: 1293–1302, 1981.
75. Rao NS, Williams AW: Normal and ischaemic jejunal mucosa of mice. Scanning electron microscopic study. *J Microsc* 15: 219–224, 1972.
76. Kenworthy R, Allen WD: Influence of diet and bacteria on small intestine morphology, with special reference to early weaning and *E. Coli. J Comp Path* 76: 291–296, 1966.
77. Barker IK, Beveridge I: Development of villus atrophy in the small intestine of sheep infected with *Trichostrongylis rugatus. Vet Parasitol* 13: 67–75, 1983.
78. Takeuchi A, Zeller JA: Scanning electron microscopic observations on the surface of the normal and spirochaete-infested colonic mucosa of the rhesus monkey. *J Ultrastruct Res* 40: 313–324, 1972.
79. Anderson JH, Withers RH: Scanning electron microscopic studies of irradiated rat intestinal mucosa. *Scanning Electron Microscopy/1973 IITRI* 6: 566–572, 1973.
80. Hard RC: Intestinal crypt lesions in neonatal thymectomized hamsters. *Nature* (Lond) 204: 455–457, 1964.
81. Watterson RL, Brown DA, Bartha A: Reduced endocrine activity of the adrenal cortex of hypophysectomized chick embryo as indicated by modified development of the duodenum. *Anat Rec* 133: 347–348, 1959.
82. Romanoff AL: The Avian Embryo, New York: Macmillan Co p 431–531, 1960.
83. Schumacher S: Die entwicklung der glandulas oesophageae des huhnes. *Zeit f Mikr-Anat Forsch* 5: 1–22, 1926.
84. Ivey WD, Edgar SA: Histogenesis of the esophagus and crop of the chicken, turkey, guinea fowl and pigeon with special reference to ciliated epithelium. *Anat Rec* 114: 189–212, 1952.
85. Van Alten PJ, Fennell RA: Histogenesis and histochemistry of the digestive tract of chick embryos. *Anat Rec* 127: 677–695, 1957.
86. Kurohmaru M, Nishida T, Mochizuki K: Morphological and histological studies on the gastric mucosa of the Japanese field vole, *Microtus montebelli montebelli. Jpn J Vet Sci* 43: 887–899, 1981.
87. Hoover WH, Mannings CL, Sherrin HE: Observations on digestion in the golden hamster. *J Anim Sci* 28: 349–352, 1969.
88. Sakata T, Tamate H: Light and electron microscopic observation of the forestomach mucosa in the golden hamster. *Tohoku J Agric Res* 27: 25–39, 1976.
89. Klein RM, McKenzie JC: The role of cell renewal in the ontogeny of the intestine. I. Cell proliferation patterns in adult, fetal and neonatal intestine. *J Ped Gastroenterol Nutr* 2: 10–43, 1983.
90. Gustafson T, Wolpert L: Cellular movement and contact in sea urchin morphogenesis. *Biol Rec* 42: 442–498, 1967.
91. Tsai LJ, Overton J: The relation between villus formation and the pattern of extracellular fibers as seen by scanning microscopy. *Dev Biol* 52: 61–73, 1976.
92. Gould SJ: Ontogeny and Phylogeny. Cambridge, Mass: Belknap Press.

Author's address:
Dr. F.N. Low
Department of Anatomy
Louisiana State University
Medical Center
1901 Perdido Street
New Orleans, LA 70112
USA

CHAPTER 2

The esophagus: Normal ultrastructure and pathological patterns

K. GEBOES, J. MEBIS, & V. DESMET

1. Introduction

The esophageal wall is composed of the tissular layers common to the digestive tube in general: the mucous membrane (tunica mucosa), the submucosa (tunica submucosa), the muscularis externa (tunica muscularis), and the adventitia-serosa (tunica adventitia-serosa). A remarkable deviating histology, however, reflects the specific functions of the esophagus in comparison with the more distal gastrointestinal tract. The mucous membrane is composed of an epithelial membrane, a supporting layer of connective tissue (the lamina propria), and a thin abluminal layer of smooth muscle (the muscularis mucosae). During embryonic development, the esophagus (unlike the rest of the digestive tube), has stepwise replaced the typical simple cylindrical epithelium of entodermal origin by a multilayered squamous epithelium. As a consequence, the luminal side of the esophagus is lined by a non-keratinizing stratified squamous epithelium, except for a short segment at the gastroesophageal junction lined by columnar epithelium.

The squamous epithelium comprises a basal cell layer (germinative layer or stratum basale), a prickle cell layer, also called stratum intermedium (spinosum), a polygonal or intermediate layer, and a stratum superficiale which can be divided into a functional and a superficial layer. In the normal esophageal epithelium, multipotential basal cells that have the capacity to divide are located in the most basal layer forming a palisade of cells. They are responsible for the renewal of the more superficial layers (1,2).

The lower border of the epithelium is irregular because of the presence of stromal papillae and ridges, interdigitating with the epithelium. The lamina propria is separated from the submucosa by the lamina muscularis mucosae. The smooth muscle cells of the lamina muscularis mucosae are arranged longitudinally, except for the proximal and distal segments, where they have a rather spiral arrangement. The tunica submucosa is composed of loose areolar connective tissue. Both mucosal and submucosal glands occur in the esophagus. In the submucosa a well-developed vascular network (Heller's plexus) can be found together with a submucous nervous plexus. Ganglia are mainly localized near the inner circular coat of the muscularis propria, in contrast with what is seen in the other segments of the gastrointestinal tract where the submucous plexus can be divided into Meissner's plexus (near the lamina muscularis mucosae) and Henle's plexus (near the inner circular muscle coat).

The tunica muscularis propria is composed of striated muscle fibers in the uppermost esophagus. The number of these fibers decreases sharply below 6 cm to 8 cm where they are gradually replaced by smooth muscle. The muscularis is usually divided into an inner circular (stratum circulare) and an outer longitudinal (stratum longitudinale) muscle coat. However, individual muscle fibers may be arranged spirally, thus forming a connected system. In the tunica muscularis, the myenteric plexus of Auerbach is found between the inner and the outer coat. The outermost layer of the esophageal wall, the tunica adventitia, is a connective tissue sheath contain-

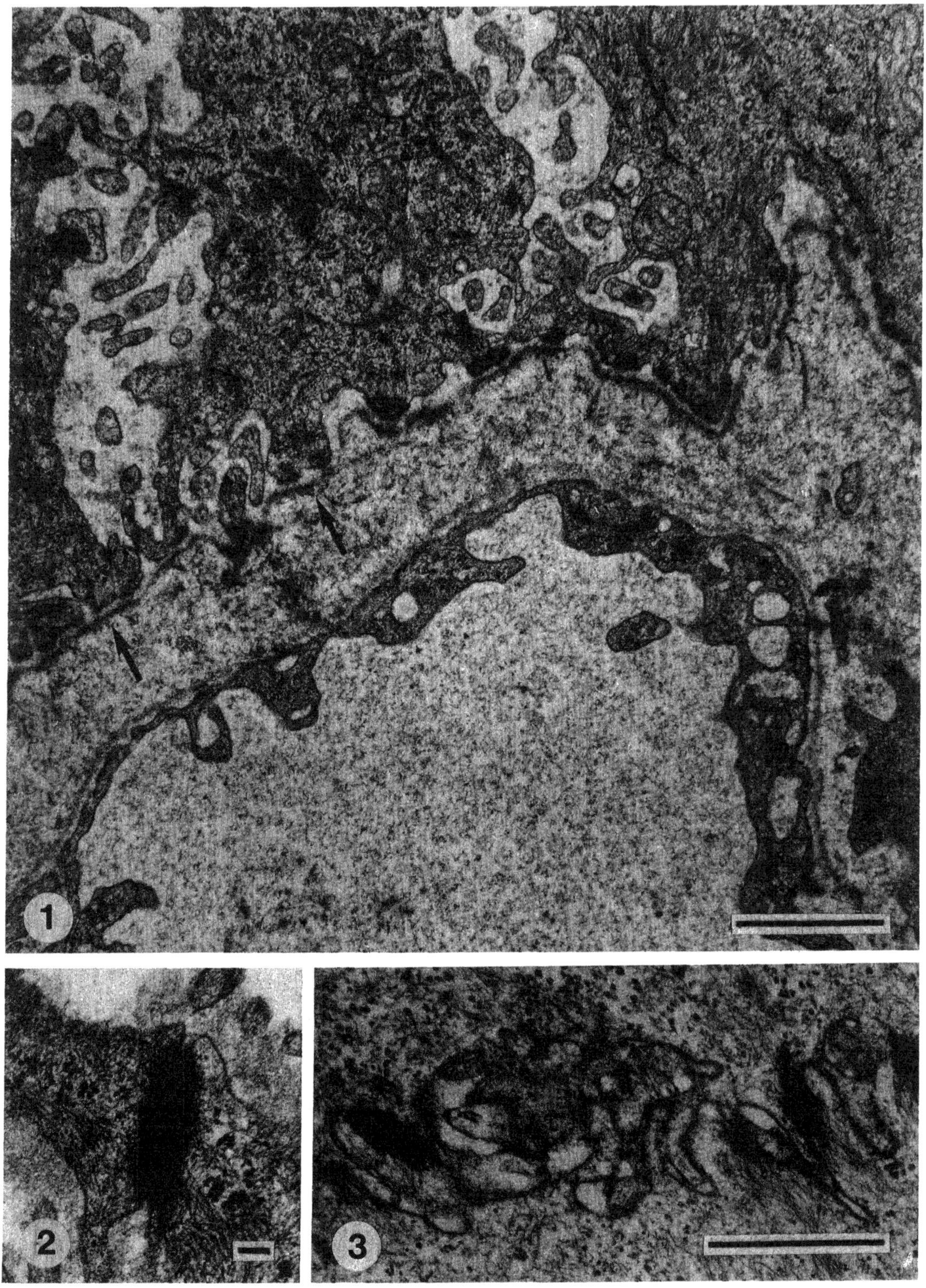
1
2
3

ing vascular and nervous plexuses as well as an important network of elastic fibers. The short abdominal segment is covered with a serosal coat. The ultrastructure of the esophagus, especially the epithelium, has been studied by several groups. Yet various elements are still lacking and further study is needed of the normal esophagus as well as its pathological conditions.

2. Mucous membrane

2.1. Epithelium

2.1.1. Squamous epithelium

The *basal lamina* separates the epithelium from the lamina propria, and closely follows the contours of the basal poles of the palisading cells of the germinative layer. Ultrastructurally it appears as a continuous, apparently homogeneous, moderately electron-dense (lamina densa) basal lamina (40–60 nm thick), separated from the basal cells by a clear space, the lamina lucida (40–50 nm thick); anchoring fibrils extend perpendicularly from the lamina densa over a short distance into the lamina propria (Fig. 2-1).

The *basal cell layer* mostly contains small, columnar or fusiform cells, oriented with their long axis perpendicular to the basal lamina. The basal cell has an elongated nucleus, which fills much of the cell volume; one or more nucleoli are visible. The cytoplasm contains the usual cell organelles in relatively small quantities: a few, mainly infranuclear mitochondria, a moderate number of free ribosomes and polysomes, some scattered cisternae of rough-surfaced endoplasmic reticulum, a small supranuclear Golgi apparatus, and some lysosomes. Lipid droplets may occur near the inferior pole of the nucleus. Glycogen is not present. The cytoskeleton is represented by fine bundles of tonofilaments (keratin intermediate filaments); microfilaments or microtubules have not been reported. The cell membrane shows the usual trilaminar fine structure.

The cell border facing the basal lamina has an irregular course. Numerous fingerlike extensions protrude in the underlying connective tissue. Occasionally small invaginations and pinocytotic vesicles occur along the inferior cell border. Typical hemidesmosomes are thought to attach the basal cells to the underlying basal lamina (Fig. 2-1). The lateral cell borders are facing adjacent epithelial cells. Their irregularly interdigitating cell membranes are interconnected within a distinct intercellular space through a few typical desmosomes. Both desmosomes and hemidesmosomes appear to be interconnected by fine bundles of tonofilaments, inserting into the electron-dense cytoplasmic plates of these adhering junctions (Figs. 2-2, 2-3). Only a few gap junctions are found as communicating junctions between the basal cells.

As the cells leaving the germinal layer disconnect from the basal lamina, they become larger, polyhedral, and subsequently more flattened and constitute the multilayered prickle cells. Together with the palisading basal cell layer, the basal portion of the prickle cell layer represents a basal zone characterized by common basophilia on light microscopy. The basal zone normally comprises 10% to 15% of the total epithelial thickness but may be greater in the distal esophagus (2). However, unlike the columnar basal cell, the polyhedral *prickle cell* does not rest on the basal lamina. Nuclear indentations become deeper and more numerous. The increased cytoplasmic volume reveals the same set of organelles, mainly in a perinuclear disposition, but clusters of free ribosomes are more numerous, the Golgi apparatus becomes more prominent, and glycogen deposits appear as electron-dense perinuclear fields or single rosettes. The distribution of glycogen

←

Figure 2-1. The basal lamina (lamina densa) (arrow) separates the basal epithelial cells from the lamina propria connective tissue. Hemidesmosomes are clearly visible (arrowhead). In the stromal papilla, a blood vessel occurs near the basal lamina. Bar = 1 μm. (From *Gastrointest Endosc* 26: 31, 1980, with permission).

Figure 2-2. A desmosome is composed of nine lines: five electron-dense and four electron-lucid. The two outer thick lines are the attachment plaques for cytoplasmic tonofilaments. Bar = 0.1 μm.

Figure 2-3. Bundles of tonofilaments are inserted into the cytoplasmic plates of the desmosomes, as illustrated on this figure from the prickle cell layer. Bar = 1 μm.

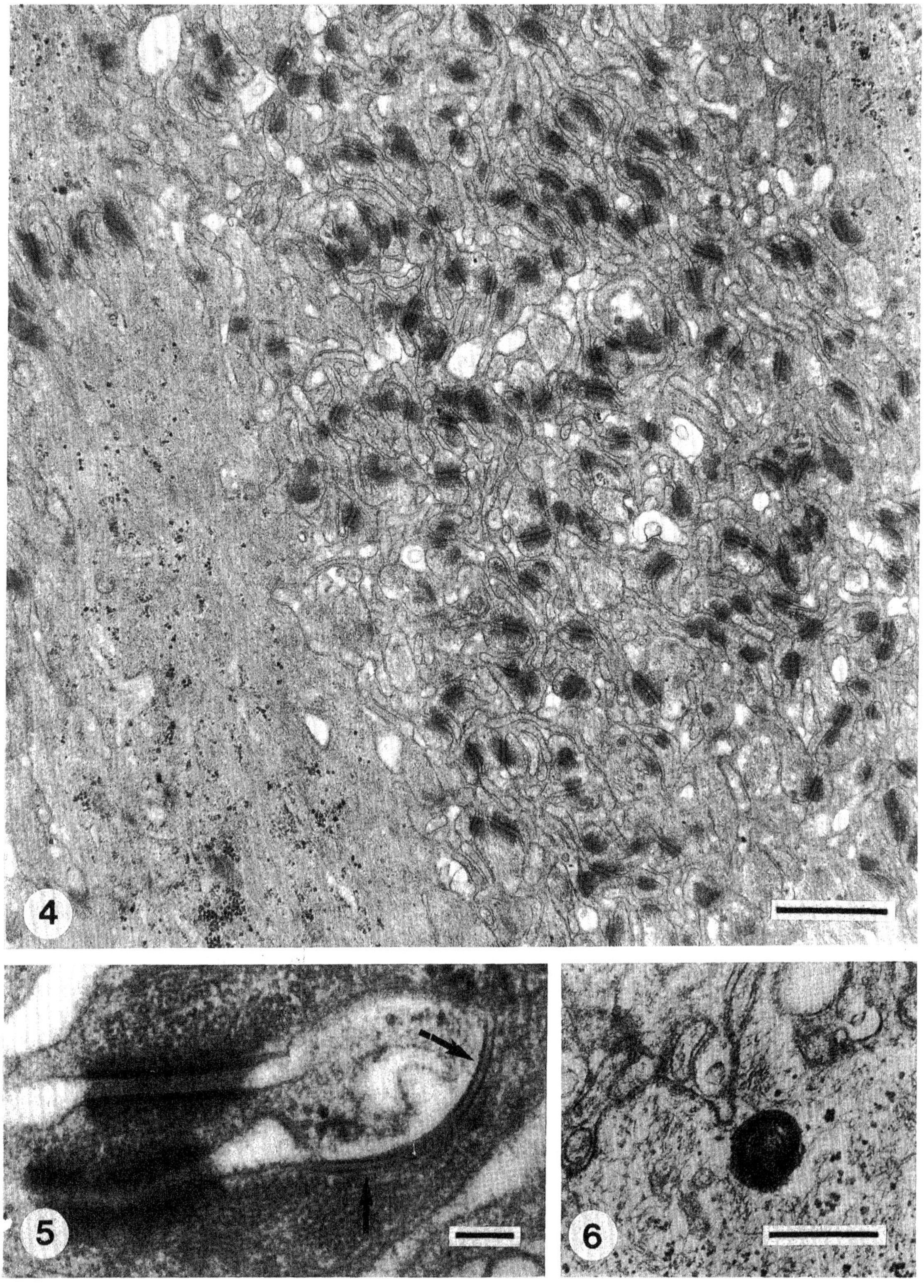
4
5
6

within the squamous epithelium can partly be explained by its role as a source of energy (3). In the cell periphery increased numbers of thick bundles of tonofilaments appear oriented perpendicular to the cell surface. The apposed cell membranes delineate elaborate interdigitating cytoplasmic folds and microvillous processes, richly supplied with desmosomes. In sections parallel to the intercellular space a labyrinth of desmosome-studded, branching cell processes may be observed as a desmosome field (Fig. 2-4). Microvillous processes of prickle cells often present gap junctions (Fig. 2-5). Where junctional structures are present, the cell surfaces of adjacent cells are in proximity; elsewhere, they tend to enclose a distinct intercellular space.

Of special interest is the presence of membrane-coating granules in the more flattened upper prickle cells. These are round, oval, or irregular membrane-bound structures, usually with a granular matrix, measuring 200–300 nm in diameter (Fig. 2-6). A fine internal periodicity, as in the rat esophageal epithelium, has not been established in man. Membrane-coating granules are seen in the region of the Golgi apparatus, towards the apical surface of the cell, and even in contact with the cell membrane. Electron-dense material resembling the matrix of these granules is seen in the intercellular space (4, 5, 6).

In the superficial layers, the cells gradually become more flattened, and consequently are oriented parallel to the surface (Fig. 2-7). Superficial cells retain their nuclei and cytoplasmic organelles. Some mitochondrial degeneration may be apparent. The Golgi apparatus is still abundant and membrane-coating granules are seen at the cell periphery. Microtubules are occasionally seen. The cellular processes become simpler and fewer, and desmosomes are attached to the cell body rather than to cell processes. Gap junctions become less frequent and the intercellular space is more narrow (7). Tight junctions have not been described in the normal squamous esophageal epithelium.

The two or three most superficial layers are slightly different from the rest of the superficial layers. The cells are markedly flattened. The nuclei appear shrunken and more or less deeply indented. The cytoplasm is essentially occupied by fine filaments with a few interspersed ribosomes and glycogen rosettes. Cell organelles, although present, are few in number and show more degeneration and disorganisation. Membrane-coating granules are still present. Lipid droplets are more frequent than in other layers. The cell membrane is particularly prominent due to a significant increase in thickness and staining intensity of its inner cytoplasmic leaflet. It is, however, much simpler with fewer, shorter processes, especially on the most superficial cells. The edges of these cells tend to overlap (Fig. 2-8). Between the edges, desmosomes may be present. The luminal free surface presents many finger-like projections, representing sectioned folds or microplicae of the plasmalemma (Fig. 2-6). Some of these possess a very fine filamentous coat. The intracellular matrix of the microplicae consists of a disordered array of fine filaments in a cytoplasmic matrix of variable electron density. No terminal web is present and therefore the microplicae differ from microvilli.

In the scanning electron microscope the normal esophageal luminal surface is seen covered by polygonal squamous cells (Fig. 2-9). Their cell surfaces display a characteristic network of densely packed ridgelike folds (microplicae) and villi of variable height and length (Fig. 2-10). These ridges follow straight or winding paths and often branch. The functional significance of the microplicae is not exactly known. It is assumed that they represent remnants of plasmalemmal folds which once provided for intercellular interdigitation and adhesion. These may be nonfunctional or provide protection by reducing the surface area

←

Figure 2-4. Desmosomes are numerous in the stratum intermedium of the esophageal epithelium. They can occur on the cell surface or on cell processes. In areas where numerous branching cell processes occur, "desmosome fields" may be observed. Bar = 1 μm.

Figure 2-5. In the prickle cell layer, junctional structures are composed of adhering desmosomes and communicating gap junctions (arrows). Bar = 0.1 μm.

Figure 2-6. Membrane-coating granules are round or oval inclusions which are found at the cell periphery and may be important for cell cohesion. Bar = 0.5 μm.

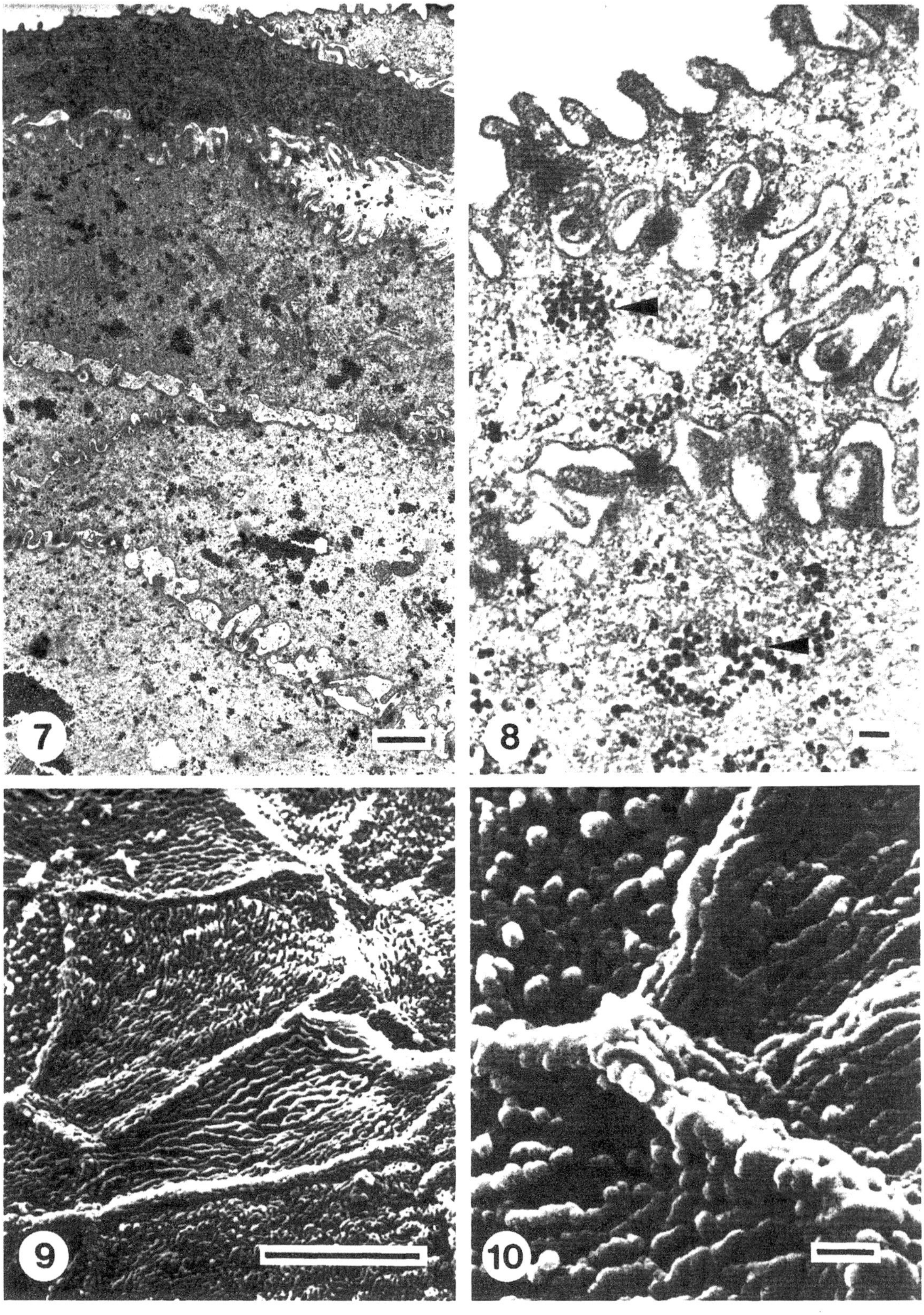
7
8
9
10

of contact and by holding a layer of lubricating mucus (9, 10). The borders of the closely contiguous superficial cells are lined by well-developed intercellular ridges. At the esophagogastric junction the squamous epithelium abuts in a jagged manner upon the columnar epithelium. Macroscopically this junction is observed as the Z-line (8).

2.1.2. *Specialized cells*

A variety of specialized cell types such as lymphocytes and Langerhans cells are seen within the squamous epithelium of the esophagus (11). Langerhans cells are mainly found in the middle and superficial layers of the squamous epithelium. Cells of this type were previously identified in the skin and various squamous epithelia, including buccal mucosa, and the female genital tract. They have a clear cytoplasm and are often irregular or dendritic in outline with cytoplasmic protrusions extending between the surrounding squamous cells. Langerhans cells are characterized by the complete absence of desmosomes and tonofilaments, which are a prominent feature of the normal squamous cells, and by the presence of the typical Langerhans or Birbeck granule. This is a rod-shaped inclusion up to 460 nm in length, having a characteristic linear, laminated internal structure. In addition to these organelles, there are moderate numbers of free ribosomes, profiles of granular endoplasmic reticulum, a well-developed Golgi apparatus, few mitochondria, and moderate numbers of membranelined electron-dense bodies (12, 13). Langerhans cells belong to a family of morphologically similar dendritic mononuclear cells that reside in lymphoid tissue, where they play a critical role in antigen presentation (11). A second clear cell often found in the human esophagus is the typical intraepithelial lymphocyte (14). Like Langerhans cells, the lymphocytes lack tonofilaments and desmosomes. Their cytoplasmic organelles are, in general, less well organized, and their nucleus tends to have a prominent heterochromatin component (Fig. 2-11). In some cases it is difficult to make a firm distinction on the basis of morphology between lymphocytes and Langerhans cells. The majority of the intraepithelial lymphocytes express the OKT_3^+, OKT_8^+, $OKIA^+$ phenotype and can therefore be considered to be suppressor/cytotoxic T lymphocytes. A minority are of the helper/inducer subset. These lymphocytes belong to the GALT (Gut-associated lymphoepithelial tissue) and have a function similar to that of the interepithelial lymphocytes in the intestine (11). About 4% to 8% of esophagi possess intraepithelial melanocytes. Melanocytes are found mainly in the basal cell layer. They also lack desmosomes and tonofilaments, and contain a variable number of melanosomes in different stages of development. About 28% of the esophagi contain unevenly distributed argyrophilic cells in the deeper epithelial strata representing neuroendocrine or APUD cells. The related amphicrine cells, which stain for mucus as well as for argyrophilic or argentaffin granules, are also seen here (14).

In cats and monkeys, the existence of free-ending intraepithelial nerve fibres has been demonstrated at the ultrastructural level in the upper third of the esophagus (15).

2.1.3. *Histochemisty*

In biopsies treated with osmium tetroxide ferrocyanide, a cell surface coat can be seen, with filaments 200–300 Å long, originating from the cell membrane (16). Neutral mucosubstances can be identified in the cell surface coat of normal

←

Figure 2-7. As the cells differentiate and move towards the lumen, they become markedly flattened, as illustrated in this picture of the superficial layers. Bar = 1 μm.

Figure 2-8. The luminal free surface of the superficial cells presents irregular cell processes, representing sectioned folds. No terminal web is present. Adjacent cells are connected by desmosomes. Glycogen can be observed in the cytoplasm (arrowheads). Bar = 0.1 μm.

Figure 2-9. The surface of the normal esophageal epithelium as seen by SEM is covered by polygonal closely contiguous cells with well-developed intercellular ridges. Bar = 10 μm.

Figure 2-10. The surface of the individual epithelial cell shows numerous folds and villi following straight or winding paths (SEM). Bar = 1 μm.

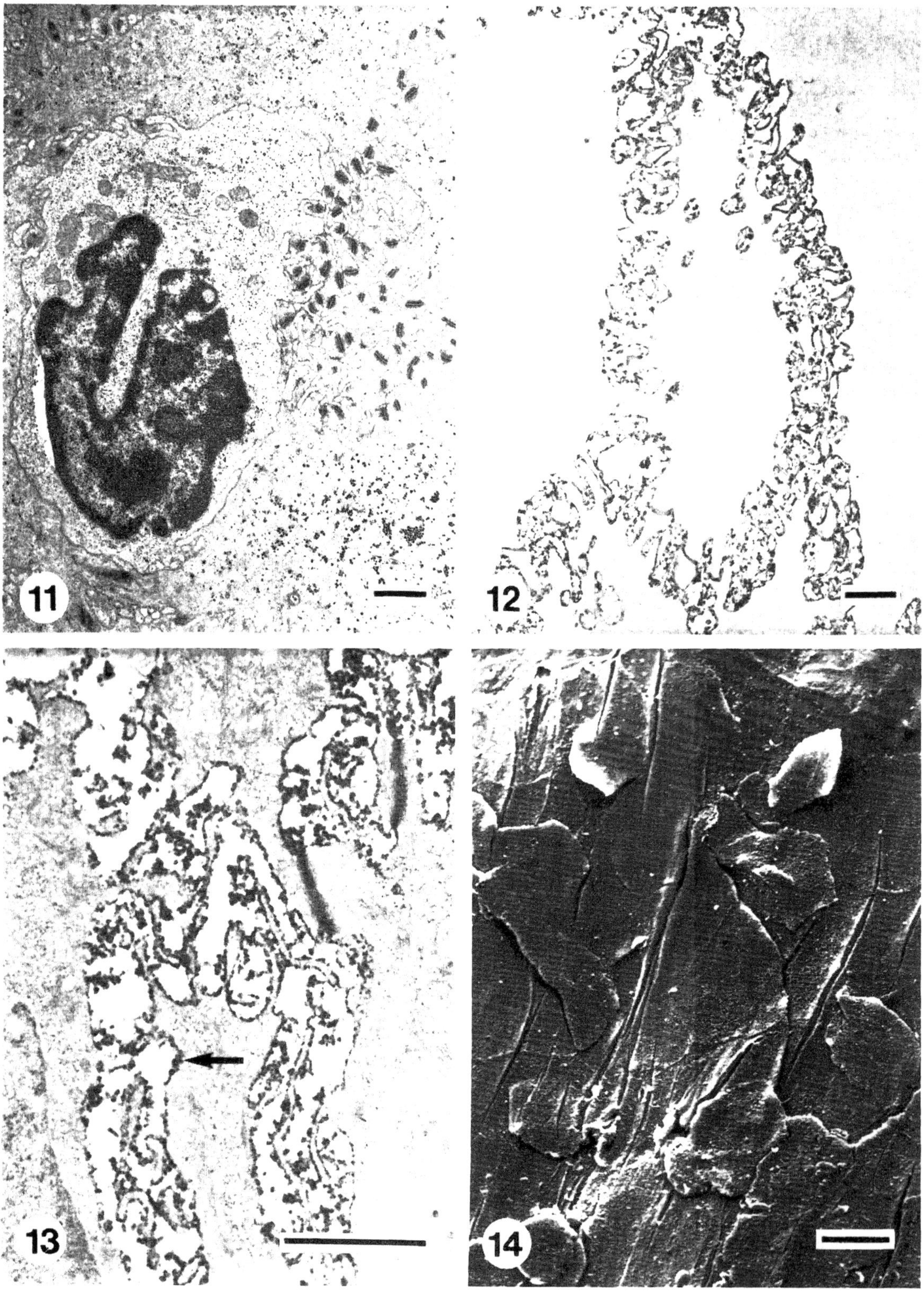
11
12
13
14

human esophageal epithelial cells by light microscopy and electron microscopy. Acid mucosubstances are also present on all epithelial cells but in larger amounts on the superficial cells. This combination of acid and neutral mucosubstances is compatible with the known composition of cell coat granules. Concanavalin A receptors (indicating the presence of α-D-mannosyl and α-D-glycosyl residues) can be found on the epithelial surface, in the perinuclear space, and in the rough endoplasmic reticulum.

Membrane-coating granules contain neutral and acid mucosubstances and carbohydrates. When these are secreted into the intercellular space, the carbohydrate moiety may contribute to the intercellular debris. Similar granules are found in the epidermis but in the esophagus they are few in number. In the skin these granules are sometimes called Odland bodies. After excretion they establish a barrier to water loss and mediate cell cohesion. In the esophagus the intercellular space of the most superficial of the functional layers contains neutral and sialic acid-rich mucopolysaccharides. The esophageal epithelial cell coat and the intercellular material, therefore, have the characteristics to provide a protective function on the cell surface. The mucus may be held in place by the fingerprintlike microplicae seen on the surface cells by scanning electron microscopy (17, 18).

Esophageal epithelial cells have little endogenous peroxidase activity. Neutral fat droplets are present in all layers, especially in the functional layer. No unsaturated lipids or fatty acids have been demonstrated yet. Much glycogen is present in the cytoplasm of the prickle and functional cell layers. Glycogen can also be found in the intercellular space (19).

Immuno-electron microscopy allows the demonstration of MHC class I antigens in the cell surface coat of normal esophageal epithelial cells (Figs. 2-12, 2-13). MHC class II antigens can not be demonstrated on epithelial cells in normal conditions. They are, however, expressed on intraepithelial lymphocytes and Langerhans cells. Other epithelial cell markers, such as Human Milk Fat Globule Antigen are also expressed on the superficial layers of the esophageal epithelium.

2.1.4. *Pathology*

2.1.4.1. Reflux esophagitis. In patients with mild esophagitis, scanning electron microscopy reveals an uneven mucosal surface due to different stages of desquamation. The individual cellular surface shows less well-defined ridges and whorls. In more severe cases, sloughing of the superficial cells and distinct widening of the intercellular spaces can be seen (Fig. 2-14). Ultimately this leads to the development of ulcerative lesions with a complete loss of the lining epithelial cells (20).

The epithelial changes found with transmission electron microscopy are either related to the intercellular spaces or to the epithelial cells themselves or both. In the inflamed esophageal mucosa, the intercellular space becomes widened (Fig. 2-15). This is most marked in the prickle cell layer. Particulate material accumulates in the intercellular space, either as free clumps or smeared against the cell membrane. Much of the debris is a neutral carbohydrate, although glycogen or neutral mucosubstances derived from the epithelial cells may also be present. Intrusive cells are regularly found in the intercellular space. Lymphocytes and neutrophils predominate, but eosinophils and mast cells are encountered (21) as well as Langerhans cells (Fig. 2-16).

Some of the epithelial changes are related to the intercellular phenomena. The most remarkable of these is the increase in distance between desmosomes on epithelial cells. The epithelial cell

←

Figure 2-11. Lymphocytes are regularly seen within the normal esophageal epithelium. They have a clear cytoplasm and lack desmosomes which are normally present between epithelial cells, as illustrated by the desmosome field in this figure. Bar = 1 μm.
Figure 2-12. Immuno-electron microscopy allows the cellular demonstration and localization of antigens. This biopsy was stained for MHC class I antigens and clearly shows a positive membranous staining of the esophageal epithelial cells. Bar = 1 μm.
Figure 2-13. MHC class I antigen is expressed as a dark electron-dense staining on the membrane of the esophageal cells (arrows). (Immunostain with counterstain). Bar = 1 μm.
Figure 2-14. Reflux esophagitis is characterized by an increased loss of superficial epithelial cells. The mucosal surface becomes uneven, and the intercellular spaces are widened (SEM). Bar = 30 μm.

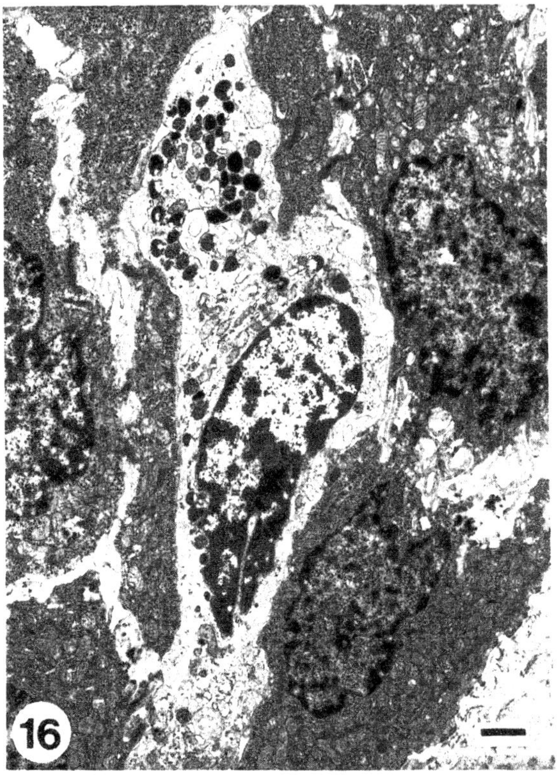

Figure 2-15. Reflux esophagitis. Basal layer. The intercellular space is considerably widened and filled with particulate material. Bar = 10 μm.

Figure 2-16. Reflux esophagitis: An intrusive mast cell between epithelial cells. Note the lack of desmosomes along the epithelial intrusive cell interface. Bar = 10 μm.

membranes remain generally intact. They may be stretched or show ruffling. Pinocytotic vesicles are more common and long strands of tonofibrils appear in the cytoplasm of adjoining cells. Desmosomes and gap junctions are almost normal. Various features of cellular damage, including edema, mitochondrial damage, and the occurrence of membrane whorls are seen. The endoplasmic reticulum shows either dilation or the appearance of granular material within the cisternae, and cytoplasmic glycogen is lost. A decrease in the number of membrane-coated granules is noted. Between the most superficial cells, tight junctions are formed. These cells normally tend to overlap, but in the inflamed esophagus the cell borders just meet and are joined by multiple desmosomes and tight junctions.

In cases with more severe inflammation, the Golgi apparatus is dilated. There is a marked loss of membrane-coating granules and parakeratotic changes may occur. Either the whole cell is involved or only a few keratohyaline granules are found. The basal lamina shows changes ranging from thickening to loss of continuity. There is an increase in the anchoring fibrils and in the number of hemidesmosomes. The cellular staining of neutral and acid mucosubstances remains virtually unaltered, whereas significant differences in the enzyme activity of membrane-coating granules are noted. Acid phosphatase activity is decreased but β-glucuronidase activity is increased (18).

When human esophageal biopsies are incubated in vitro with saline, 0.1 N HCl, and various gastrointestinal fluids, severe epithelial cell damage is observed with gastric juice, pepsin, duodenal juice, and bile acids. No morphological change is produced by saline, and HCl alone causes little damage. Bile acids are damaging only at high acid levels (22,23). The superficial layer appears to be able to withstand the attack better than the prickle and basal cells. Membrane-coating granules are secreted in the presence of hydrochloric acid but not of enzymes.

In acute experimental conditions in rabbits, the epithelial lesions are most marked in the middle layers and characterized by progressive swelling and rupture of the cells, resulting in the formation of vesicular spaces (24).

These various ultrastructural observations confirm the toxic influence of a wide variety of components of upper gastrointestinal fluids upon the esophageal mucosa. The esophagus contains a poorly-developed protective system in the most superficial epithelial layers, as shown by the formation of tight junctions and the ability of these cells to withstand in vitro incubation. This is also confirmed by animal experiments. Esophagitis is associated with an increase in permeability as a result of widening of the intercellular spaces and a loss of membrane-coating granules and with an increase in number and types of intrusive cells.

2.1.4.2. Infectious esophagitis. Occasional biopsies of *herpes simplex* and concomitant herpes-monilial esophagitis have been examined by electron microscopy. Epithelial cells infected by herpes contained numerous intranuclear viral capsids approximately 100 nm in diameter. Occasional virions were present in the cytoplasm. The nuclear membrane showed thickening and duplications. Focal thickenings in the cell membrane were seen in areas where the short projections of the luminal plasmalemma (microplicae) were associated with clusters of encapsulated virus particles. The cell wall of the monilia was separated from the cell membrane of the infected epithelial cell by a lucent space. Monilia could be seen together with herpes within individual epithelial cells (25, 26, 27). Herpes virus particles were also identified in biopsies from a patient with a squamous cell carcinoma (28, 59).

2.1.4.3. Epithelial dysplasia. Because of the high incidence of locally-recurrent esophageal carcinoma, even after apparently curative surgery, the mucosa has been examined in order to detect early dysplastic lesions. Using scanning electron microscopy, it has been found that esophageal tumor surfaces are markedly deranged. The few recognizable squamous cells have a grossly deficient microplical pattern and no intercellular cement. They appear more rounded and cell borders are indistinct (29).

Dysplastic mucosa and the mucosa adjacent to tumors display a microplical pattern which is less dense and more unfolded than normal. The microridges are irregular and pleomorphic. The intercellular segments are greatly reduced in number, and there are varying degrees of detachment and desquamation of the superficial cells (30). Transmission electron microscopy of dysplastic changes reveals a heterogeneous picture characterized by hypertrophy and hyperchromasia of the nucleus with deep invaginations of the nuclear membrane and swelling of the mitochondria. In the cytoplasmic matrix, myelinlike structures appear. The number of free ribosomes is increased and tonofibrillar keratohyaline complexes are formed. Parakeratosis, characterized by excessive accumulation of keratin filaments and nuclear maturation reaching the stage of pyknosis may occur. Increased numbers of cytoplasmic processes may be seen (31, 32). Various types of malignant epithelial esophageal tumors have been studied by transmission electron microscopy. No features diagnostic for carcinoma of the esophagus emerged from these studies. Neurosecretory granules can be demonstrated in small cell carcinoma of the oat cell type but are not present in all cases. Undifferentiated, non-oat cell type small cell carcinoma may be derived from primitive esophageal mucosal cells and can be diagnosed by electron microscopy (33, 34, 35).

2.1.5. Junctional epithelium

2.1.5.1. Normal junctional epithelium. The mucosa along the luminal border at the gastroesophageal junction and in the cryptlike mucosal glands is lined by columnar epithelial cells which resemble the gastric epithelial cells. Intestinal-type goblet and absorptive cells and Paneth cells are absent from this junctional layer. The apical surface of the junctional epithelial cells is characterized by the presence of only a few rudimentary microvilli, the fine structure of which resembles that of other alimentary epithelial cells. The terminal web is virtually nonexistent. The lateral plasma membranes show frequent infoldings. Although intercellular spaces of significant size are not uncommon, adjacent cells are usually in close contact with each other. Special attachment zones including tight junctions and various numbers of desmosomes are common. The basal plasma membrane is close to a continuous basal lamina. The usual cell organelles, including an elaborate Golgi apparatus, abundant rough endoplasmic reticulum, and mitochondria, are present in the cytoplasm. The supranuclear cytoplasm is filled with mucous granules (36).

2.1.5.2. Barrett's metaplasia. A large body of clinical and experimental evidence supports the concept that Barrett's metaplasia is usually an acquired condition mainly due to gastroduodenal reflux. Scanning electron microscopy reveals a mosaic pattern of architectural forms that are normally encountered in the stomach and small bowel. It ranges from the rather flat foveolar form of the stomach through various types of rudimentary villi to true small bowel-type villi. These forms are seen in intimate admixture. In areas of dysplasia the villar structures tend to be quite distorted. Inflammation can also distort the villi, which makes this feature diagnostically less useful (37).

Light microscopy and transmission electron microscopy reveal that Barrett's epithelium is an intimate admixture of all the various cell, gland, and architectural types found in the normal stomach and small bowel.

2.2. The lamina propria

The lamina propria consists of loose areolar connective tissue containing a network of collagenous and elastic fibers and very few fibroblasts. Scattered lymphocytes and solitary lymph follicles, preferentially located along the excretory ducts of the submucosal glands can be found. The latter can be considered to be lymphoepithelial complexes. Small mucosal cardia-type glands can occur. They are not always present. When present, they are scattered throughout the esophagus and not only at either end of the tube. They are composed of mucin-secreting cells comparable to the junctional cells, although occasionally serous cells have been noted.

SEM of the lamina propria at the epithelial stromal interface shows complex topographical patterns consisting of various numbers of simple and branched papillae and bladelike stromal folds. The papillae have helical contours, probably because of the spiral blood vessels within them. There is a great variation in papillary length and patterns between individuals and this results in unexpected complexity (38, 39). Transmission electron microscopy reveals that the blood vessels in the papillar extensions are separated from the basal epithelial cells by a thin layer of connective tissue. The endothelium may be fenestrated. Immunohistochemistry reveals the presence of factor VIII related antigen, of IgG and of MHC class II antigens on endothelial cells. Therefore, at least some of these vessels have characteristics of the postcapillary venules described in lymph nodes, and may be involved in immune mechanisms (40, 41).

The looplike capillaries of the superficial papillar layers of the lamina propria drain into a network of veins just below the squamous epithelium. These veins are visible as interconnecting linear ridges in the interpapillary zones on SEM. Fine longitudinal veins are normally present in the lamina propria of the distal esophagus. They originate from submucosal veins at the esophagogastric junction and run for approximately 3–4 cm upward. There again they pierce the muscularis mucosae and are reconnected to the submucosal veins.

2.2.1. Reflux esophagitis

In patients with reflux esophagitis, cellular infiltrates consisting of neutrophilic granulocytes, eosinophilic granulocytes, or both, and of lymphocytes are currently seen. Scanning electron microscopy of the connective tissue-epithelial junction shows intrusive cells (lymphocytes and granulocytes) in various stages of passage through the basal lamina. Transmission electron microscopy confirms discontinuity of the basal lamina as well as of the endothelium lining the capillaries in the lamina propriae papillae (39, 40, 21).

2.2.2. Esophageal varices

In biopsies from patients with esophageal varices, the subepithelial vessels may appear extremely dilated. Electron microscopy confirms the absence of an endothelial lining in such cases and demonstrates that red blood cells are in direct communication with flattened squamous cells, showing evidence of degeneration but still identifiable by tonofilaments and desmosomes. Rupture and bleeding of esophageal varices may be due to erosion of the variceal wall due to reflux esophagitis. The critical point seems near the area where the longitudinal lamina propria veins connect to the submucosal plexus (42, 43).

2.3. Muscularis mucosae

The muscularis mucosae is composed of smooth muscle fibers, oriented in a longitudinal direction. The muscle bundles may be interrupted by intercalated elastic tendons. The thickness of the muscularis mucosae increases distally, reaching up to 300 μm at the level of the cardia.

3. Tunica submucosa

3.1. Connective tissue

The submucosa is a broad layer of loose connective tissue which contains blood vessels, nerves, tubulo alveolar glands, lymph follicles, scattered lymphocytes, and groups of fat cells.

3.2. Submucosal glands

The submucosal esophageal glands are mucin-secreting glands, composed of lobules each consisting of acinar and ductal structures. The ducts are formed by cuboidal epithelium that becomes stratified squamous as it reaches the mucosa. These glands are considered a continuation of the minor salivary glands of the mouth and pharynx. The principal acinar cells are columnar mucous cells. They have a pyramidal shape. Dark and clear cells reflecting alternating activity have been distinguished. They are closely packed together and connected by the usual junctional complexes. The bulging luminal surface bears short microvilli. The nucleus is pushed to the periphery of the cell by the accumulation of pale mucus granules. The cytoplasm contains a well-organized granular endoplasmic reticulum, an elaborate Golgi system, mitochondria, microtubular inclusions, and membrane-limited microfibrillar bundles. Besides the predominant mucous cells, occasional subsidiary secretory cells with similar cytoplasmic characteristics are seen as well as typical oncocytes. The latter are found in small numbers, occasionally between duct and acinus. They have numerous closely-packed mitochondria, but no unusual surface features. They are connected to adjacent cells by junctional complexes and desmosomes. They rest on the basal lamina without basal specializations, and their free surfaces have few or no microvilli. The acini also contain myoepithelial cells. These are flattened, spindle-shaped cells, connected by desmosomes to the overlying epithelial cells and resting upon a continuous basal lamina (5).

The lining of the duct ranges from flattened, cuboidal epithelium at the junction of acinus and duct through two layers to a stratified pattern. Myoepithelial cells are present only in those ducts which have a simple epithelial lining. The main duct cells have short irregular microvilli on their luminal surface and are joined together in a conventional manner. They do not appear to contribute to mucus secretion (5).

The innervation of the submucosal glands appears to be by indirect contact. Neural elements are found in the periacinar connective tissue but do not enter the epithelium.

4. Tunica muscularis

4.1. Electron microscopy of the normal pattern

In the human esophagus the tunica muscularis is composed of smooth and striated muscle fibers. The transitional zone between the two types of muscle fibers is spread throughout the middle third of the length of the muscularis externa. Whereas, in some animals, the transitional zone is seen to occur within the inner circular and outer longitudinal muscles at the same level, in man, smooth muscle fibers are observed more proximally in the inner circular layer. Striation appears in the proximal esophagus at a later stage than the somite period. This suggests that striation is the result of a primary differentiation of the esophageal muscle itself and not from a pharyngeal extension. The ultrastructure and organization of striated muscle cells in the human esophagus are largely similar to those found in striated skeletal muscle. Differences in the arrangement, the amount of fibrosis, and motor innervation between striated esophageal and skeletal muscle fibers have been described in the guinea pig and marmoset. The amount of fibrosis is also much higher in the human esophagus.

In man, two histochemically distinct populations of striated muscle fibers can be discerned: slow-twitch and fast-twitch fibers in a proportion of two to one. These fibers are evenly distributed close to the pharynx and more distally (45). The smooth muscle fibers of the human esophagus are organized into continuous sheets of loosely arranged cells. Three types of myo-myal junctions can be observed: nexuses (or gap junctions) allowing rapid intercellular communications, interdigitations, and intermediate junctions or close contacts.

The individual cells are surrounded by a basal lamina. In the sarcoplasm, around the nucleus are mitochondria, a few elements of granular endoplasmic reticulum and some free ribosomes, a small Golgi apparatus, glycogen particle, and occasional lipid droplets. The remainder is occupied by myofilaments. At the transitional zone, striated and smooth muscle fibers are closely intermingled (Fig. 2-17). They are separated from each other by a small gap containing collagen fibers, ground substance, and sometimes neurovascular elements. Specialized contacts between smooth and striated muscle fibers are neither observed in man nor in the guinea pig and the marmoset. Yet both cell types can come in close contact at certain points of their surface. They are arranged together in such a way that they run a similar course and belong to the same construction system (36, 44, 45).

Interstitial cells of Cajal are present in the human esophagus wherever the musculature contains smooth muscle cells. These cells are considered to be intercalated anatomically and functionally between autonomic nerves and effector cells. They are abundant in the esophageal body, even more numerous in the lower esophageal sphincter, but very rare in the gastric portion of the lower esophageal sphincter. The majority of the interstitial cells are located at the periphery of the muscle bundles in both the inner and outer layer, and they do not accompany the nerve plexuses. Several types of contacts and junctions exist between the interstitial cells of Cajal and between the interstitial cells and Schwann cells or nerve endings. The interstitial cells of Cajal probably play an important role in esophageal motility (46) and may transmit nerve impulses to the muscle cells.

Smooth muscle cells, interstitial cells, and nerves form bundles, which constitute anatomical units separated by connective tissue septa, containing blood vessels and nerves.

4.2. Pathology

4.2.1. Achalasia

Various smooth muscle cell changes are described in achalasia of the esophagus. The ultrastructural alterations are, however, nonspecific and neither pronounced nor consistent. In the junctional zone between the dilated and distally-narrowed segment of the esophagus, cellular atrophy as well as cellular hypertrophy has been described. In individual cells a detachment of myofilaments from the surface membranes was observed. Various other ultrastructural alterations have been described in both achalasia and diffuse esophageal spasm. The smooth muscle changes include: filament disarray, mottling of the fiber density, thick and long cytoplasmic dense bodies, long dense plaques, and decrease of nexus junctions. Statistical analysis of the number of muscle cells suggested that the gross thickening of the muscular wall of the esophagus in achalasia is due to hyperplasia and not to hypertrophy of muscle cells (47). In achalasia, the interstitial cells of Cajal and the nerve endings in the lower esophageal sphincter are also damaged in a diffuse way. These alterations are more pronounced in older patients. On the contrary, the ultrastructural changes (increase in smooth endoplasmic reticulum and mitochondria) observed in biopsies from patients with a hypertensive sphincter suggest an enhanced activity of the interstitial cells (48).

4.3. Scleroderma

In scleroderma, fibrosis with extensive replacement of the muscle fibers of the esophageal wall is the most characteristic abnormality. Ultrastruc-

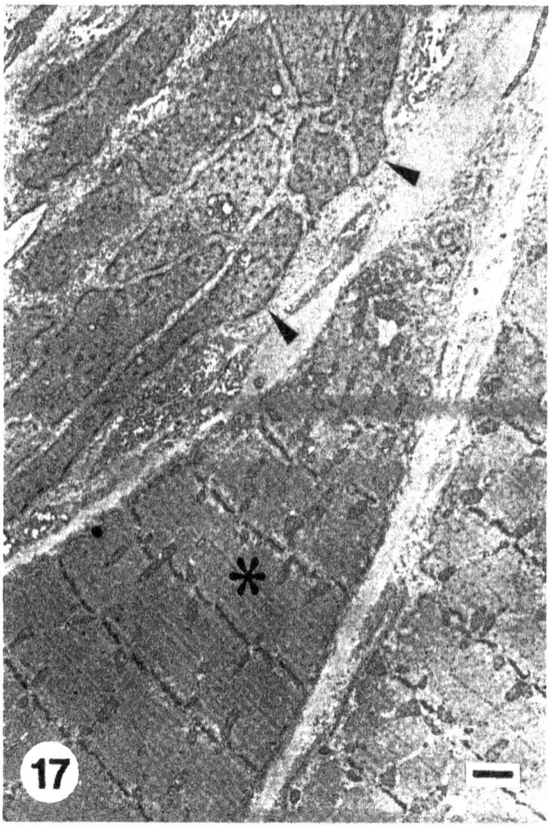

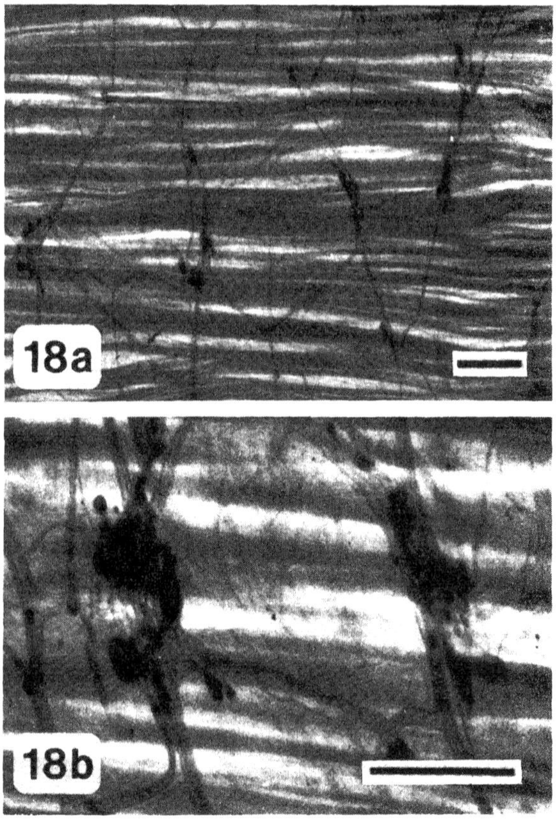

Figure 2-17. The tunica muscularis is composed of striated muscle cells (*) in the pars cervicalis and of smooth muscle cells (arrowheads) in the distal esophagus. At the transition zone (illustrated on this photograph), both cell types are closely intermingled and run a similar course. Bar = 10 μm.

Figure 2-18. The myenteric plexus is composed of a polygonal network of nerve bundles. (a) The ganglia can be located at a bifurcation. Various types can be distinguished such as parafascicular and intrafascicular. In some ganglia, individual cells are lying outside the main mass of the ganglion. Cholinesterase stain. Bar = 1 mm. (b) Cholinesterase stain. Bar = 0.1 μm.

tural muscular changes include thickened dense bodies and plaques and increased cell organelles in smooth muscle cells. In addition, the muscle capillaries show thickening and lamination of the capillary basal lamina and swollen endothelial cells. Some capillaries may be surrounded by dense material, and the lumen may be partially occluded with platelet aggregates (49).

5. Innervation

5.1. Microanatomy of the intramural nerve plexus

Histochemically-stained stretch preparations of the human esophagus allow a precise description of the intramural nerve plexuses. Both the submucosal and the myenteric plexuses have a similar anatomy in the newborn and in the adult. The submucosal ganglionated plexus is almost entirely located near the inner coat of the muscularis propria. The nerves are sparse and form an irregular, polygonal network, containing a limited number of small ganglia. The myenteric plexus shows a more elaborate, irregular, polygonal network of nerve bundles. The ganglionic density in the myenteric plexus is slightly variable. The density is low in the uppermost segment and increases distally. In the smooth muscle part of the adult human esophagus, the density amounts to about 25($\pm$6) ganglia/cm^2. In the lower esophageal sphincter area, it decreases again. The ganglia are usually found at a bifurcation of nerve bundles. Occasionally they are localized within the bundle, at a distance from a bifurcation. Not all bifurcations contain ganglia.

Shape and size of the ganglia are variable. In the proximal (striated muscle) part, the ganglia are usually big and loosely arranged. In the distal (smooth muscle) part, the ganglia are more compact. The number of cells per ganglion is also variable, allowing a distinction between paucicellular and polycellular ganglia. Solitary ganglionic cells can be observed either at a bifurcation or within the course of a bundle (50). Some ganglia are located outside the main course of a nerve bundle and are therefore called "parafascicular" ganglia, whereas others are located within a bundle, and these are called "intrafascicular" (51). Intermediate forms can also be observed. In many ganglia, one or more nerve cells can be observed, lying outside the main mass of the ganglion, but connected to it by a relatively long, fine stalk (Fig. 2-18).

The ultrastructure of the ganglia and nerve bundles of the human esophagus has not yet been described extensively. The ganglionic cells of the autonomic nerve plexus are usually large, with pale, round nuclei and a rather uniform cytoplasm. They contain numerous ribosomes with scattered granular endoplasmic reticulum and many small mitochondria. Various cell types have been recognized on the basis of the cellular affinity for silver and other characteristics (52). The ganglionic cells are surrounded by a connective tissue capsule. Bundles of unmyelinated nerve fibers enter and leave the individual ganglion formations. These bundles are composed of axons suspended in the conventional way within Schwann cells. Specialized structures have been described in animals (53) and humans (50).

5.2. Pathology

Various neurological lesions have been described in achalasia of the esophagus. These include a loss of argyrophil cells, the presence of degenerating ganglion cells containing Lewy bodies in the myenteric plexus, and lesions of the extra-esophageal vagus (52, 54, 55). Studies of the pathology of various other motility disorders of the esophagus are scanty, at least with regard to local lesions of the nervous system. Ganglionic cell damage has been reported in scleroderma (56) but was not confirmed (57). The myenteric plexus has been reported as normal and abnormal in idiopathic intestinal obstruction (58), and ganglionic cells remain invariably present in diffuse esophageal spasm and idiopathic muscular hypertrophy.

References

1. Geboes K, Desmet V: Histology of the esophagus. *Front Gastrointest Res* 3: 1–17, 1978.
2. Weinstein WM, Bogoch ER, Bowes KL: The normal human esophageal mucosa: A histological reappraisal. *Gastroenterology* 68: 40–44, 1975.
3. Falin LI: Glycogen in the epithelium of mucous membranes and skin and its significance. *Acta Anat* 46: 244–276, 1961.
4. Hopwood D, Logan KR, Bouchier IAD: The electron microscopy of normal human oesophageal epithelium. *Virchows Arch* 26: 345–358, 1978.
5. Al Yassin TM, Toner PG: Fine structure of squamous epithelium and submucosal glands of human oesophagus. *J Anat* 123: 705–721, 1977.
6. Themann H, Ottenjann R, Roggel R, Micderer SE: Ultrastruktur der Speisröhrenschleimhaut. *Z Gastroenterol* 9: 171–184, 1971.
7. Logan KR, Hopwood D, Milne G: Cellular junctions in human oesophageal epithelium. *J Path* 126: 157–163, 1978.
8. Siew S, Goldstein ML: Scanning electron microscopy of mucosal biopsies of the human upper gastrointestinal tract. *Scan Electron Microsc* 4: 173–181, 1981.
9. Squier CA: Membrane coating granules in nonkeratinizing oral epithelium. *J Ultrastruc Res* 60: 212–220, 1977.
10. Andrews PM: Microplicae: Characteristic ridgelike folds of the plasmalemma. *J Cell Biol* 68: 420–429, 1976.
11. Geboes K, De Wolf-Peeters C, Rutgeerts P, Janssens J, Vantrappen G, Desmet V: Lymphocytes and Langerhans cells in the human oesophageal epithelium. *Virchows Arch* 401:45–55, 1983.
12. Böck P: Fine structure of Langerhans cells in the stratified epithelia of the esophagus and stomach of mice. *Z Zellforch* 147: 237–247, 1974.
13. Al Jassin TM, Toner PG: Langerhans cells in the human oesophagus. *J Anat* 122: 435–445, 1976.
14. Tateishi R, Taniguchi H, Wada A, Horai T, Taniguchi K: Argyrophil cells and melanocytes in esophageal mucosa. *Arch Pathol* 98: 87–89, 1974.
15. Robles-Chillida AM, Rodrigo J, Mayo I, Arnedo A, Gomez A: Ultrastructure of free-ending nerve fibres in oesophageal epithelium. *J Anat* 133: 227–233, 1981.
16. Logan KR, Hopwood D, Milne G: Ultrastructural demonstration of cell coat on the cell surfaces of normal human oesophageal epithelium. *Histochem J* 9: 495–504, 1977.

17. Hopwood D, Logan KR, Coghill G, Bouchier IAD: Histochemical studies of mucosubstances and lipids in normal human oesophageal epithelium. *Histochem J* 9: 153–161, 1977.
18. Hopwood D, Logan KR, Milne G, Bouchier IAD: Concanavalin A receptors in normal and inflamed oesophageal epithelium. *Histochemistry* 57: 255–263, 1978.
19. Hopwood D, Logan KR, Milne G: Mucosubstances in the normal human oesophageal epithelium. *Histochemistry* 54: 67–74, 1977.
20. Dilly PN, Mallinson CN: Ultrastructural characteristics of the oesophageal mucosa in man. *Gut* 16: 841–842, 1975.
21. Hopwood D, Milne G, Logan KR: Electron microscopic changes in human oesophageal epithelium in oesophagitis. *J Path* 129: 161–167, 1979.
22. Bateson MC, Hopwood D, Milne G, Bouchier IAD: Oesophageal epithelial ultrastructure after incubation with gastrointestinal fluids and their components. *J Path* 133: 33–51, 1981.
23. Hopwood D, Bateson MC, Milne G, Bouchier IAD: Effects of bile acids and hydrogen ion on the fine structure of oesophageal epithelium. *Gut* 22: 306–311, 1981.
24. Salo JA, Lehto VP, Kwilaakso E: Morphological alterations in experimental esophagitis. *Dig Dis Sci* 28: 440–448, 1983.
25. Mc Kay JS, Day DW: *Herpes simplex* oesophagitis. Histopathology 7: 409–420, 1983.
26. Samuels Mirra S, Bryan JA, Buth WC, Miles ML: Concomitant Herpesmonilial esophagitis: Case report with ultrastructural study. *Hum Path* 13: 760–763, 1982.
27. Buerrig KF, Borchard F, Feiden W, Pfitzer P: Herpes oesophagitis. Electron microscopical findings. *Virchows Arch* (A) 404: 177–185, 1984.
28. Spence IM: Electron microscopic evidence of herpes virus in association with oesophageal carcinoma. *S Afr Med J* 68: 103–105, 1985.
29. Robinson KM, Maistry L, Evers P: Surface features of normal and neoplastic human esophageal cells in vivo and in vitro. *Scan Electron Microsc* 2: 213–222, 1981.
30. Goran DA, Shields HM, Bates ML, Zuckerman GR, De Schrijver-Keeskemeti K: Esophageal dysplasia. Assessment by light microscopy and scanning electron microscopy. *Gastroenterology* 86: 39–50, 1984.
31. Robinson KM, Gregory MA: Transmission electron microscopy of human oesophageal carcinomas. *J Path* 135: 97–109, 1981.
32. Zhang J, Li WQ, Wu En: A preliminary report on the electron microscopy of human squamous cell esophageal carcinoma. *Clin Med J* 96: 371–378, 1983.
33. Johnson FE, Clawson MC, Bashiti HM, Silverberg AB, Broun GO: Small cell undifferentiated carcinoma of the esophagus. *Cancer* 53: 1746–1751, 1984.
34. Kishida H, Sodemoto Y, Ushigome S, Kubota S, Kataba Y: Non-oat cell small cell carcinoma of the esophagus. Report of a case with ultrastructural observation. *Acta Pathol J* 33: 403–413, 1983.
35. Burke JA, Ranchod M: Hemangiopericytoma of the esophagus. *Human Pathol* 12: 96–100, 1981.
36. Desmet VJ, Tytgat GN: Histology and electron microscopy. *In*: Diseases of the Esophagus, G. Vantrappen, J Hellemans (eds). Berlin, Heidelberg, New York: Springer-Verlag 17–39, 1974.
37. Thompson JJ, Zinsser KR, Enterlinc HT: Barrett's metaplasia and adenocarcinoma of the esophagus and gastroesophageal junction. *Hum Pathol* 14: 42–61, 1983.
38. Warfel KA, Hull MT: Scanning electron microscopic study of the epithelial-mesenchymal junction of the esophagus. *Scan Electron Microsc* 2: 697–701, 1984.
39. Warfel KA, Hull MT: Scanning electron microscopic study of human esophageal stromal papillae. *Lab Invest* 50: 65A, 1984.
40. Geboes K, Desmet V, Vantrappen G, Mebis J: Vascular changes in the esophageal mucosa. *Gastrointest Endosc* 26: 29–32, 1980.
41. Geboes K: Morphological aspects of Crohn's disease. Thesis, Leuven, University, 1984.
42. Spence RAJ, Sloan JM, Johnson GW, Greenfield A: Oesophageal mucosal changes in patients with varices. *Gut* 24: 1024–1029, 1983.
43. Noda T: Angioarchitectural study of esophageal varices. *Virchows Arch* 404: 381–392, 1984.
44. Whitmore I: The ultrastructure of oesophageal striated muscle in the guinea pig and marmoset. *Cell Tissue Res* 234: 365–376, 1983.
45. Whitmore I: Oesophageal striated muscle arrangement and histochemical fibre types in guinea pig, marmoset, macaque and man. *J Anat* 134: 685–695, 1982.
46. Faussone-Pellegrini MS, Cortesini C: Ultrastructural features and localization of the interstitial cells of Cajal in the smooth muscle coat of human esophagus. *J Submicrosc cytol* 17: 187–197, 1985.
47. Friesen DL, Henderson RD, Hanna W: Ultrastructure of the esophageal muscle in achalasia and diffuse esophageal spasm. *Am J Clin Pathol* 79: 319–325, 1983.
48. Faussone-Pellegrini MS, Cortesini C: The muscle coat of the lower esophageal sphincter in patients with achalasia and hypertensive sphincter. An electron microscopic study. *J Submicrosc Cytol* 17: 673–685, 1985.
49. Russell ML, Friesen D, Henderson RD, Hanna WM: Ultrastructure of the esophagus in scleroderma. *Arthritis Rheum* 25: 1117–1123, 1982.
50. Mebis J, Geboes K, Desmet V: Intrinsic innervation of the gut. FGWO Kontaktgroep "Sekreties en Motiliteit van het spijsverteringsstelsel", 1985.
51. Christensen J, Robinson BA: Anatomy of the myenteric plexus of the opossum esophagus. *Gastroenterology* 83: 1033–1042, 1982.
52. Smith B: The neurological lesion in achalasia of the cardia. *Gut* 11: 388–391, 1970.
53. Rodrigo J, de Felipe J, Robles-Chillida EM, Perez-Anton JA, Mayo I, Gomez A: Sensory vagal nature and anatomical access paths to esophagus laminar nerve endings in myenteric ganglia. Determination by surgical degeneration methods. *Acta Anat* 112: 47–57, 1982.
54. Cassella RR, Ellis EH, Brown AL: Fine structure changes in achalasia of the esophagus: Vagus nerves. *Am J Path* 46: 279–288, 1965.
55. Qualman SJ, Haupt HM, Jang P, Hamilton SR: Esophageal Lewy bodies associated with ganglion cell loss in achalasia. *Gastroenterology* 87: 848–856, 1984.

56. Goetz RH: The pathology of progressive systemic sclerosis with special reference to changes in the viscera. *Clin Proc* 4: 337–392, 1946.
57. Treacy WL, Baggenstoss AH, Slocumb EM: Scleroderma of the esophagus. *Ann Intern Med* 59: 351–356, 1963.
58. Schuffler MD, Pope CE: Esophageal motor dysfunction in idiopathic intestinal pseudoobstruction. *Gastroenterology* 70: 677–682, 1976.
59. Rabeneck L, Boyko WJ, Melean DM, McLeod WA, Wong KK: Unusual esophageal ulcers containing enveloped viruslike particles in homosexual men. *Gastroenterology* 90: 1882–1889, 1986.

Author's address:
K. Geboes
Department of Medical Research
Laboratory of Histochemistry and Cytochemistry
U.Z. St. Rafael K.U. Leuven
Minderbroedersstraat 12
B-3000 Leuven
Belgium

CHAPTER 3

Fine structure of gastric glands

HERBERT F. HELANDER

1. Introduction

The main topic of this review will be the ultrastructure of the mammalian gastric glands, with emphasis on man. For a broader overview the reader should consult earlier articles (1, 2, 3), where emphasis is sometimes put on other data and conflicting opinions might be presented. For information on the submammalian gastric mucosa, the reader is referred to an earlier review (1).

The first publication on the fine structure of the gastric epithelial cells appeared in 1951. It dealt with the parietal cells, which secrete HCl, and, for the first time, the microvilli of its secretory canaliculi were described (4). Ever since then research efforts in this field have been dominated by studies on the parietal cells, and therefore much of the present review will deal with morphological aspects of HCl secretion.

2. Structure of the mucosa

The gastric glands are tightly packed, slightly coiled tubules between ½ and 1 mm in length. In the adult man there are about 35 million such glands (5).

There are three types of *gastric glands*: the cardiac glands, the oxyntic glands, and the pyloric glands. In man the oxyntic glands occupy about 75% of the mucosal surface. The glands open up into the gastric pits — foveolae gastricae — which are funnellike depressions on the surface of the mucosa. In the cardiac and pyloric gland areas these pits are fairly deep; in the oxyntic gland area they reach only about 100–200 μm down into the mucosa.

3. Surface mucous cells (Fig. 3-1)

The entire surface of the gastric mucosa is covered by a single layer of surface mucous cells. These cells are rather tall — 20–40 μm — and they are seldom wider than 6 μm. The gastric pits are covered by similar, but more cuboidal cells. In the rat the mean volume of the surface and foveolar mucous cells has been calculated to be 683 μm^3 (6).

The cells on the mucosal surface are separated by wide intercellular spaces, which are presumably of importance for the transepithelial transport of water and electrolytes (7). The presence of Na^+, K^+-ATPase along the basolateral — but not the apical — cell membrane supports this hypothesis (8). The intercellular spaces are sealed off from the lumen by tight junctions. Using freeze-fracture techniques, Meyer et al. (9) recently demonstrated differences in the structural configuration of the junctional complexes between the surface mucous cells and the oxyntic gland cells in the dog. Parallel investigations where lanthanum nitrate was used as a tracer in the extracellular space indicated that the tight junctions at the mucosal surface were less permeable than those in the glands.

On the apical cell surface the plasma mem-

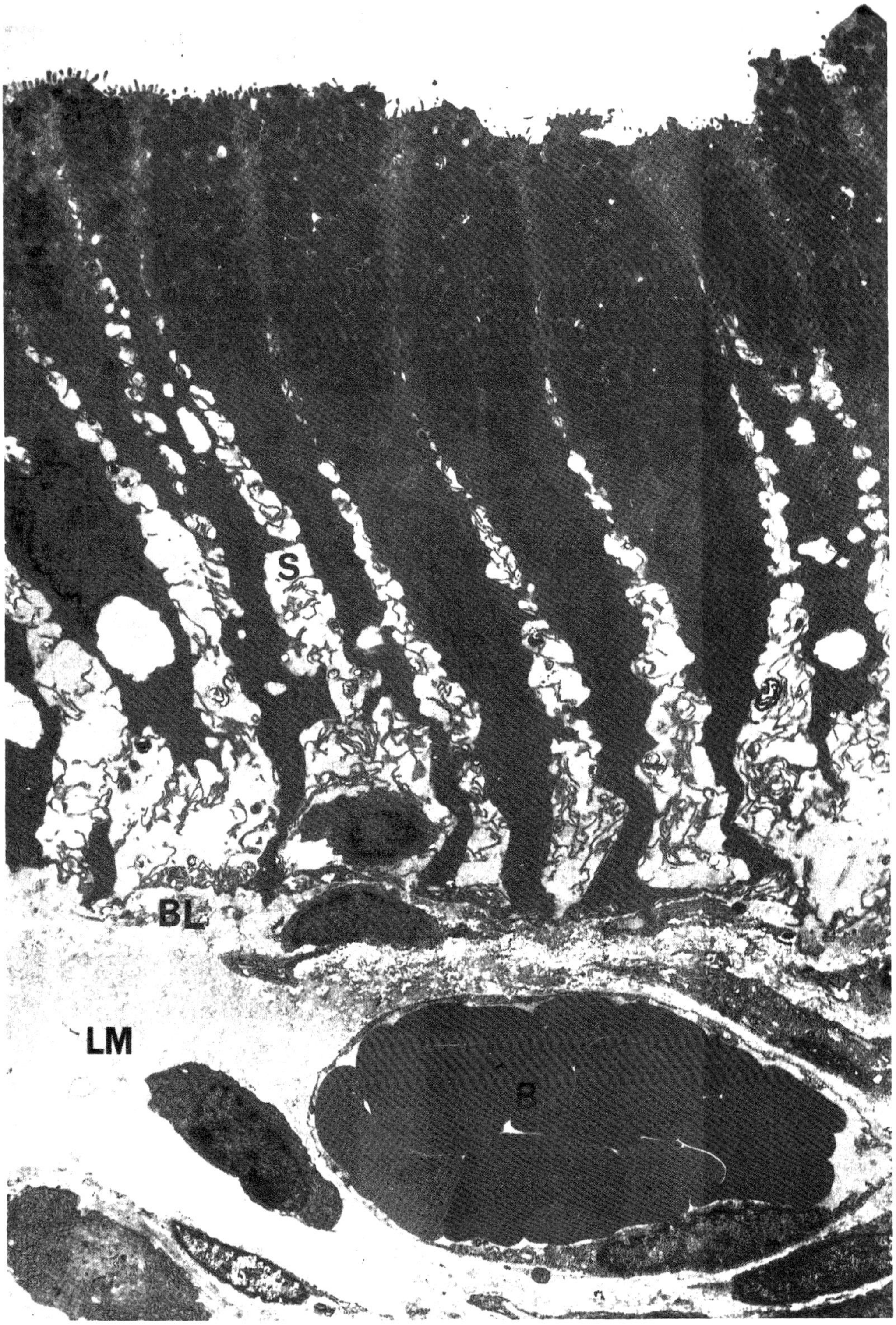
S
BL
LM
B

brane forms a few short microvilli. Beneath this membrane there is a variable number of fairly dense secretory granules, some of which are biphasic. These granules contain mucus and their ultrahistochemical properties have been investigated in detail by Spicer and coworkers (10, 11). The granules are secreted by exocytosis, and thus a mucous coat several hundred microns thick is formed which protects the gastric mucosa against mechanical and chemical injury (12).

A very important observation in this context is that there exists a pH gradient within the mucous coat; immediately above the surface mucous cells the pH is normally always close to neutrality, whereas the more superficial layers of the mucous coat assume the pH of the gastric cavity. Thus, during acid secretion the pH in the mucous coat ranges from about 7 to about 2. Secretion of bicarbonate from the surface mucous cells is responsible for maintaining the neutral pH at the cell surface, and if this secretion is inhibited, e.g., by indometacin or acetazoleamide, the gradient disappears, and superficial erosions rapidly develop (12).

A morphological correlate of this bicarbonate secretion appears in the form of large activities of carbonic anhydrase, located in the apical cytoplasmic matrix and the cores of the microvilli (13).

Secretion from the surface mucous cells is stimulated by food and vagomimetic drugs. Zalewsky et al. (7), using intra-arterial infusion of acetylcholine in dogs, obtained a copious secretion of alkaline mucus. In these dogs, there was a decrease in the amount of mucous secretory granules in the surface cells and an enlargement of the intercellular spaces between the interfoveolar cells, indicating fluid movement.

The Golgi apparatus of the surface mucous cells is usually found close to the nucleus. Electron histochemical investigations have revealed carbohydrate material in the Golgi apparatus as well (10, 11).

There are only a few, mainly elongated, mitochondria and the endoplasmic reticulum is sparse. There is an abundance of filaments, many of them attached to desmosomes; it has been suggested that the system of intermediate filaments provides strength to resist mechanical wear and tear. Other filaments, contractile in nature, could also participate in the mechanism that generates the motion responsible for the migration of the epithelial cells (14).

Fat droplets are often observed in suckling animals, indicating absorbtion of milk lipids (15). The nucleus usually contains a prominent, centrally located nucleolus.

The surface mucous cells are formed by mitoses in immature mucous cells of the gland neck. By injecting ^{3}H-thymidine, the DNA in mitotic cells can be labelled and these cells can be identified by autoradiography (16, 17). The surface mucous cells are shed into the gastric cavity, and their turnover time has been estimated to be between 2 and 8 days (1, 16). By scanning electron microscopy of human gastric biopsies from healthy volunteers, Mackercher et al. (18) calculated the fraction of dying or dead surface mucous cells to be 2.4%.

Extrusion of the surface mucous cells is preceded by degeneration with loss of electron density of the cytoplasmic matrix and of the nucleus. In addition, there is massive exocytotic release of mucous, while the basolateral plasma membranes, including the tight and gap junctions, remain well preserved until the final steps of extrusion (17, 19).

4. Cardiac glands

The cardiac glands are found in a narrow zone surrounding the cardia. In comparison with the oxyntic glands, the cardiac glands are more coiled, and their lumen is often more dilated. The main cell of these glands, the cardiac gland cell,

←

Figure 3-1. Mucous cells at the surface of the human gastric oxyntic mucosa. These tall, columnar cells are separated by wide intercellular spaces (S), where much of the transepithelial water and electrolyte transport presumably takes place. Tight junctions seal off the intercellular spaces from the gastric cavity. The basal end of the cells taper off into a thin pedicle, which stands on the basal lamina (BL). Mucous secretory granules occupy the supranuclear region of the cytoplasm. B, blood capillary; LM, lamina propria. X 4,000.

is quite similar in structure to the mucous neck cells with numerous mucous secretory granules. In addition to a few parietal cells and endocrine cells, a special serous cell has been described, which is characterized by numerous relatively dense, non-mucous secretory granules (20).

5. Oxyntic glands

5.1. Parietal cells (Figs. 3-2–3-4)

In man the number of parietal (oxyntic) cells has been estimated at about 1,000 million (21, 22). In

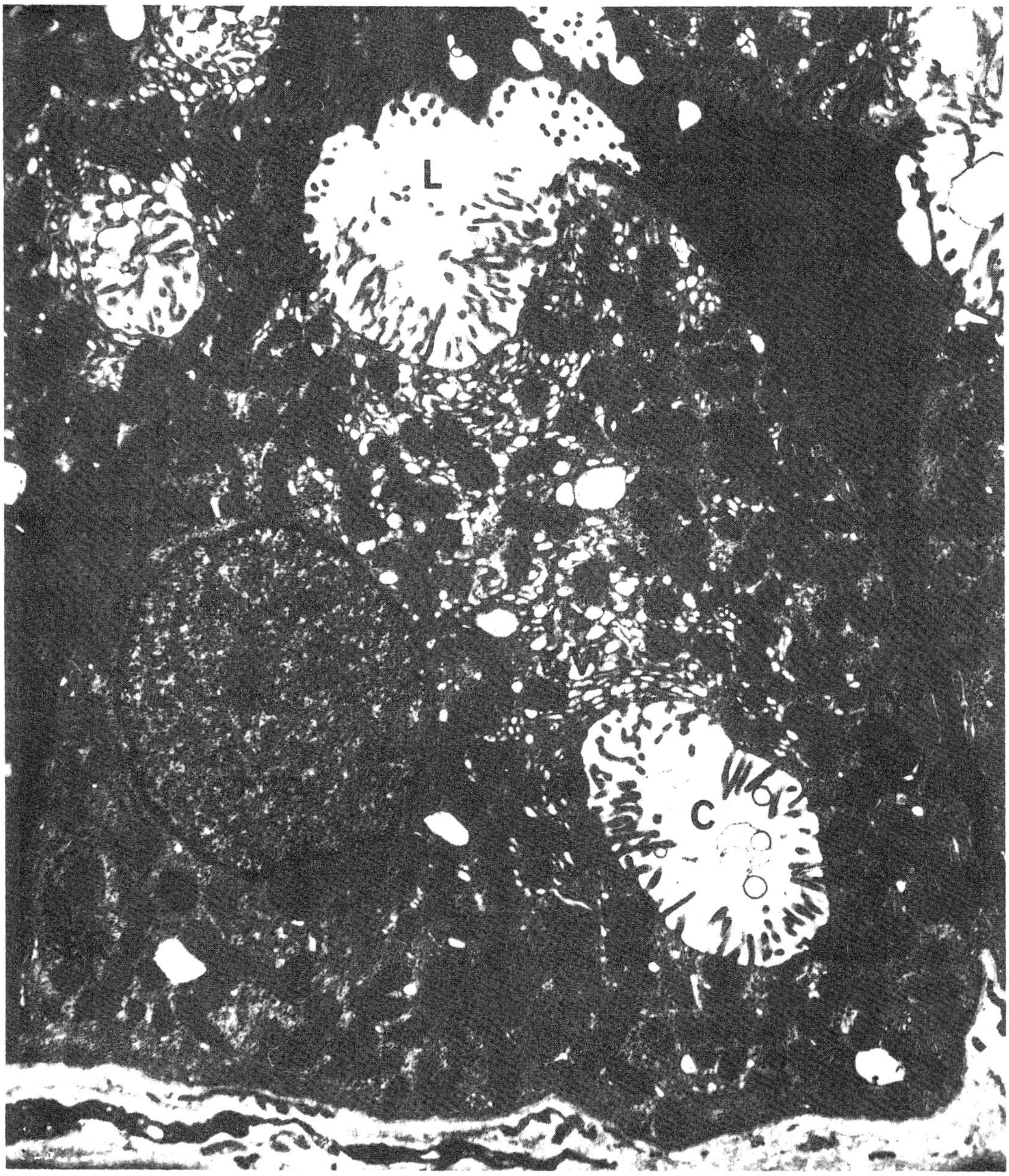

Figure 3-2. Parietal cell from a nonstimulated person. The cytoplasm contains numerous mitochondria (M) and tubulo-vesicles (V). A cross-cut secretory canaliculus (C) is seen; microvilli project from its walls. Microvilli are also seen on the apical cell surface which faces the gland lumen (L). ER, rough endoplasmic reticulum; LY, lysosome; N, nucleus, T, tight junction. X 8,000.

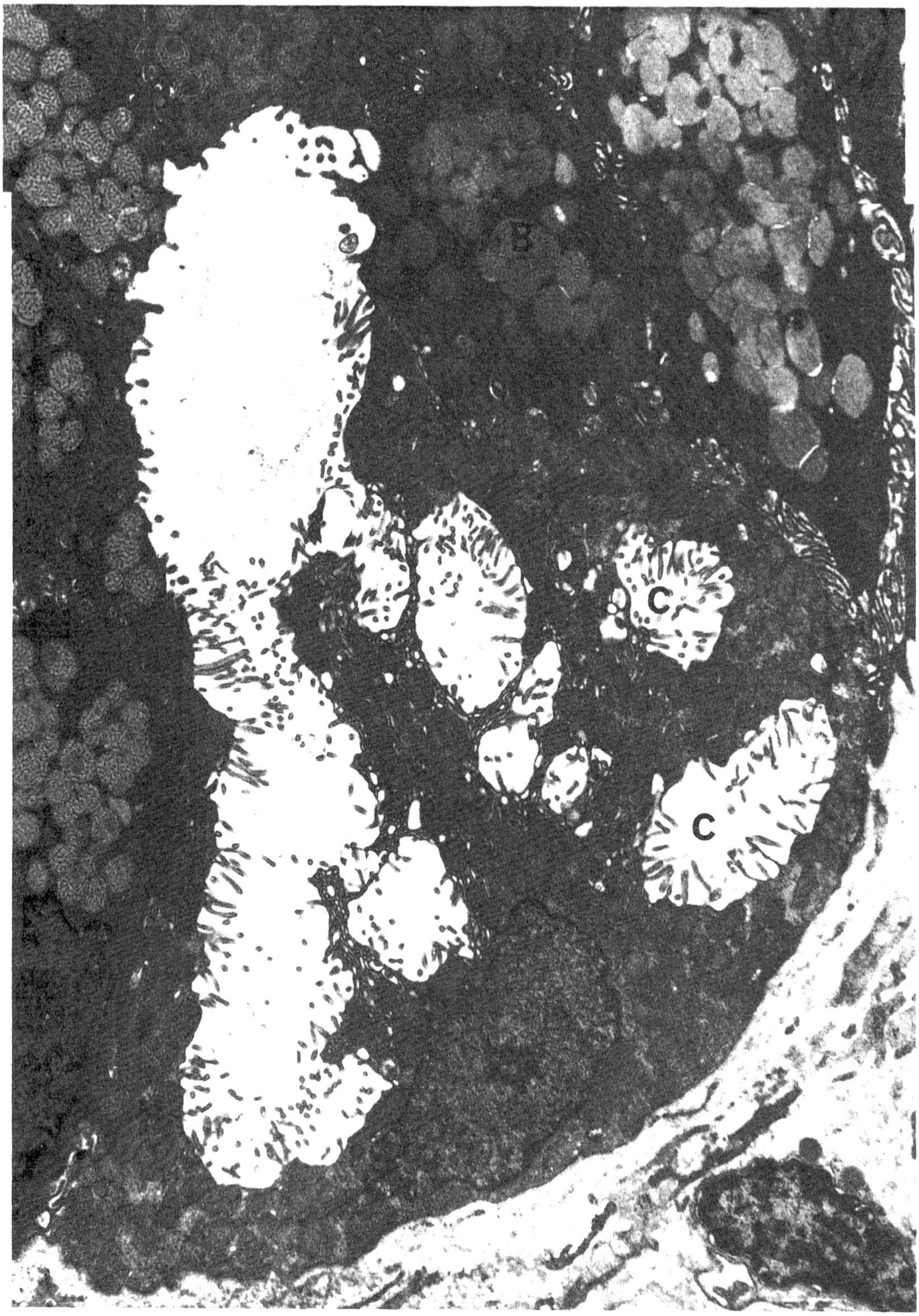

Figure 3-3. Human parietal cell during maximal acid secretion. In comparison with Figure 3-2 there is an expansion of the secretory canaliculi (C) and an increase of the number of microvilli. The extra membrane material required for this transformation is presumably derived from the tubulo-vesicles which are much less frequent than in the resting parietal cells. Several mucous neck cells also face the gland lumen; some of their secretory granules are biphasic (B). X 7,600.

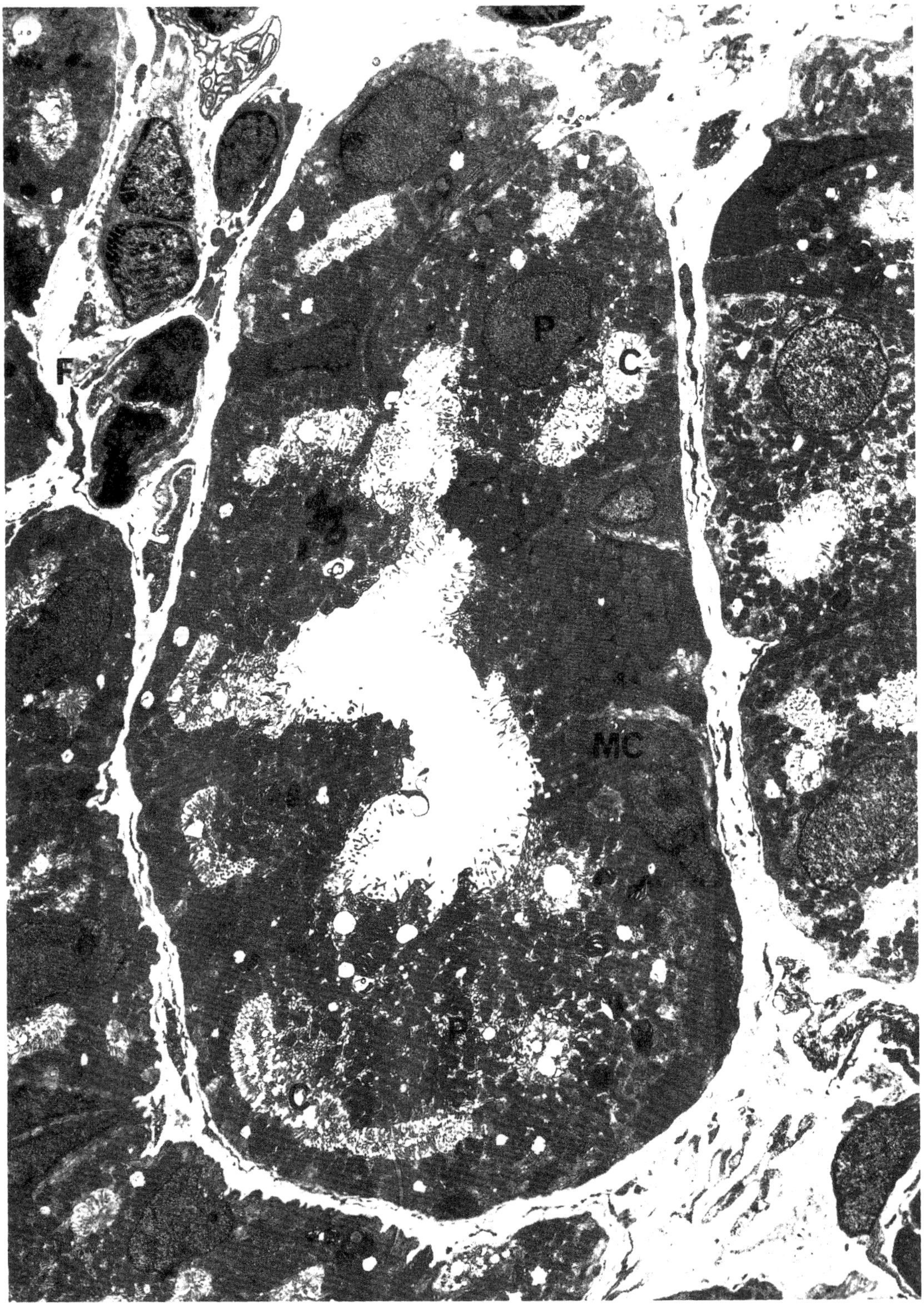

Figure 3-4. Oblique section through midsection of human oxyntic gland during secretory rest. In this region the parietal cells (P) dominate, but there are also a few mucous neck cells (MC). In the lamina propria there are nerve fibers (F) and connective tissue cells. C, secretory canaliculus. X 3,100.

duodenal ulcer patients with hypersecretion of acid, there is sometimes an increased number of parietal cells (23). The great majority of the parietal cells are found in the oxyntic glands; in the cardiac and in the pyloric glands they are quite scarce. The mean volume of the human parietal cells has been calculated to be between 1,800 and 5,500 μm^3 (24, 25, 26). Estimates of the mean volume of the various cells in the rat gastric epithelium have shown that the parietal cells are much larger than the other types of cells (6).

Most of the parietal cells are found in the midportion of the oxyntic glands, where many of them occupy a truly parietal position, with short, narrow intercellular channels connecting them with the main gland lumen. In the sections, some of the parietal cells therefore appear to be isolated from the gland tubules.

There are three characteristic features of the parietal cells:

(i) the mitochondria
(ii) the tubulo-vesicles
(iii) the secretory canaliculi

The mitochondria are observed mainly along the basal cell surface and close to the secretory canaliculi. They are round or oval, 0.5 by 1 μm in size, and they occupy roughly one-third of the cytoplasmic volume (24, 25, 27). This is a much higher figure than has been found for other types of epithelial cells. For comparison, it should be mentioned that the corresponding figure for rat zymogen cells is 5% (28), for human hepatocytes about 19% (29), but for human myocardial cells it is 33% (30). The high proportion of mitochondria in the parietal cells indicates a very high rate of oxidative metabolism. The closely spaced mitochondrial cristae also point in this direction. No morphological alterations have been found in the mitochondria as a result of normal changes in parietal cell function. Stachura et al. (31) described the presence of condensed mitochondria in dogs given omeprazole with or without histamine; such changes have not been observed in man or in other species.

The cytoplasmic tubulo-vesicles are round or elongated, smooth-surfaced structures, about 0.2 μm in cross-section. However, their size and shape vary considerably with different types of fixatives (32). As a rule, their content is clear, but small particles have been found in the tubulo-vesicles of the human parietal cells (3). Most of the tubulo-vesicles are observed in supranuclear cytoplasm and close to the secretory canaliculi, and, in man, Aase et al. (25) estimated that they occupied about 11% of the parietal cell volume during secretory rest. Following maximal pentagastrin stimulation, the volume fraction of the tubulo-vesicles fell to about 2%. Using a slightly different method, Frexinos et al. (33) observed a reduction of similar magnitude: about 80% of the tubulo-vesicles disappeared during maximal acid stimulation. During prolonged fasting (20–30 days) almost all of the tubulo-vesicles were lost (34).

Earlier investigators have demonstrated histochemical differences between the tubulo-vesicles and the regular endoplasmic reticulum (35), and more recent studies in the rabbit have shown the presence of H^+, K^+-ATPase at the tubulovesicular membrane. This enzyme, which is responsible for the final step of HCl production, is also present during stimulation of acid secretion at the secretory surface facing the canaliculi and the gland lumen (36). It has frequently been discussed whether HCl is produced and stored in the tubulo-vesicles, but there are no methods available to test this hypothesis.

The secretory canaliculi were first described by Golgi in 1893 in the rabbit (37). Over the years, many unsuccessful attempts were carried out using indicators to try to demonstrate the presence of acid in the secretion from the parietal cells. The final proof did not come until 1979, when Sachs and coworkers observed an acid shift in the fluorescent dye acridine orange following stimulation of gastric glands isolated from rabbits (38). From their start deep in the cytoplasm, the canaliculi follow a tortuous course until they empty out onto the apical cell surface which faces the gland lumen. No other normal mammalian cell type exhibits cytoplasmic canaliculi of a similar kind. As a result of these canaliculi, the secretory surface — which lines the canaliculi and the apical cell surface — becomes very large, an effect which is reinforced by the presence of numerous microvilli projecting from these surfaces.

The exterior surface of the secretory membrane is covered by a thin glycocalyx consisting of neutral mucosubstances (39, 40, 41); this layer might afford the parietal cells protection against the very strong (0.16 molar) HCl that is produced

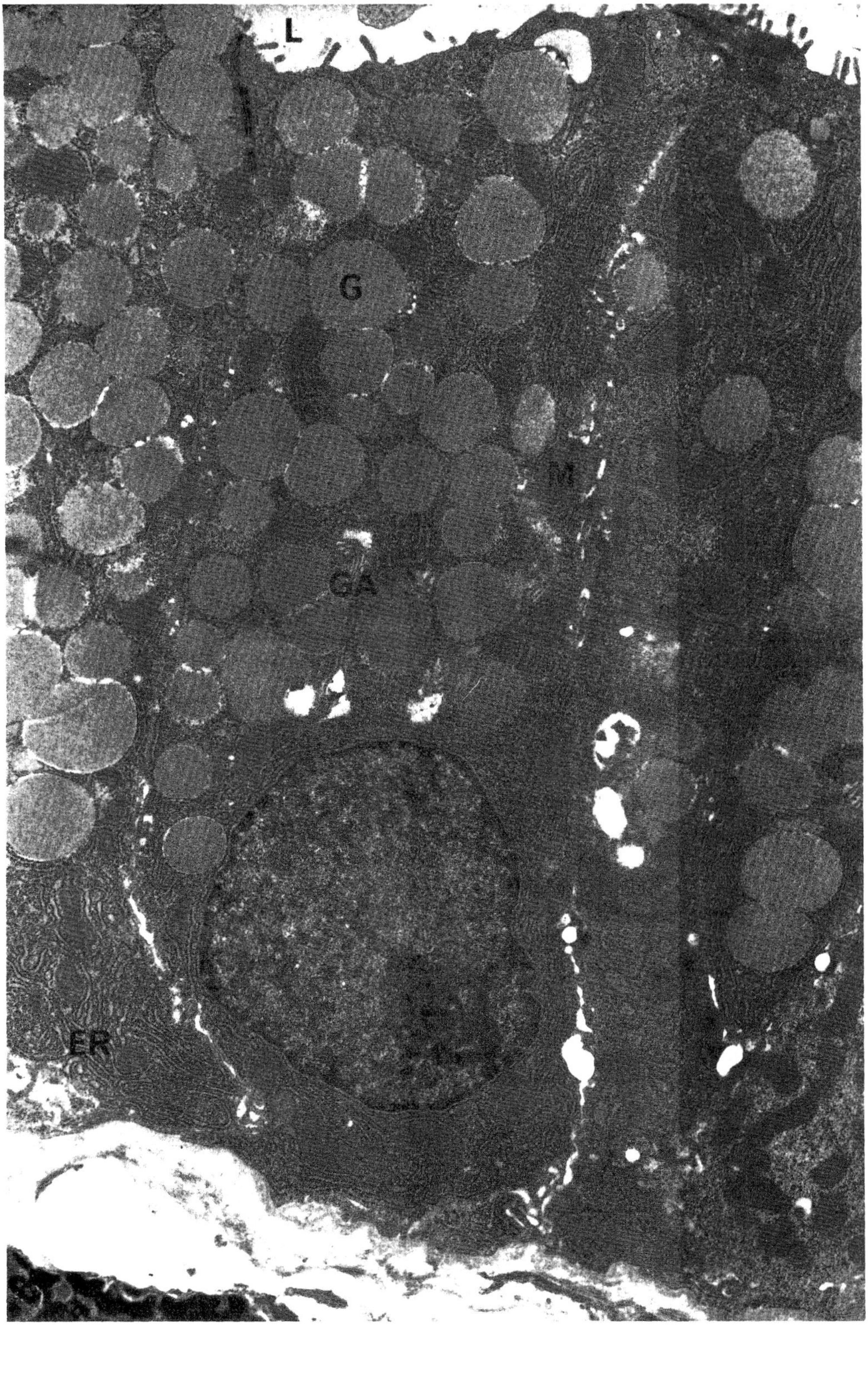
L
G
M
GA
ER

immuno-electron microscopy of human gastric biopsies in the perinuclear membrane, the tubulovesicular membrane, and the rough endoplasmic reticulum of these cells. It has been speculated that the neck parietal cells might be responsible for the production of intrinsic factor (64), but no firm data are available.

5.2 Zymogenic cells (Figs. 3-5, 3-6)

The zymogenic (chief) cells are confined to the oxyntic gland area, and they are the dominating cell type in the basal part of the glands. The zymogenic cells are much smaller than the parietal cells; available data from rats indicate a mean cell volume of about 500 μm^3 (65), i.e., less than half of the mean parietal cell volume in this species (see also 6).

The zymogenic cells are usually shaped like truncated pyramids or cubes. The luminal surface is provided with a few microvilli and is covered by a glycocalyx containing nonreducing terminal galactose residues (42). During stimulation the luminal surface area increases, presumably as an effect of incorporation of zymogen granule membrane during exocytosis (65). The most characteristic features of the zymogenic cells are the secretory granules and the rough endoplasmic reticulum. The secretory granules — often referred to as zymogen granules — occupy much of the supranuclear cytoplasm. These granules are round or oval with a homogeneous, moderately dense content. A more clear content, often in combination with fragmentation of the surface membrane of the granules, in most cases indicates poor fixation. During fasting conditions in the rat, the zymogen granules have an average diameter of about 0.8 μm, and they occupy about 30% of the cytoplasm. Stimulation to secretion results in smaller and fewer zymogen granules, whereas inhibition of secretion tends to increase the size and proportion of these granules (65). Secretion takes place by exocytosis (1). With immunohistochemical methods, pepsinogen can be demonstrated in the zymogen granules (66, 67). In some species, mucosubstances are also detected in the zymogen granules (10; see also 1).

The basal half of the zymogenic cells is dominated by a very extensive rough endoplasmic reticulum. This and the well-developed Golgi apparatus are reminiscent of other protein exporting cells.

In contrast to the parietal cells, the mitochondria are few, only 5% of the cytoplasm is taken up by these organelles. The nucleus is found in the basal part of the cell. In man, binucleate zymogen cells constitute 1% of the total number of zymogen cells; in the rat, binucleate zymogenic cells are roughly twice as large as mononucleate ones (65).

Zymogenic cells in the midportion of the oxyntic glands frequently display smaller and denser secretory granules than those in the bottom of the glands, and it has been proposed that the zymogen cells with dense granules might represent immature cells (10, 68). Extensive studies in rodents have demonstrated that most zymogenic cells in the adult gastric mucosa arise from mitotic replication of pre-existing zymogenic cells (16, 61, 68); only a small proportion are derived from mitotic stem cells or immature mucous neck cells. This proportion increases under abnormal conditions, such as in healing gastric ulcers (16, 69). For normal, proliferating zymogenic cells in mice, the average cell cycle duration has been estimated at about two months (70); for the zymogen cells in the golden hamster, the life span is 200 ± 100 days (57).

5.3. Mucous neck cells (Fig. 3-3)

These cells are found in large numbers in the neck and isthmus regions of all gastric glands. In the rat their mean volume has been calculated to be 180 μm^3, which is much less than the other exocrine cell types of the gastric mucosa (6). Their shape varies considerably: cuboidal, round, and pear-shaped forms are frequently seen.

The apical part of the cytoplasm contains secretory granules, which in general are some-

←

Figure 3-5. Zymogenic cells from human oxyntic mucosa during secretory rest. There is an abundance of rough endoplasmic reticulum (ER) and secretory granules (G). On the apical cell surface there are a few microvilli projecting into the gland lumen (L). GA, Golgi apparatus; M, mitochondrion. X 11,000.

what larger and less dense than those of the surface mucous cells. Many of the granules are biphasic, containing regions of different densities (10). Extensive ultrahistochemical studies have been carried out to characterize the mucosubstances in these cells (10, 11, 71). The secretory granules, in addition, display pepsinogen immunoreactivity (66). Mucous neck cells and more immature cells of the neck region often undergo mitosis, and it has been demonstrated that the dividing cells differentiate into epithelial cells of other types (1). Transitional types between the

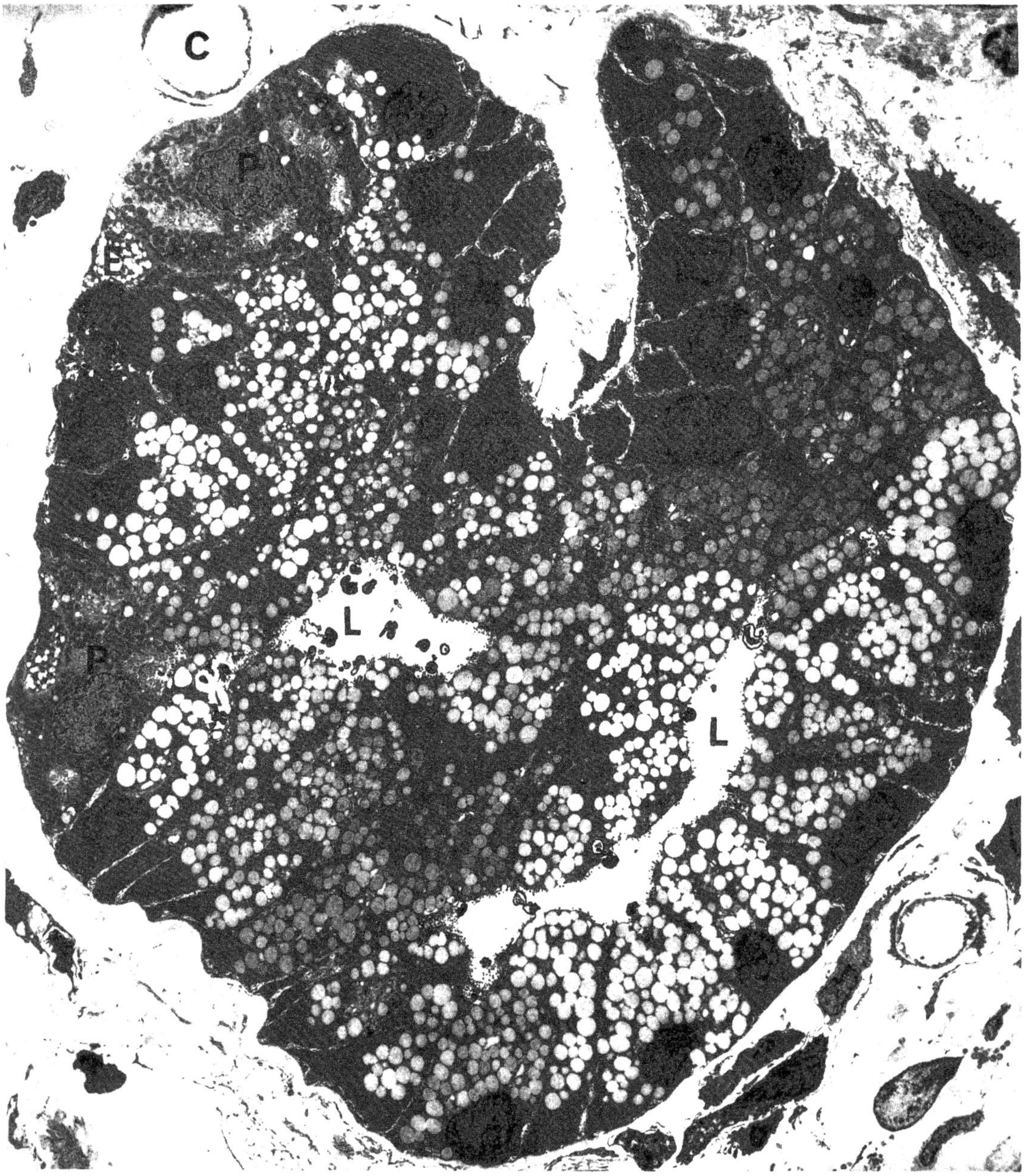

Figure 3-6. Survey of bottom of human oxyntic gland during maximal acid secretion. The gland lumen (L) has here divided into two branches. The great majority of the cells are zymogenic cells containing a large number of secretory granules. Two parietal cells (P) are also observed, seemingly without contact with the main gland lumen. A few endocrine cells (E) are seen at the periphery of the gland. C, capillary. X 1,900.

mucous neck cells and other exocrine cells are often observed (10, 72; see also 1).

5.4. Endocrine cells (Fig. 3-6)

In man, the gastric mucosa contains several types of endocrine cells. However, since these cells will be dealt with extensively in Chapter 13 of this book, only some quantitative data will be given here.

The endocrine cells of the human oxyntic mucosa constitute only 0.3% of the total mucosal volume, but approximately 4% of the total number of epithelial cells (73). The most common types of endocrine cells are the ECL cells and D cells, which in man comprise 40% to 60% and 10% to 20% of the endocrine cell population, respectively (74). In rats, Alumets et al. (75) found that 65% of the endocrine cells were ECL-cells and 25% A-like cells. In the antrum, the G cells dominate (in man 40% to 60% of the endocrine cell population) over the EC cells (20% to 40%), the D cells (10% to 30%), and the P and D_1 cells (~5%) (74). Arnold et al. calculated that the ratio of G/D-cells in healthy subjects is 4.4 (76). In normal rats, Reinhardt et al. (6) estimated that the G cells occupied 0.3% of the antral mucosal volume. In man, Oberholzer et al. (77) found the corresponding figure to be 1.6% and the numerical density of the G cells to be 13,104 cells per mm^3 of antral mucosa. The total number of G cells in normal human stomachs has been estimated at between 11 and 42 million (78).

The endocrine cells are small: in the rat, the mean cell profile area for the parietal cells is about 120 μm^2 (54), while the corresponding figure for the ECL cells is only 26 μm^2, for the A-like cells 31 μm^2, and for the G cells 58 μm^2 (75). Assuming that the G cells are spheres (they are in reality elongated and polyhedral), this figure for the mean cell profile area of the rat G cell would correspond to a mean cell volume of about 325 μm^2. In healthy persons, Oberholzer et al. (54) calculated a mean G cell volume of 1,221 μm^3. This fourfold difference between the rat and man cannot be explained solely by methodological differences between the studies.

The close proximity between the somatostatin (D) cells and the parietal cells has been regarded as an indication of a regulatory function, whereby the local release of somatostatin would inhibit parietal cell secretion (79, 80). However, since the D cells only constitute a very small proportion of the epithelial cells, it seems doubtful that such direct inhibition would play any significant role.

6. Pyloric glands

In the antral mucosa the gastric pits often penetrate more than half of the epithelial layer. Both the mucosal surface and the pits are covered by surface mucous cells. The short pyloric glands consist mainly of mucous cells similar to the mucous neck cells (10, 81, 82, 83). In addition to mucus, the secretory granules also contain pepsinogen (67) and lysozyme (84). Parietal cells are encountered, but rarely; zymogenic cells are absent. The endocrine cells have been discussed above.

7. Other epithelial cells

The fibrillo-vesicular cells are pear-shaped cells observed interspersed between the mucous cells of the gastric groove, which is the boundary between the forestomach and the glandular stomach of the rat. On rare occasions these cells are also found in other parts of the gastrointestinal tract. They are characterized by long, thick microvilli which project from the apical cell surface. Each microvillus contains a prominent bundle of microfilaments, which extends deep into the cytoplasm. There is also a system of 0.1 μm-wide interconnected vesicles or tubules in the cytoplasm (20). The role of these cells remains unknown; tentatively, they have been attributed a receptor function (1).

Lymphocytes are frequently seen within the epithelial layer, as well as a few mast cells. Ciliated cells have also been found in the human gastric mucosa (85).

8. Lamina propria and morphological aspects of stimulation

The tissue between the gastric glands — referred to as the lamina propria — contains connective tissue cells, smooth muscle cells, leukocytes, lymphocytes, plasma cells, and mast cells. Occasionally, particularly in the antrum, small accumulations of lymphoid cells are encountered. In human oxyntic mucosa the lamina propria occupies about one-third of the mucosal volume (43).

The blood capillaries of the lamina propria arise from arteries which divide around the base of the glands. Scanning electron microscopical studies of corrosion casts from rats have revealed two to four capillaries around each gastric gland (86). These capillaries, which are fenestrated, form a honeycomb network as they approach the basal aspect of the surface mucous cells. Infrequent, straight collecting venules, 20–50 μm in luminal diameter, drain the blood into the veins of the submucosa. Gannon et al. (87) pointed out that this microvascular architecture would expose the surface layers of the mucosa to the alkaline tide that results from HCl production in the subjacent parietal cells. This should promote HCO_3^- secretion from the surface mucous cells (see Section 3) when HCl secretion takes place.

The nerve fibers of the gastric mucosa are unmyelinated but are often seen in close contact with Schwann cells. Electron microscopy of rat oxyntic mucosa has revealed three types of nerve endings, which, based on the structure of the synaptic vesicles, were characterized as cholinergic, adrenergic, and peptidergic nerves. Throughout the mucosa the cholinergic nerves were predominant as compared with adrenergic nerves; the peptidergic nerves were rare (88). The Schwann cells and the nerve fibers are surrounded by a basal lamina, similar to that surrounding the glands and the capillaries. Usually, there is a distance between the nerve fibers and the target cells — gland cells or capillaries — of at least 100 nm (1). However, closer contact has been observed in a few cases (1, 89).

In this context, it should be recalled that impulses can pass between the epithelial cells of the gastric mucosa, since the cells are electrically coupled (90). It is assumed that the gap junctions that are present between gastric epithelial cells (91) play an important role in this communication.

Kiseleva and Bersimbaev (92) attempted to locate ^{3}H-pentagastrin given to rats. Light and electron microscopical autoradiography disclosed a selective site in the peptic and the endocrine cells of the oxyntic mucosa. The parietal cells were not labelled in this study.

Similar autoradiographic studies have been carried out to localize muscarinic acetylcholine receptors in the rat gastric mucosa. These investigations demonstrated the presence of ^{3}H-quinuclinidyl benzilate, a potent muscarine antagonist, at the basolateral plasma membrane of the parietal cells and at the basal plasma membrane of the blood capillaries (93).

The location of injected ^{14}C-histamine in rats has been studied by light microscopic autoradiography. The labelling of the gastric mucosa was not pronounced. In the oxyntic area, the heaviest uptake was in the muscularis mucosae and the artery and arteriole walls (94). In contrast, there was a preferential uptake of ^{3}H-metiamide and ^{3}H-cimetidine in the parietal cell cytoplasm.

References

1. Helander HF: The cells of the gastric mucosa. *Internat Rev Cytol* 70: 217–298, 1981.
2. Ito S: Functional gastric morphology. In: Physiology of the Gastrointestinal Tract, LR Johnson, J Christensen, (eds), New York: Raven Press, p 517–550, 1981.
3. Rubin W, Ross LL, Sleisenger MH, Jeffries GH: The normal human gastric epithelia. *Lab Invest* 19: 598–626, 1968.
4. Dalton AJ, Kahler H, Lloyd BJ: The structure of the free surface of a series of epithelial cell types in the mouse as revealed by the electron microscope. *Anat Rec* 111: 67–77, 1951.
5. Bloom W, Fawett DW: A Textbook of Histology. Philadelphia: Saunders, 1969.
6. Reinhart WH, Müller O, Halter F: Influence of long-term 16, 16-dimethyl prostaglandin E_2 treatment on the rat gastrointestinal mucosa. *Gastroenterology* 85: 1003–1010, 1983.
7. Zalewsky CA, Moody FG, Allen M, Davis EK: Stimulation of canine gastric mucus secretion with intraarterial acetylcholine chloride. *Gastroenterology* 85: 1067–1075, 1983.
8. Helander HF, Durbin RP: Localization of ouabain binding sites in frog gastric mucosa. *Am J Physiol* 243: G505–G510, 1982.
9. Meyer RA, McGinley D, Posalaky Z: The gastric mucosal barrier of intercellular junctions in the dog. *J Ultrastruct Res* 86: 192–201, 1984.
10. Sato A, Spicer SS: Ultrastructural cytochemistry of complex carbohydrates of gastric epithelium in the guinea

pig. *Am J Anat* 159: 307–329, 1980.
11. Spicer SS, Katsuyama T, Sannes PL: Ultrastructural carbohydrate cytochemistry of gastric epithelium. *Histochem J* 10: 309–331, 1978.
12. Flemström G, Turnberg LA: Gastroduodenal defense mechanisms. *Clin Gastroenterol* 13: 327–354, 1984.
13. Sugai N, Ito S: Carbonic anhydrase, ultrastructural localization in the mouse gastric mucosa and improvements in the technique. *J Histochem Cytochem* 28: 511–525, 1980.
14. Dabiké M, Koenig CS: Intermediate filaments of the cytoskeleton in glandular cells of the rat fundic mucosa: Immunofluorescence and electron microscopy study. *Anat Rec* 207: 297–308, 1983.
15. Helander HF, Olivecrona T: Lipolysis and lipid absorption in the stomach of the suckling rat. *Gastroenterology* 59: 22–35, 1970.
16. Willems G: Cellular kinetics and gastric mucosal disease. In: Magen und Magenkrankheiten W Domschke, KG Wormsley (eds), Stuttgart: Thieme, p 195–204, 1981.
17. Tatsumi H, Fujita H, Tamura S: Electron microscopic studies on the physiological cell loss in the gastric mucosa of the golden hamster. *Cell Tiss Res* 239: 343–347, 1985.
18. Mackercher PA, Ivey KJ, Baskin WN, Krause WJ: A scanning electron microscopic study of normal human oxyntic mucosa using blunt dissection and freeze fracture. *Dig Dis* 23: 449–459, 1978.
19. Kataoka K, Takeoka Y, Hirano S: Electron microscopic observations of surface mucous cells in the mouse gastric mucosa during physiological degeneration and extrusion. *Arch Histol Jap* 48: 327–339, 1985.
20. Wattel W, Greuze JJ: The cells of the rat gastric groove and cardia. *Cell Tiss Res* 186: 375–391, 1978.
21. Cox AJ: Stomach size and its relation to chronic peptic ulcer. *Arch Pathol* 54: 407–422, 1952.
22. Naij SR, Bajaj SC, Goyal RK, Gupta DN, Chuttani HK: Parietal cell mass in healthy human stomach. *Gastroenterology* 61: 682, 1971.
23. Baron JH: Pathophysiology of gastric and pepsinogen secretion. In Magen und Magenkrankheiten W Domschke, KG Wormsley (eds), Stuttgart: Thieme, p 131–149, 1981.
24. Helander HF, Vanderwee M, Leth R, Olbe L: Quantitative electron microscopical studies of normal human oxyntic mucosa. *Gastroenterology* 84: 1184, 1983.
25. Aase S, Dahl E, Roland M, Hars R: Morphometric studies of parietal cells during basal conditions and during stimulation with pentagastrin in healthy subjects. *Scand J Gastroent* 18: 913–923, 1983.
26. Hogben CA, Kent TH, Woodward PA, Sill AJ: Quantitative histology of the gastric mucosa: Man, dog, cat, guinea pig, and frog. *Gastroenterology* 67: 1143–1154, 1974.
27. Aase S, Roland M, Liavåg I, Dahl E: Morphometric studies of human parietal cells before and 7 days after proximal gastric vagotomy. *Scandinav J Gastroenterol* 19: 697–706, 1984.
28. Helander HF: Ultrastructure and function of gastric parietal cells in the rat during development. *Gastroenterology* 56: 35–52, 1969.
29. Oberholzer M, Rohr H-P: Liver morphometry. In: Morphometry in Diagnostic Pathology, JPA Baak, J Oort (eds), Berlin: Springer, p 126–131, 1983.
30. Fleischer M, Warmuth H: Ultrastructural morphometric analysis on normal loaded and hypertrophied human myocardial left ventricles. In: Myocardial Biopsy. Diagnostic Significance. HD Bolte (ed), Berlin: Springer, p 49–52, 1980.
31. Stachura J, Konturek SJ, Ciezkowsky M, Dabros W, Zakrzewska, Konturek J: Comparison of the effect of omeprazole — a substituted benzimidazole — and ranitidine — a potent H_2-receptor antagonist — on histamine-induced gastric acid secretion and the ultrastructure of the canine parietal cells. *Hepato-gastroenterol* 30: 205–210, 1983.
32. Vial JD, Orrego H: Effects of molar concentration of fixative upon the fine structure of oxyntic cells. *Protoplasma* 61: 164–180, 1966.
33. Frexinos J, Carballido M, Louis A, Ribet A: Effects of pentagastrin stimulation on human parietal cells. *Dig Dis* 16: 1065–1074, 1971.
34. Morozov IA, Kovanova LA, Vodolagin VD, Minayilenko: Ultrastructure of the parietal cells of the stomach and their functional activity. *Biull Eksp Biol Med* 79: 14–18, 1975.
35. Rubin W: Enzyme cytochemistry of gastric parietal cells at the fine structure level. *J Cell Biol* 42: 332–338, 1969.
36. Smolka A, Helander HF, Sachs G: Monoclonal antibodies against gastric H^+, K^+-ATPase. *Am J Physiol* 245: G589–G596, 1983.
37. Golgi C: Sur la fine organisation des glandes peptiques des mammifères, *Arch Biol* 19: 448–453, 1893.
38. Berglindh T, DiBona DR, Ito S, Sachs G: Probes of parietal cell function. *Am J Physiol* 238: G165–176, 1980.
39. Sedar AW: Electron microscopic demonstration of polysaccharides associated with the acid-secreting cells of the stomach after "inert dehydration." *J Ultrastruct Res* 28: 112–124, 1969.
40. Sato A, Spicer SS: Ultrastructural cytochemistry of complex carbohydrates of gastric epithelium in the guinea pig. *Am J Anat* 159: 307–329, 1981.
41. Vial JD: The secretory membrane of the parietal cell. A look at an unusual membrane behaviour. *Arch Biol Med Exp* 15: 475–480, 1982.
42. Albeidi FM, Barsotti P, Mingazzini P, Marinozzi V: Visualization of the secretory canaliculi of human parietal cells with a peroxidase-labelled peanut lectin. *Cell Tiss Res* 239: 447–450, 1985.
43. Helander HF, Leth R, Olbe L: Stereological investigations on normal human gastric mucosa. *Anat Rec* 216; 373–380, 1986.
44. Tarnawski A, Ivey KJ, Krause WJ, Sherman D, Burks M, Hewett J: Quantitative analysis of human parietal cells after pentagastrin. Correlation with gastric potential difference. *Lab Invest* 42: 420–426, 1980.
45. Black JA, Forte TM, Forte JG: Structure of oxyntic cell membranes during conditions of rest and secretion of HCl as revealed by freeze-fracture. *Anat Rec* 196: 163–172, 1980.
46. Sato A, Spicer SS, Tashian RE: Ultrastructural localization of carbonic anhydrase in gastric parietal cells with the

immunoglobulin-enzyme bridge method. *Histochem J* 12: 651–659, 1980.
47. Gonzales A, Garrido J, Vial JD: Epidermal growth factor inhibits cytoskeleton related changes in the surface of parietal cells. *J Cell Biol* 88: 198–114, 1981.
48. Helander HF: Parietal cell structure during inhibition of acid secretion. *Scand J Gastroenterol* (suppl) 101: 21–26, 1984.
49. Helander HF, Sundell GW: Ultrastructure of inhibited parietal cells in the rat. *Gastroenterology* 76: 1064–1071, 1984.
50. Helander HF, Ramsay C-H, Regårdh C-G: Localisation of omeprazole and metabolites in the mouse. *Scand J Gastroenterol* (suppl) 108: 95–104, 1985.
51. Black JA, Forte TM, Forte JG: Inhibition of HCl secretion and the effects on ultrastructure and electrical resistance in isolated piglet gastric mucosa. *Gastroenterology* 81: 509–519, 1981.
52. Gibert AJ, Hersey SJ: Morphometric analysis of parietal cell membrane transformations in isolated gastric glands. *J Membr Biol* 67: 113–124, 1982.
53. Helander HF, Johansson C, Blom H, Uribe A: Trophic actions of E_2 prostaglandins in the rat gastrointestinal mucosa. *Gastroenterology* 89: 1393–1399, 1985.
54. Helander HF: Stereological changes in rat parietal cells after vagotomy and antrectomy. Gastroenterology 71: 1010–1018, 1976.
55. Aase S, Roland M, Liavåg I, Dahl E: Stereological analysis of human parietal cells before and 6 months after vagotomy. *Scand J Gastroenterol* 19: 697–706, 1984.
56. Ragins H, Wincze F, Liu SM, Dittbrenner M: The origin and survival of gastric parietal cells in the mouse. *Anat Rec* 162: 99–109, 1968.
57. Hattori T, Fujita S: Tritiated thymidine autoradiographic study on cellular migration in the gastric gland of the golden hamster. *Cell Tiss Res* 172: 171–184, 1976.
58. Lipkin M: Proliferation and differentiation of mucus containing cells in normal and diseased mucosa. *Biol et Gastro-entérol* 5: 550c, 1972.
59. Aigne G, Poynter D: H_2 receptor antagonists: Ultrastructure of canine parietal cells after long-term treatment with ranitidine. *Scand J Gastroent* 16 (suppl) 70: 143–154, 1981.
60. Morozov IA: Ultrastructural topography of parietal cells of the gastric mucosa. *Biull Eksper Biol Med* 82: 1390–1394, 1976.
61. Tamura S, Fujita H: Fine structural aspects on the renewal and development of surface mucous cells and glandular cells of the gastric body of the adult golden hamster. *Arch Histol Jpn* 46: 501–521, 1983.
62. Coulton GR, Firth JA: Cytochemical evidence for functional zonation of parietal cells within the gastric glands of the mouse. *Histochem J* 15: 1141–1150, 1983.
63. Levine JS, Nakane PK, Allen RH: Immunocytochemical localization of human intrinsic factor: The non-stimulated stomach. *Gastroenterology* 79: 493–502, 1980.
64. Gérard A: Etudes histochemiques et histophysiologiques des glycoprotéines de la muqueuse gastrique chez le chien. *Arch Biol* (Liége) 79: 1–104, 1968.
65. Helander HF: Quantitative ultrastructural studies on rat gastric zymogen cells under different physiological and experimental conditions. *Cell Tiss Res* 189: 287–303, 1978.
66. Suzuki S, Furihata C, Tsuyama S, Murata F: Immunocytochemical localization of pepsinogen in rat stomach. *Histochemistry* 79: 167–176, 1983.
67. Hirsch-Marie H, Loisillier F, Touboul JP, Burtin P: Immunochemical study and cellular localization of human pepsinogens during ontogenesis and in gastric cancers. *Lab Invest* 34: 623–632, 1976.
68. Kataoka K: Electron microscopic observations on cell proliferation and differentiation in the gastric mucosa of the mouse. *Arch Histol Jpn* 32: 251–273, 1970.
69. Helander HF: Morphological studies on the margin of gastric corpus wounds in the rat. *J Submicrosc Cytol* 15: 627–643, 1983.
70. Willems G, Galand P, Vansteenkiste Y, Zeitoun P: Cell population kinetics of zymogen and parietal cells in the stomach of mice. *Z Zellforsch* 134: 505–518, 1972.
71. Suzuki S, Tsuyama S, Murata F: Post-embedding staining of rat gastric mucous cells with lectins. *Histochemistry* 73: 563–575, 1983.
72. Suzuki S, Tsuyama S, Murata F: Cells intermediate between mucous neok cells and chief cells in rat stomach. *Cell Tiss Res* 233: 475–484, 1983.
73. Helander HF: Stereological data on parietal and endocrine cells of the gastric corpus in patients treated with omeprazole. *Gastroenterology* 88: 1416, 1985.
74. Solcia E, Capella C, Sessa F, Rindi G, Cornaggia M, Riva C: Gastric carcinoids and related endocrine growths. *Digestion* (in press).
75. Alumets J, El Munshid HA, Håkanson R, Liedberg G, Oscarson J, Rehfeld JF, Sundler F: Effect of antrum exclusion on endocrine cells of rat stomach. *J Physiol* 286: 145–155, 1979.
76. Arnold R, Hülst MV, Neuhof C, Schwarting H, Becker HD, Creutzfeldt W: Antral gastrin-producing G-cells and somatostatin-producing D-cells in different states of gastric acid secretion. *Gut* 23: 285–291, 1982.
77. Oberholzer M, Heitz PU, Kayasseh L, Gyr K, Stadler GA, Rohr HP: Immunozytochemie und morphometrie am modell der gastrinzelle der normalen menschlichen magenantrumschleimhaut und bei ulcus duodeni. *Acta Histochem* (suppl) 24: 283–288, 1981.
78. Keuppens F, Williams G, De Graef J, Woussen-Colle MC: Antral gastrin cell hyperplasia in patients with peptic ulcer. *Ann Surg* 191: 276–281, 1980.
79. Kusumoto Y, Iwanaga T, Ito S, Fujita T: Juxtaposition of somatostatin cell and parietal cell in the dog stomach. *Arch Histol Jpn* 42: 459–465, 1979.
80. Larsson L-I, Goltermann N, De Magistris L, Schwartz TW: Somatostatin cell processes as pathways for paracrine secretion. *Science* 205: 1393–1395, 1979.
81. Helander HF: Ultrastructure of secretory cells in the pyloric gland area of the mouse gastric mucosa. *J Ultrastruct Res* 10: 145–154, 1964.
82. Krause WJ, Ivey KJ, Baskin WN, MacKercher P: Ultrastructure of the human pyloric glands with emphasis on the mucous cell component. *Acta Anat* 99: 1–10, 1977.
83. Zeitoun P, Potet F, Zylberberg L: Histologie de la

muqueuse antrale. *J Fr Gastro-Entérol*, 10–18, 1969.
84. Klockars M, Reitamo S: Tissue distribution of lysozyme in man. *J Histochem Cytochem* 18: 340–000, 1975.
85. Okuda T, Ogata T: An electron microscopic study of the ciliated cells in the human gastric mucosa. *Arch Histol Jpn* 39: 149–156, 1976.
86. Hattori T: On cell proliferation and differentiation of the fundic mucosa of the golden hamster. *Cell Tiss Res* 148: 213–226, 1974.
87. Gannon B, Browning J, O'Brien P: The microvascular architecture of the glandular mucosa of rat stomach. *J Anat* 135: 667–683, 1982.
88. Nakamura M, Watanabe N, Tsukada N, Oda M, Tsuchiya M: Demonstration of the adrenergic nerves in the rat gastric mucosa — a histofluorescence and electron microscopic study in comparison with the distribution of the cholinergic nerves. *Okajimas Folia Anat Jpn* 59: 65–86, 1982.
89. Kyösola K, Rechardt L, Veijola L, Waris T, Pentillä O: Innervation of the human gastric wall. *J Anat* 131: 453–470, 1980.
90. Blum AL, Hirschowitz BI, Helander HF, Sachs G: Electrical coupling of isolated cells of *Necturus* gastric mucosa. *Biochim Biophys Acta* 241: 261–272, 1971.
91. Peracchia C: Gap junctions. Structural changes after uncoupling procedures. *J Cell Biol* 72: 628–641, 1977.
92. Kiseleva EV, Bersimbaev RI: Studies on localization of ^{3}H-pentagastrin in the region of fundal glands in the rat gastric mucosa using light and electron autoradiography. *Tsitologia* 17: 726–728, 1975.
93. Nakamura M, Oda M, Yonei Y, Tsukada N, Watanabe N, Komatsu H, Tsuchiya M: Demonstration of the localization of muscarinic acetylcholine receptors in the gastric mucosa — light and electron microscopic autoradiographic studies using ^{3}H-quinuclidinyl benzilate. *Acta Histochem Cytochem* 17: 297–309, 1984.
94. Cross SAM: Localization of histamine and histamine H_2-recptor antagonists in the gastric mucosa. *Histochem J* 9: 619–644, 1977.

Author's address:
Hässle Gastrointestinal Research Laboratories
Department of Biology
S-431 83 Mölndal, Sweden

CHAPTER 4

Ultrastructural aspects of the turnover of stomach mucosal epithelium

HARUYUKI TATSUMI & HISAO FUJITA

1. Introduction

The epithelia of the gastric body and fundus are comprised of gastric pits (foveolae) and fundic glands. The transitional portion between the pit and the gland, corresponding to the isthmus and neck region, is the generative cell zone (1,2), where mitosis takes place. Most cells generated there migrate upwards along the gastric pit to the surface of the stomach to become surface mucous cells and are finally disposed of into the gastric lumen; whereas a few cells in the generative zone move downwards to differentiate into mucous neck cells, parietal cells, and chief cells, and eventually cell death occurs at the base of the gland, though their turnover is much slower than that of surface mucous cells.

Bizzozero (3) first pointed out the occurrence of mitosis at the bottom of the gastric pit, corresponding to the generative cell zone, and concluded that the cells at the bottom are the source of the surface cells. Noting the constant occurrence of mitosis in the gastrointestinal tract, Leblond and Stevens (4) and Stevens and Leblond (5) examined the mitotic activity of the mucosa and first suggested that there should be a loss of cells balancing the production of newly formed cells under physiological conditions. These suggestions were further confirmed with convincing data obtained by autoradiography using tritiated thymidine (6), which elucidated the renewal rate and the upward migration of the surface mucous cells. This means that the surface mucous cells are arranged in chronological order along the pit, and they have a life span that loses a few days. Therefore, detailed observations along the gastric pit should give us some clues to understanding the cell renewal phenomenon (including differentiation, migration, and degeneration) and some protective mechanisms of the tissue.

The surface and pits of the stomach mucosa are lined with surface mucous cells, which are varied in morphology according to cell position in the pit. Before getting onto the main subject matter of this chapter, we will discuss the differences in structure and depth of the pit among animal species. The portion from the uppermost parietal cell to the uppermost mucous neck cell is referred to as the isthmic region, and immature cells are found here. The isthmic region is obscure in some animals, and there is no isthmic region in dogs (7) and golden hamsters (2), because the boundary between the surface mucous cells and mucous neck cells coincides with the uppermost parietal cells in these animals (8). In mice and rats, the depth of the pit is shorter than that in hamsters, which is rather similar to that in human. The depth of the pit is reflected in the renewal rate of the surface mucous cells: one day in mice (9), three days in rats (5), four days in hamsters (1), and four to six days in human (10). Therefore, hamsters are often used in experimental studies investigating stomach structure as models of the human stomach (1,2,11,12,13,14).

The following descriptions are based mainly on our data on the hamster stomach.

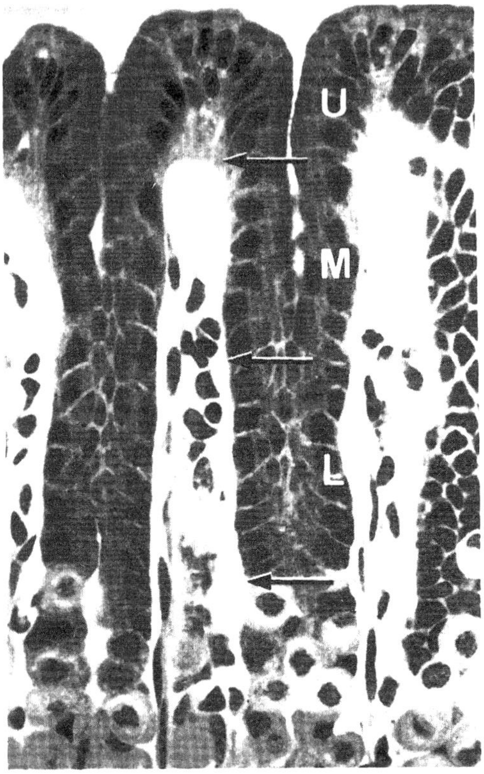

Figure 4-1. Gastric pits and glands of the golden hamster. Periodic acid-Schiff (PAS)-hematoxylin staining. The gastric pit is subdivided into upper (U), middle (M) and lower (L) portions. X 500.

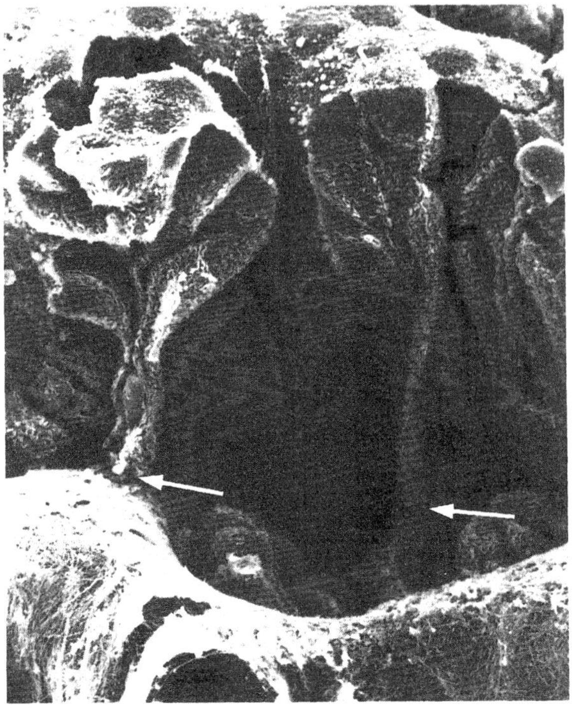

Figure 4-2. Scanning electron micrograph of the surface of the gastric wall. Some surface mucous cells are removed mechanically from the underlying dense connective tissue layer, which is continuous downwards to the fibrous stromal sheaths encasing each gastric pit. The center of the interfoveolar fibrous layer (dense connective tissue layer) between the pits bulges towards the gastric lumen as a ridge (arrows) on which the basal parts of the tall funnel-shaped cells converge. X 2,100.

2. Immature and mature surface mucous cells

For convenience in describing the morphological changes of the cell, the gastric pit of the hamster is subdivided into three parts: the upper, middle, and lower portions (Fig. 4-1). The surface mucous cells of the upper portion facing the lumen of the stomach are called *covering mucous cells* or *interfoveolar cells*. Between the pits, the interfoveolar cells are located on a connective tissue layer, which we call the interfoveolar ridge of the fibrous tissue layer (Fig. 4-2). Its precise structure will be described later in this chapter.

Immature surface mucous cells (Figs. 4-1, 4-3, 4-4) are situated in the lower portion of the pit and are characterized by an abundance of free ribosomes and a scarcity of mucous granules. The nucleus is located in the basal half of the cell, and the small Golgi complex is localized in the supranuclear region. The rough endoplasmic reticulum is also very poorly developed. Some of these cells are labelled two hours after a pulse injection of ^{3}H-TdR, and the mitotic figures of the immature surface mucous cells are sometimes observed in this area. It is difficult to find an undifferentiated cell without any kind of secretory granule in the lower gastric pit. Cells in the gastric gland below the pit also sometimes show the ability to take up ^{3}H-TdR and go through mitosis. The other characteristic feature of this generative cell zone is the fact that huge phagosomes are occasionally seen in the immature cells (12). This finding means that some cells of this zone might be degenerated.

Cells migrate from the lower to the upper portion of the pit, towards the surface of the gastric wall, and gradually maturate, showing an increased number of mucous granules, which are

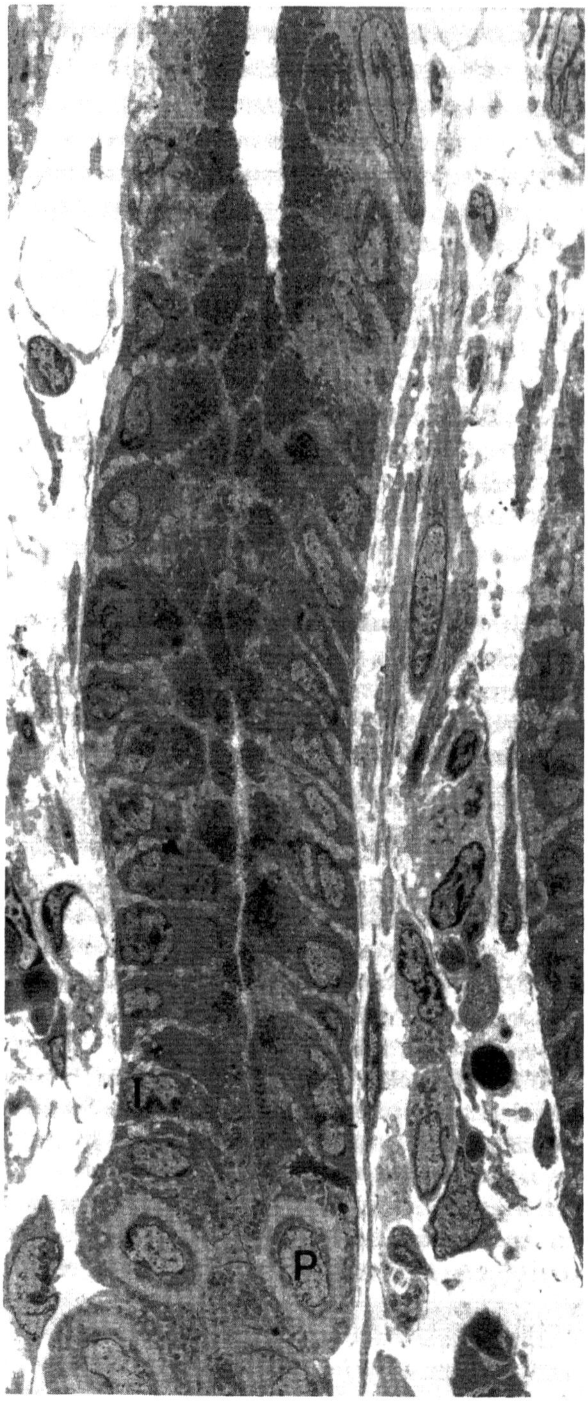

Figure 4-3. Low magnification electron micrograph of an area from the lower to middle portion of the gastric pit. Secretory granules in surface mucous cells increase gradually in number as these cells migrate upwards. In the lamina propria, there lie many kinds of connective tissue cells, small capillaries, and bundles of the smooth muscular fibers running perpendicularly to the lamina muscularis mucosae. I, immature surface mucous cell; P, parietal cell. X 1,500.

round or oval in shape (400–800 nm) and packed densely together in the apical cytoplasm (Fig. 4-5). A few mucous granules can also be seen at the supranuclear region, corresponding to the Golgi area. These are immature granules, round in shape, and lower in electron density than mature granules in the apical cytoplasm.

These surface mucous cells are well known to be of a simple columnar type, but each real contour is far from columnar and is markedly changed during migration from the middle portion of the pit to the surface of the gastric lumen. The cells, arranged in a tube at the lower and middle portion of the gastric pit, are rectangular in shape in a longitudinal section along the gastric pit and are sectorial in shape in a transverse section across the pit. As a whole, the shape of this cell is truncated or wedge-shaped, showing a narrow apex about 2 by 5 μm in diameter, a wide base about 5 by 13 μm in diameter, and a low height of 15–20 μm. In contrast, the interfoveolar cells are tall and funnellike, with an extremely narrow basal part of cytoplasm (Fig. 4-2).

3. Interfoveolar surface mucous cells

Near the top of the pit, the surface mucous cells become slender and funnellike and show well-developed interdigitations and basal infoldings (Fig. 4-6). The covering mucous cells, making up the surface of the gastric wall, ultimately have a rather wide apex, about 6 by 9 μm in diameter, a narrow base, about 1 by 3 μm in diameter, and a tall height of 30–40 μm. Though the basal part of the cell is very small at the interfoveolar ridge between the pits (Fig. 4-2), the complete detachment of the basal plasma membrane from the basal lamina is never seen (Figs. 4-6, 4-7).

3.1. Dislocation of the Golgi apparatus in the interfoveolar cell

Along with the aforementioned changes in cell contour, the nucleus moves upward to the apical half of the cytoplasm just beneath the packed mucous granules, assuming an elongated form, while the Golgi apparatus — a few mucous granules and lysosomes, which are usually located in the supranuclear cytoplasm — becomes situated lateral to or under the nucleus (13, 14, 15, 16). The mucous granules in the infranuclear cytoplasm of the tall funnel-shaped cells are visible by periodic acid-Schiff staining. When viewed with the electron microscope, they appear to be immature secretory granules localized near the Golgi apparatus, for they are more round in shape and relatively electron lucent in comparison with the apical mature ones. Some of the immature mucous granules in the infranuclear cytoplasm contain heterogeneously electron-dense materials and are occasionally associated with dense bodies and small vesicles.

As mentioned above, the reverse relationship in localization between the nucleus and Golgi apparatus is seen in tall funnel-shaped cells at the interfoveolar portion. This kind of finding in the stomach of domestic fowl was reported by Golgi himself (15, 16) in 1909 using the silver impregnation method and later by D'Agata (17) and Hibbard (18), but it wasn't until the advent of the electron microscope that it was rediscovered and reinvestigated (13, 14). In the stomach surface mucous cell, the infranuclear Golgi apparatus is small and compact and is surrounded by many small vesicles, immature granules, and lysosomes. This phenomenon is intimately related with the degeneration of the cell, in which the nucleus moves upward. A similar finding has been reported in intestinal epithelial cells damaged by ^{32}P administration (19).

3.2. Lysosomes and lipid droplets

Lysosomes in the tall funnel-shaped cell are of greater number and size, and the activities of acid

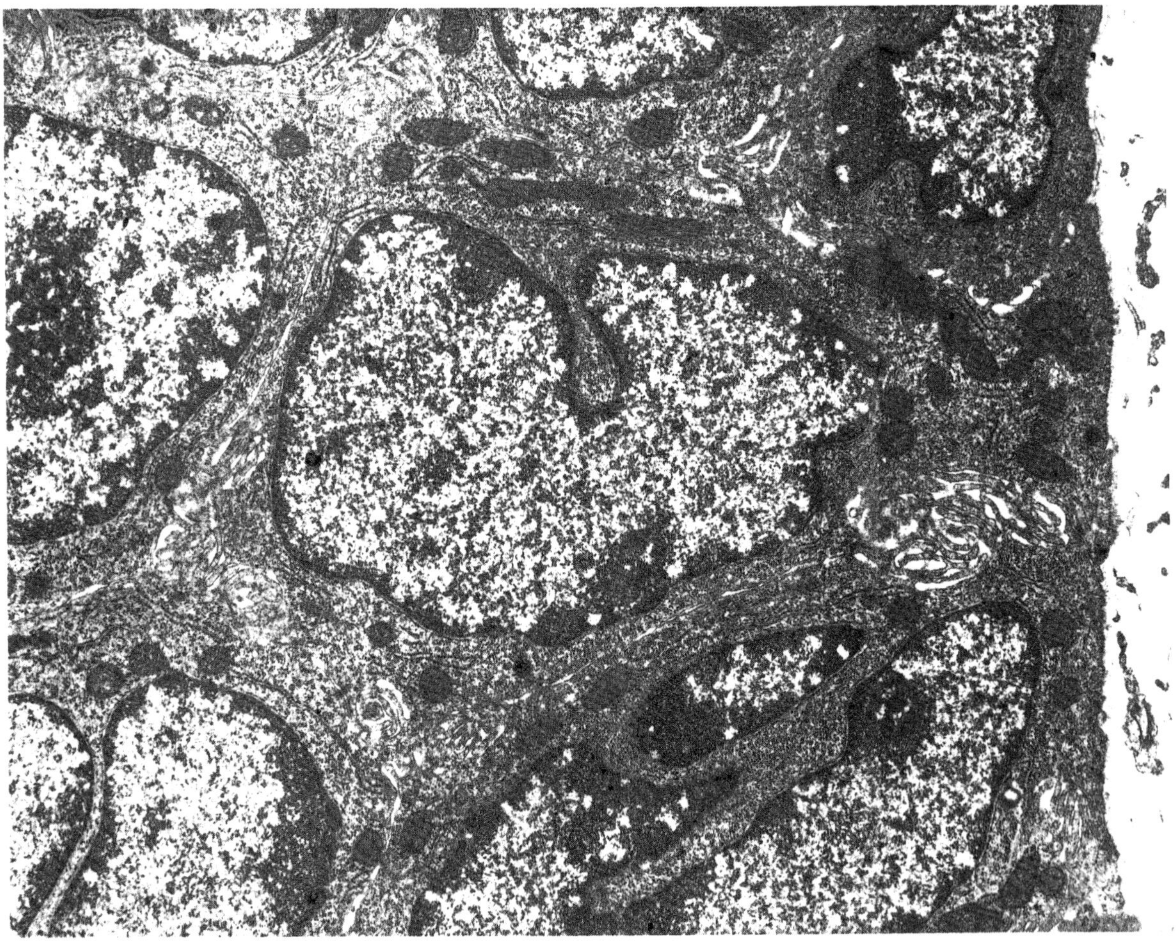

Figure 4-4. Immature surface mucous cells in the lower portion of the gastric pit. The cytoplasms of these cells contain abundant free ribosomes. X 11,000.

phosphatase are restricted within these organelles (13, 14, 20, 21) (Fig. 4-8). They often house rough endoplasmic reticulum, myelinlike structures, and lipid droplets (13, 14). In addition, mucous granules which are to be disposed of, are also seen in the lysosomes. This phenomenon is regarded as crinophagy and was reported in the mammotrophs of the anterior pituitary gland by Smith and Farquhar (22), who postulated that the appearance of lipid droplets in the lysosome marks the last stage of growth of secondary lysosomes. We maintain that the production and transportation of the secretory granules are disordered in the funnel-shaped cells.

It is not likely that the increase in size and number of the lysosomes in the interfoveolar cells

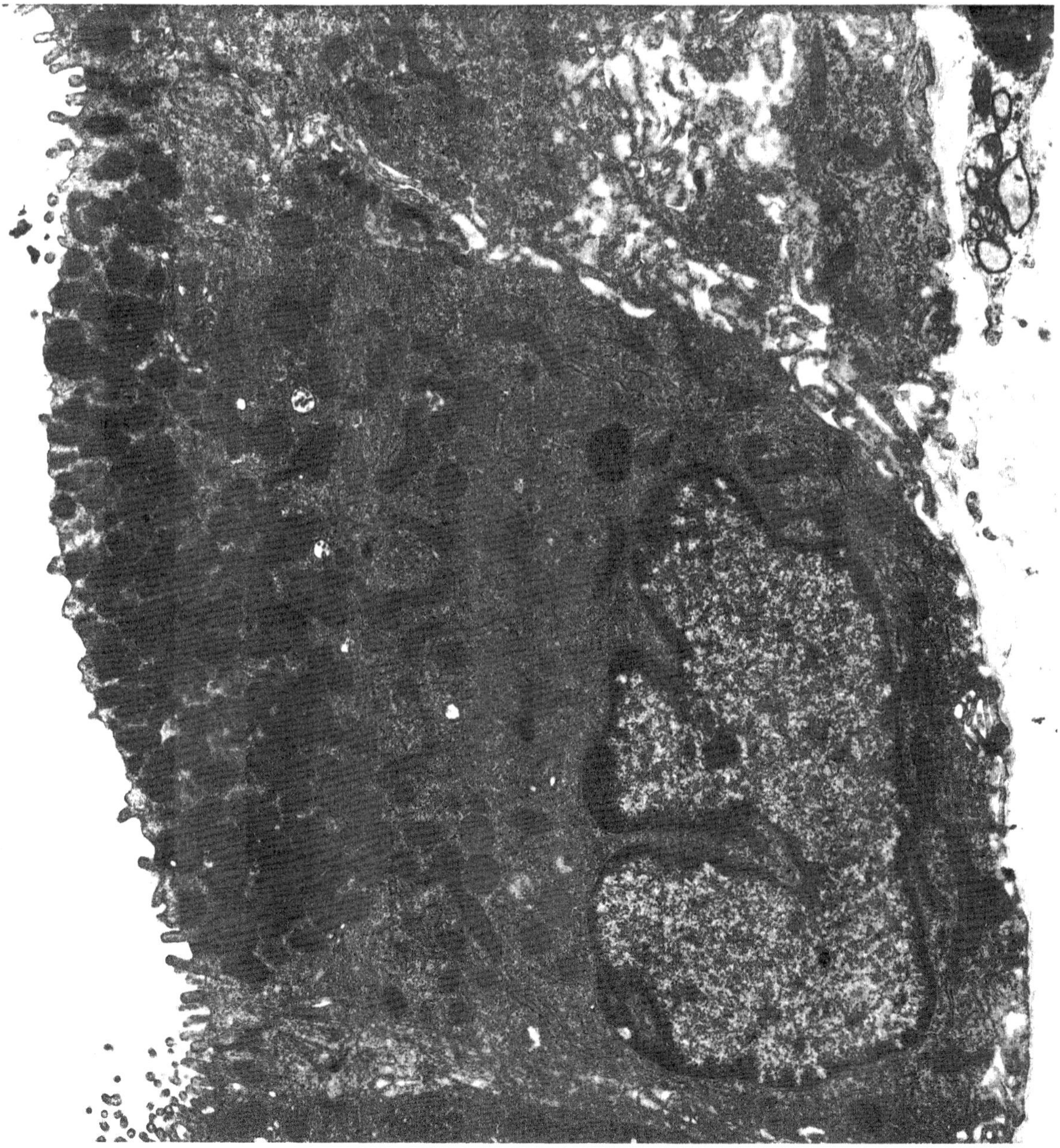

Figure 4-5. Mature surface mucous cell at the middle portion of the gastric pit. The apical cytoplasm is filled with numerous secretory granules. The nucleus is located in the basal half region of the cell. X 9,500.

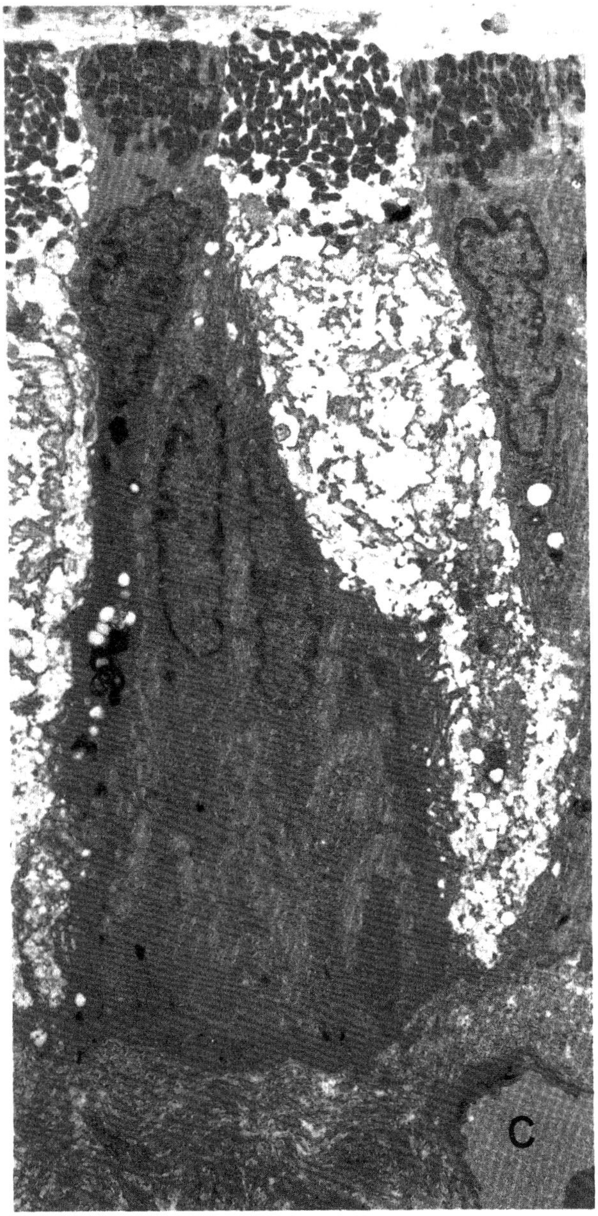

Figure 4-6. Surface mucous cells at the interfoveolar region. Some of them show degeneration in situ. These cells are tall and funnellike with many lysosomes and lipid droplets, most of which are situated in the infranuclear region. In the upper region of the lamina propria, corresponding to the interfoveolar ridge, are densely packed collagen fibrils and a blood capillary (C). X 4,000.

causes the autolysis of the cell itself, since the acid phosphatase activities are confined to primary as well as secondary lysosomes, and they do not diffuse into the cytosol. Besides autophagy, phagocytotic activity of the interfoveolar cells is reported under pathological conditions by Yeomans (23) and Morris and Harding (24), and in the pyloric region large secondary lysosomes are frequently encountered (25). So there is a possibility that the interfoveolar cell may take up the remaining structures of the adjacent ruptured cells or the entire degenerated cells under physiological conditions.

Lipid droplets are usually seen in the basal part of the tall funnel-shaped cells (Fig 4-6). This is ascertained by observations of cross-sections of the gastric wall at the level of the basal part of the interfoveolar cells (14). These lipid droplets are uncommon in the surface mucous cells in the middle and lower portions of the pit (Figs. 4-3, 4-4, 4-5). In suckling rats, the covering surface mucous cells contain a lot of lipid droplets, which are thought to be derived from food in the stomach cavity through absorption (26). On the other hand, the lipid droplets in the basal cytoplasm of the interfoveolar cells seem to be produced from secondary lysosomes, since they increase in number with the accumulation of lysosomes and/or with morphological changes in the cell. As for some lipids in the gastric mucosa, intriguing papers have been published that prostaglandin has a cytoprotective effect on the gastric mucosa (27–34) and that the mucosal surface of the stomach has a hydrophobic lining that may be attributed to the surface-active phospholipid known to be present in both gastric mucosa and juice (35, 36). Lichtenberger et al. (35) showed that the concentration of surface-active phospholipids is increased by the administration of prostaglandin and suggested that the hydrophobic lining of the gastric mucosa may play a role in mucosal protection. A plausible idea that may be deduced from these data is that the lipid droplets in the interfoveolar cell might have a cytoprotective function by giving a hydrophobic property to the epithelium at the time of the cell rupture.

3.3. Keratin filaments

In the cytoplasm of the interfoveolar cell, bundles of randomly running intermediate filaments are fairly well developed as compared with those of the cells located in the lower portion of the gastric pit (Fig. 4-9), and are sometimes inserted into the dense plaque of the desmosomes (13, 14). Intermediate filaments are thought to have an important role in the mechanical integration of various cytoplasmic organelles (37, 38), and Dabiké et al. (39, 40) showed that the surface mucous cells in the rat have an abundant prekeratinlike material. Taking these findings together with our observations, we would like to suggest that immature surface mucous cells in the lower portion of the gastric pit become specialized as they migrate into the covering epithelium, as do keratinocytes of the epidermis, and thereby gain some strength or hardiness against the mechanical stimuli generated by food passage. So these changes would constitute differentiation into a protective, covering epithelium.

3.4. Carbonic anhydrase activity

Carbonic anhydrase, which plays an important role in the production of gastric acid (41, 42) and maintenance of the acid-base balance (43), is localized not only in the parietal cells (41) but also in the surface mucous cells (44–46). This fact implies that the surface mucous cells would be equipped with a special mechanism to defend themselves from hydrochloric acid. In fact, surface mucous cells need a large of supply of carbon dioxide, a substrate of carbonic anhydrase, to

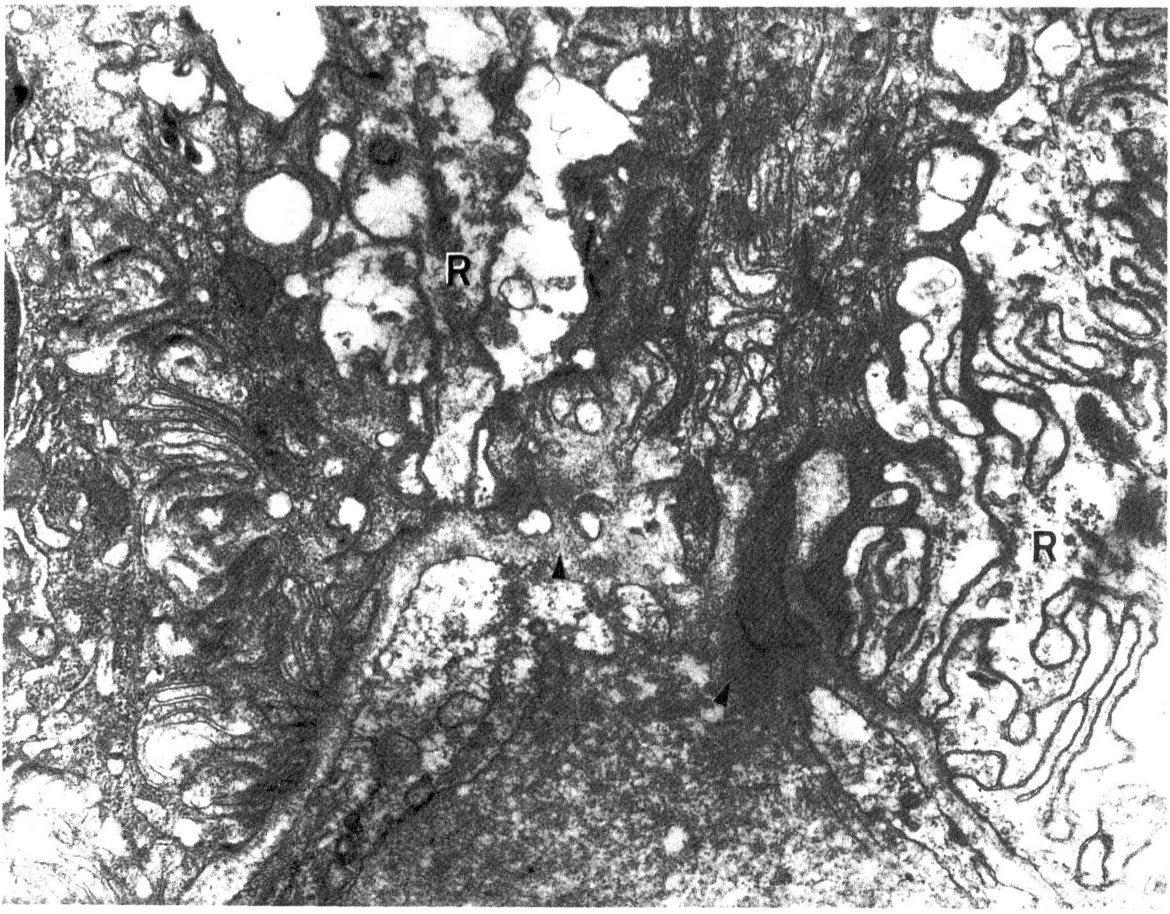

Figure 4-7. Basal parts of the interfoveolar cells showing well-developed infoldings and interdigitations of the plasma membranes. The basal plasma membrane is not detached from the basal lamina (arrow heads) which is thickened at the interfoveolar ridge. In spite of the discharge of the contents of the cell into the gastric lumen, the basal and lateral plasma membranes of the ruptured cells (R) remain behind. X 18,000.

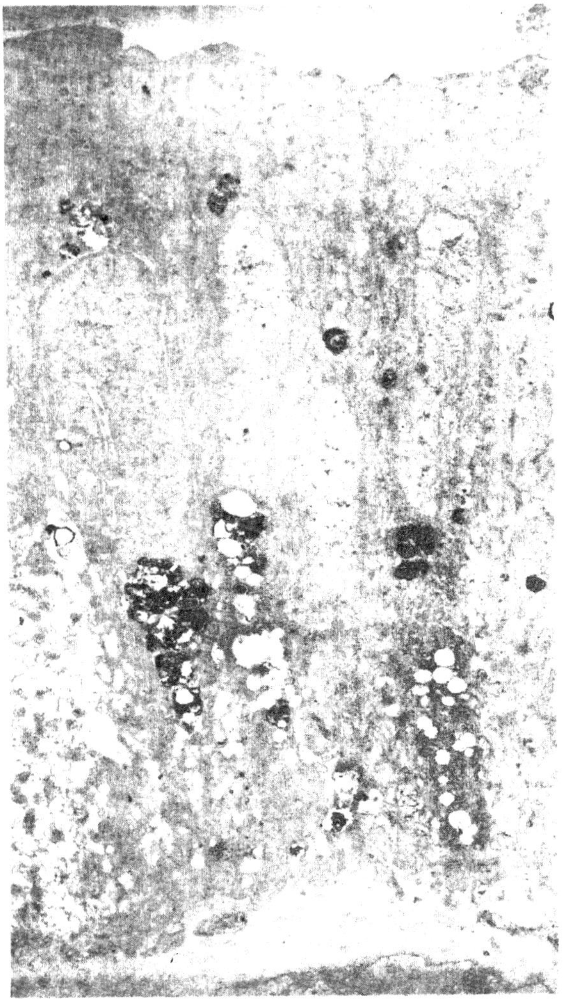

Figure 4-8. Demonstration of acid phosphatase activity in the funnel-shaped cell at the interfoveolar ridge without counter staining. Most of the secondary lysosomes showing reaction products are situated under the nucleus. X 3,800.

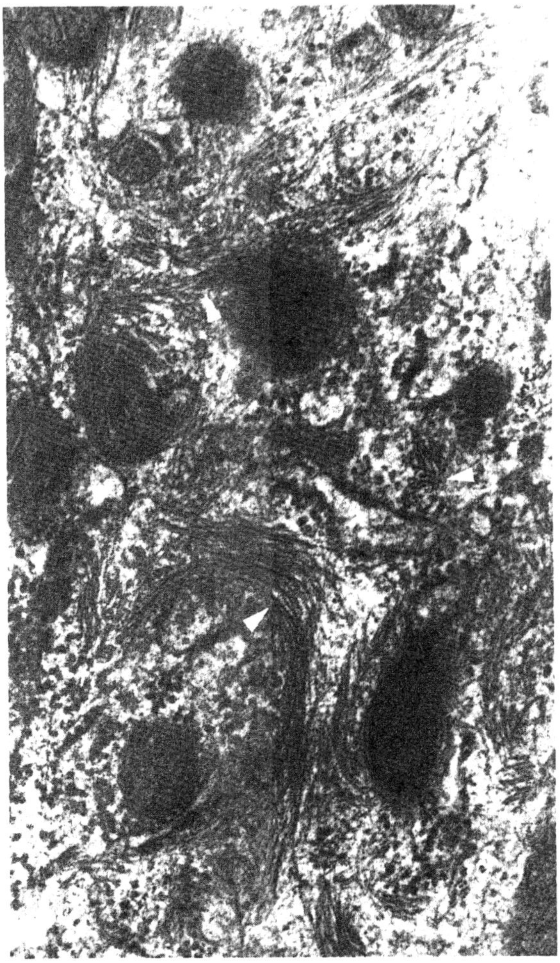

Figure 4-9. Abundant intermediate filaments (arrowheads) in the cytoplasm of the interfoveolar cell. X 41,000.

protect themselves from acid; otherwise, ulcerations and perforations occur (47).

4. Physiological cell loss

Interfoveolar cells show discernible degenerative changes, manifested by the low electron density of their cytoplasmic matrix and the highly contrasted chromatin (Fig. 4-6). The interfoveolar cells lose their apical plasma membrane and discharge their contents into the gastric lumen, but the lateral and basal plasma membranes always remain as a boundary between the gastric lumen and the underlying connective tissue (13, 14, 21), where the basal lamina is rather thickened and wavy (Fig. 4-7). It has been reported that the zonula occludens remains intact, even with degeneration induced by ethanol application (48, 49).

4.1. Mode of cell loss

Two alternative mechanisms have been proposed for physiological cell loss: (a) extrusion of individual foveolar cells (50) and (b) in situ degeneration (51). "Extrusion" was used originally as a mode of cell loss in the intestinal villi by Leblond and

Stevens (4). By extrusion, they meant the shedding of intact cells, which appear to die soon thereafter in the gastric lumen. On the other hand, Koelliker (quoted in 52) thought that the tops of the cells broke off, though the precise mechanism was not elucidated. Pfeiffer (54) described that an initial sign of "in situ" cell degeneration was a small cavitation in the convex surface of the cell, and said that this type of degeneration was the normal mode, calling it "in situ" degeneration. Harding and Morris (50) showed the image of extrusion by scanning and transmission electron microscopy. However, it is difficult to deduce, from our findings, that the shedding cell detaches from the basal lamina and the surrounding cells with loss of the junctional complex. We think that the cells, undergoing in situ degenerative changes as previously described, disrupt their apical plasma membranes without detaching from the epithelial lining and discharge their intracellular contents into the gastric lumen, leaving the basolateral plasma membranes behind (Figs. 4-6, 4-7).

4.2. Basolateral membrane and basal lamina

The most common site of physiological cell loss is the interfoveolar ridge. Since at that portion the basal part of the cell is so small and the basolateral plasma membranes remain after rupturing of the apical plasma membrane under the physiological state (Fig. 4-7), the cell defect would be replenished immediately by surrounding cells without making direct communication between the gastric lumen and the underlying connective tissue. The remaining basal plasma membrane is thought to be important even after cell rupture, because if the basal lamina under the basal plasma membrane is damaged, the recovery of the continuous epithelial lining would be retarded. This is supported by a report that the intact state of the basal lamina is a critical factor in restoring epithelial integrity in the case of erosion caused by acid (53).

4.3. Secretory activity and mode of secretion

The secretory activity of the surface mucous cells has a strong bearing on the intake of foodstuffs, which is accompanies by cell loss. Grant et al. (52) stated that loss of cells and mucosal changes in some areas during digestion of meat were analogous to the changes caused by mild irritants. Zalewsky et al. (54) found that intra-arterial injection of acetyl choline induces not only a copious secretion of mucus, but also an increase in the amount of DNA in the gastric secretion. This fact means that mucous secretion is closely related to cell loss. Zalewsky and Moody (55) described that exocytosis is found especially in the sulfated glycoprotein-rich foveolar cells located in the lower to middle portion of the pit and that apical expulsion, resulting in an explosive release of the entire apical mucous package, occurs in the oldest cells at the interfoveolar portion, whose mucus predominantly stains for neutral glycoproteins. Morris et al. (56) reported that the mucosa is only partially covered by an interconnected network of mucous strands which is strongly acidic and is released by epithelial cells in the isthmic and lower foveolar regions. In addition, they noted that after topical application of an ulcerogenic agent, a thick layer of gelatinous mucus is released as a result of exfoliation and disintegration of interfoveolar surface epithelial cells. Gerard et al. (57) showed that interfoveolar cells incorporate much less intravenously injected $Na_2S^{35}O_4$ than the surface mucous cells located in the lower gastric pit. Taken together with our findings indicating dysfunction in secretory granule production, we deduced the following: the surface mucous cells in the lower to middle pit vigorously produce the mucous granules and store them in the apical cytoplasm, some of which are continuously but slowly released by exocytosis, so that the amount of mucous granules in the cell increases with migration to the interfoveolar region. As the cell comes near the interfoveolar region, changes in cell shape and cytoplasmic structure take place, resulting in dysfunction of mucous granule production. When the interfoveolar cells are endangered by food intake or an ulcerogenic agent, all of the mucous granules as well as other intracellular contents might be released into the gastric lumen through rupture of the apical membrane. The mode of secretion in the later stage is a kind of holocrine secretion (12, 13, 18, 58). All the interfoveolar cells are thought to secrete mucus in a holocrine manner.

5. Connective tissue elements around the gastric pit

5.1. Fibrous layers in the mucosa

Beneath the interfoveolar epithelium is connective tissue with sinusoid capillaries. The connective tissue between the capillary endothelium and the interfoveolar cells is relatively thick (ca. 5 μm), and is characterized by the occurrence of numerous, densely-packed collagen fibrils (12, 13). This layer is stained intensely blue by the azan method. The collagen fibrils are much more numerous and dense in the pit portion, especially in the interoveolar ridge zone, as compared with the gastric glandular portion (12, 13).

Foodstuffs just entering into the stomach which are not very soft are mixed with the gastric juice to become chyme, and then they are delivered into the duodenum. Since the stomach is subject to mechanical friction generated by the passage of solid materials, the connective tissue structure mentioned above could serve as a mechanical support against the mechanical force and might protect the underlying vascular mesh (13). This kind of connective tissue structure has been reported in human colonic mucosa by Donnellan (59), and he referred to it as a "reticular layer." This structure would also be necessary in the colon to defend against mechanical insults, because the soft contents of the colon become more solid again in becoming feces.

5.2. Smooth muscles, blood capillaries, and nerves in the mucosa

In the lamina propria mucosae, smooth muscle cells, blood capillaries, and nerve fibers are seen running from the lamina muscularis mucosae along the gastric gland and pit, to the interfoveolar ridge. According to Kaur and Potten (60) administration of noradrenalin, which inhibits contraction of the smooth muscles within the small intestine, immediately suppresses cell migration on the villus but seems to have no immediate effect on crypt cell migration. These facts suggest that the smooth muscle cells and neuronal elements of the villi are implicated in the control of cell migration and that contraction of the smooth muscles might facilitate replacement of the cell defect by the surrounding cells. This would be true of the stomach to some extent, but the architecture of the mucous epithelium of the stomach is more similar to that of the crypt than that of the villus.

As for the innervation of the stomach mucosa, sympathetic and parasympathetic nerves are thought to regulate the gastric mucosal blood flow by changing the diameter of the submucosal arteriole (61, 62), and no findings on the occurrence of arteriovenous anastomosis have been reported in the mucosa and submucosa (63). Nakamura et al. (64–66) found that the activities of acetylcholinesterase and the muscarinic acetylcholine receptors are distributed on the plasma membrane of the axon as well as on the capillary endothelium in the lamina propria mucosae.

5.3. Three-dimensional observations

Dynamic changes in cell contour and connective tissue architecture around the opening of the gastric pit to the gastric lumen are well demonstrated by scanning electron microscopy using a specimen in which the covering epithelium was mechanically removed from the underlying connective tissue after fixation (Fig. 4-2). Under the covering mucous cells, a dense fibrous connective tissue layer, mentioned above, is well recognized, and is continuous downwards with the stromal sheaths encasing each gastric pit (11, 13, 14). The middle part of the fibrous layer between the gastric pits bulges towards the gastric lumen as a ridge, and we call it an interfoveolar ridge of the fibrous tissue layer (Fig. 4-2). Observations of the specimen revealed that a succession of changes in cell shape occurs near the opening of the pit to the gastric lumen and that covering mucous cells comprising the inner surface of the stomach are all located on the interfoveolar ridge and have a characteristic tall and funnellike, with an extremely narrow basal cytoplasmic portion.

6. Discussion of factors affecting cell shape

It has been reported that when cell division or cell differentiation is blocked by some agents, the intestinal epithelial cell becomes cuboidal or squamous in shape (67), and when attacks to the gastric epithelium result in erosion or ulcer, squamous cells appear at the restitution site, where the cell loss outnumbers the cell supply from the lower gastric pit (23, 68, 69). Intake of food has also been known to decrease the total number of surface mucous cells in a pit (52, 70, 71), and the height of the interfoveolar cell in fasting animals becomes significantly greater than that of nonfasting animals (unpublished data). Therefore, cell height and the tall, funnel cell

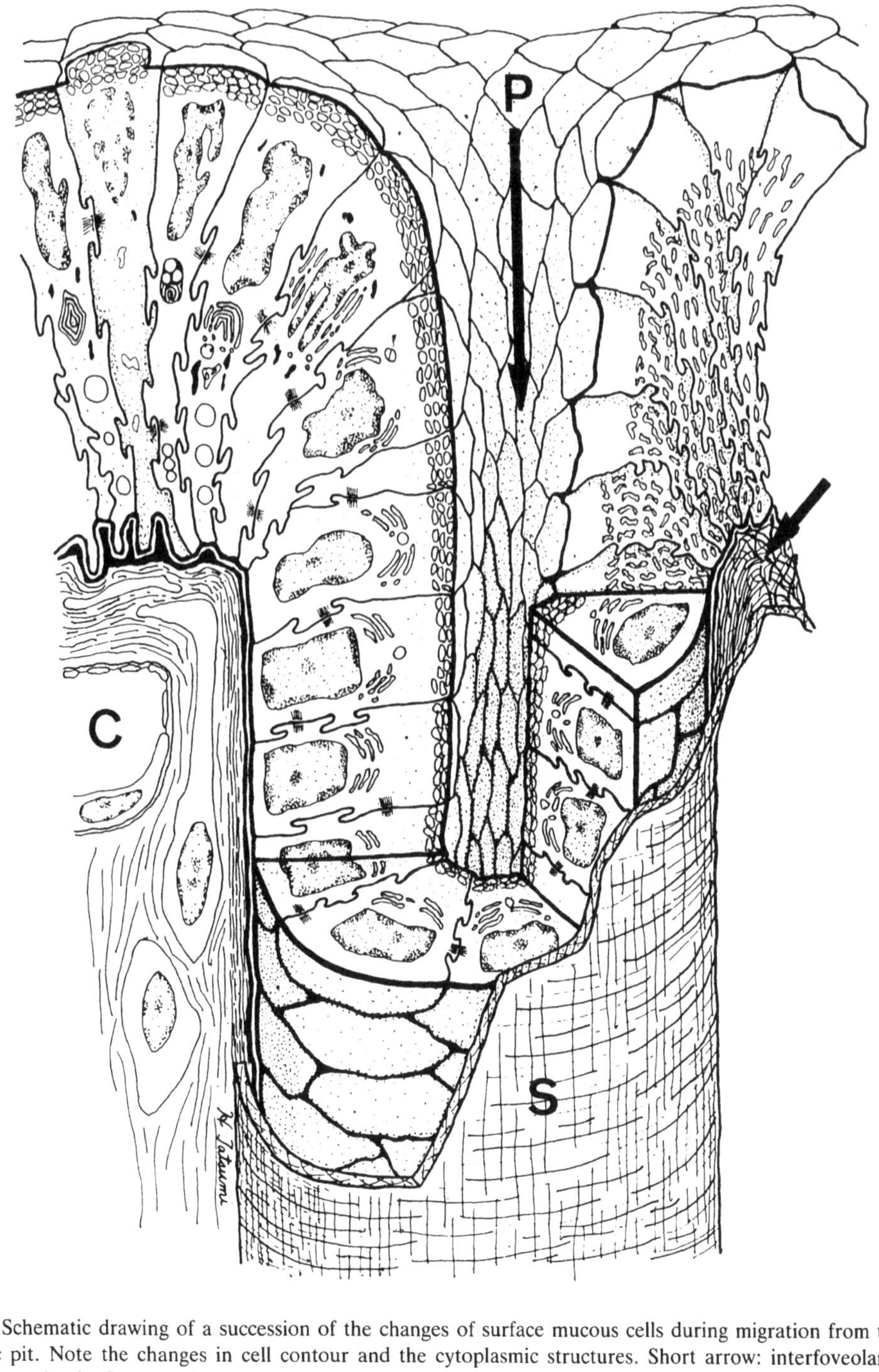

Figure 4-10. Schematic drawing of a succession of the changes of surface mucous cells during migration from the middle to the upper gastric pit. Note the changes in cell contour and the cytoplasmic structures. Short arrow: interfoveolar ridge; C, blood capillary; P, gastric pit; S, stromal sheath.

shape are considered to represent an outcome of balancing between cell loss and supply (cell division rate) in the single-layered epithelium.

The tall funnellike shape of the cell, the elongated nucleus, and organelle displacement seen in the interfoveolar cell suggest that a compressive force is generated by neighboring cells and that there is partial detachment of the basal plasma membrane from the basal lamina; but complete detachment has never been observed under physiological conditions (13, 14). The compressive force is presumably derived from the migration of cells and/or the contraction of smooth muscles running along the gastric pits and perpendicularly to the lamina muscularis mucosae (12). Mitosis has been thought to play a part in the migration force, but cell migration can occur without proliferation (72, 73). Hunt (70) reported that mitotic activities are increased progressively to a maximum at 20 to 24 hours after refeeding, which promotes cell loss. So we think that the shape of the surface mucous cell is subject to the rate of cell loss and proliferation, as well as to the surrounding connective tissue structure.

The changes of the surface mucous cells described in this chapter are summarized in Figure 4-10.

References

1. Kaku H: Dynamic analysis of the stomach epithelium — ^{3}H-thymidine autoradiographic study. *Arch Histol Jpn* 27: 223–246, 1966 (in Japanese).
2. Hattori T: On cell proliferation and differentiation of the fundic mucosa of the golden hamster. *Cell Tissue Res* 148: 213–226, 1974.
3. Bizzozero G: Uber die schlauchformigen drusen des magendarmkanals und die Beziehungen ihres Epithels zu dem Oberflachenepithel der Schleimhaut. *Archiv Mikroskop Anat* 42: 82–152, 1893.
4. Leblond CP, Stevens CE: The constant renewal of the intestinal epithelium in the albino rat. *Anat Rec* 100: 357–371, 1948.
5. Stevens CE, Leblond CP: Renewal of the mucous cells in the gastric mucosa of the rat. *Anat Rec* 115: 231–245, 1953.
6. Messier B, Leblond CP: Cell proliferation and migration as revealed by radioautography after injection of thymidine-^{3}H into male rats and mice. *Am J Anat* 106: 247–285, 1960.
7. Willems G, Vansteenkiste Y, Verbeustel S: Autoradiographic study of cell renewal in fundic mucosa of fasting dogs. *Acta Anat* 80: 23–32, 1971.
8. Wattel W, Geuze JJ, de Rooij DG: Ultrastructural and carbohydrate histochemical studies on the differentiation and renewal of mucous cells in the rat gastric fundus. *Cell Tissue Res* 176: 445–462, 1977.
9. Creamer B, Shorter RG, Bamforth J: The turnover and shedding of epithelial cells. *Gut* 2: 110–118, 1961.
10. MacDonald WC, Trier JS, Everett NB: Cell proliferation and migration in the stomach, duodenum, and rectum of man: Radioautographic studies. *Gastroenterology* 46: 405–417, 1964.
11. Hattori T, Fujita S: Tritiated thymidine autoradiographic study on cellular migration in the gastric gland of the golden hamster. *Cell Tissue Res* 172: 171–184, 1976.
12. Tamura S, Fujita H: Fine structural aspects on the renewal and development of surface mucous cells and glandular cells of the gastric body of the adult golden hamster. *Arch Histol Jpn* 46: 501–521, 1983.
13. Tatsumi H, Fujita J, Tamura S: Electron microscopic studies on the physiological cell loss in the gastric mucosa of the golden hamster. *Cell Tissue Res* 239: 343–347, 1985.
14. Tatsumi H, Ban T, Fujita H: Electron microscopic studies on morphological changes in the surface mucous cell during migration in the stomach of the golden hamster. *Arch Histol Jpn* 48: 527–539, 1985.
15. Golgi C: Sur une fine particularité de structure de l'epithélium de la muqueuse gastrique et intestinale de quelques vertébrés. *Arch Ital Biol* 51: 213–245, 1909.
16. Golgi C: Di una minuta particolarità di struttura dell'epitelio della mucosa gastrica ed intestinale di alcuni vertebrati. *Arch Sc Med* 33: 1–36, 1909.
17. d'Agata G: Sur les modifications de l'appareil réticulaire interne dans l'épithélium de la muqueuse gastrique. *Arch Ital Biol* 54: 425–428, 1911.
18. Hibbard H: The "Golgi apparatus" during development in the stomach of *Gallus domesticus*. *J Morphol* 70: 121–149, 1942.
19. Grad B, Stevens CE: Histological changes produced by a single large injection of radioactive phosphorus (P^{32}) in albino rats and in C3H mice. *Cancer Res* 10: 289–296, 1950.
20. Kataoka K, Sakano Y: Panoramic observation of the mouse gastric mucosa by superwide-field electron microscopy. *Arch Histol Jpn* 47: 209–221, 1984.
21. Kataoka K, Takeoka Y, Hirano S: Electron microscopic observations of surface mucous cells in the mouse gastric mucosa during physiological degeneration and extrusion. *Arch Histol Jpn* 48: 327–339, 1985.
22. Smith RE, Farquhar MG: Lysosome function in the regulation of the secretory process in cells of the anterior pituitary gland. *J Cell Biol* 31: 319–347, 1966.
23. Yeomans N: Electron microscopic study of the repair of aspirin-induced gastric erosions. *Am J Dig Dis* 21:

533–541, 1976.
24. Morris GP, Harding RK: Phagocytosis of cells in the gastric surface epithelium of the rat. *Cell Tissue Res* 196: 449–454, 1979.
25. Lee ER: Dynamic histology of the antral epithelium in the mouse stomach: III. Ultrastructure and renewal of pit cells. *Am J Anat* 172: 225–240, 1985.
26. Helander HF, Olivecrona T: Lipolysis and lipid absorption in the stomach of the suckling rat. *Gastroenterology* 59: 22–35, 1970.
27. Whittle BJR: Relationship between the prevention of rat gastric erosions and the inhibition of acid secretion by prostaglandins. *Eur J Pharmacol* 40: 233–239, 1976.
28. Robert A, Nezamis JE, Lancaster C, Hanchar AJ: Cytoprotection by prostaglandins in rats: Prevention of gastric necrosis produced by alcohol, HCl, NaOH, hypertonic NaCl, and thermal injury. *Gastroenterology* 77: 433–443, 1979.
29. Chaudhury TK, Robert A: Prevention by mild irritants of gastric necrosis produced in rats by sodium taurocholate. *Dig Dis Sci* 25: 830–836, 1980.
30. Kauffman GL, Reeve JJ, Grossman MI: Gastric bicarbonate secretion: Effect of topical and intravenous 16, 16-dimethyl prostaglandin E2. *Am J Physiol* 239: G44–G48, 1980.
31. Bickel M, Kauffman GL: Gastric gel mucus thickness: Effect of distention, 16, 16-dimethyl prostaglandin E2 and carbenoxolone. *Gastroenterology* 80: 770–775, 1981.
32. Cohen MM, MacDonald WC: Mechanism of aspirin injury to human gastroduodenal mucosa. *Prostaglandins Leukotrienes Med* 9: 241–255, 1982.
33. Wood LR, Dubois A: Scanning electron microscopy of the stomach during modifications of acid secretion. *Am J Physiol* 244: G475–G479, 1983.
34. Tarnawski A, Hollander D, Stachura J, Krause WJ, Gergely H: Prostaglandin protection of the gastric mucosa against alcohol injury — a dynamic time–related process. *Gastroenterology* 88: 334–352, 1985.
35. Lichtenberger LM, Graziani LA, Dial EJ, Butler BD, Hills BA: Role of surface-active phospholipids in gastric cytoprotection. *Science* 219: 1327–1329, 1983.
36. Hills BA, Butler BD, Lichtenberger LM: Gastric mucosal barrier: Hydrophobic lining to the lumen of the stomach. *Am J Physiol* 244: G561–G568, 1983.
37. Sun TT, Shih C, Green H: Keratin cytoskeletons in epithelial cells of internal organs. *Proc Natl Acad Sci USA* 76: 2813–2817, 1979.
38. Lazarides E: Intermediate filaments as mechanical integrators of cellular space. *Nature* 283: 249–256, 1980.
39. Dabiké M, Koenig CS, Vial JD: Distribution of intermediate filaments in amphibian oxyntic cells. *Cell Tissue Res* 220: 725–737, 1981.
40. Dabiké M, Koenig CS: Intermediate filaments of the cytoskeleton in glandular cells on the rat fundic mucosa: Immunofluorescence and electron microscopy study. *Anat Rec* 207: 297–308, 1983.
41. Davenport HW, Fisher RB: Carbonic anhydrase in the gastrointestinal mucosa. *J Physiol* 94: 16–17, 1939.
42. Davies RE, Edelman J: The function of carbonic anhydrase in gastric mucosa. *Biochem J* 43: lvii-lviii, 1948.
43. Davies RE, Edelman J: The function of carbonic anhydrase in the stomach. *Biochem J* 50: 190–194, 1951.
44. O'Brien P, Rosen S, Trencis-Buck L, Silen W: Distribution of carbonic anhydrase within the gastric mucosa. *Gastroenterology* 72: 870–874, 1977.
45. Sugai N, Ito S: Carbonic anhydrase, ultrastructural localization in the mouse gastric mucosa and improvements in the technique. *J Histochem Cytochem* 28: 511–525, 1980.
46. Sato A, Spicer SS, Tashian RE: Ultrastructural localization of carbonic anhydrase in gastric parietal cells with the immunoglobulin-enzyme bridge method. *Histochem J* 12: 651–659, 1980.
47. Davies RE, Longmuir NM: Production of ulcers in isolated frog gastric mucosa. *Biochem J* 42: 621–627, 1948.
48. Eastwood GL, Kirchner JP: Changes in the fine structure of mouse gastric epithelium produced by ethanol and urea. *Gastroenterology* 67: 71–84, 1974.
49. Eastwood GL, Erdmann KR: Effect of ethanol on canine gastric epithelial ultrastructure and transmucosal potential difference. *Am J Dig Dis* 23: 429–435, 1978.
50. Harding RK, Morris GP: Cell loss from normal and stressed gastric mucosae of the rat: An ultrastructural analysis. *Gastroenterology* 72: 857–863, 1977.
51. Pfeiffer CJ: Gastric surface morphology in man, monkey, and ferret: Evidence for in situ surface cell degeneration. *Exp Mol Pathol* 13: 319–328, 1970.
52. Grant R, Grossman MI, Ivy AC: Histological changes in the gastric mucosa during digestion and their relationship to mucosal growth. *Gastroenterology* 25: 218–231, 1953.
53. Black BA, Morris GP, Wallace JL: Effects of acid on the basal lamina of the rat stomach and duodenum. *Virchows Arch* (Cell Pathol) 50: 109–118, 1985.
54. Zalewsky CA, Moody FG, Allen M, Davis EK: Stimulation of canine gastric mucus secretion with intraarterial acetylcholine chloride. *Gastroenterology* 85: 1067–1075, 1983.
55. Zalewsky CA, Moody FG: Mechanisms of mucus release in exposed canine gastric mucosa. *Gastroenterology* 77: 719–729, 1979.
56. Morris GP, Harding RK, Wallace JL: A functional model for extracellular gastric mucus in the rat. *Virchows Arch* (Cell Pathol) 46: 239–251, 1984.
57. Gerard A, Lev R, Glass GBJ: Histochemical study of the mucosubstances in the canine stomach. II. The effects of histamine, gastrin, urecholine and food. *Lab Invest* 19: 29–39, 1968.
58. Helander HF: The cells of the gastric mucosa. *Int Rev Cytol* 70: 217–289, 1981.
59. Donnellan WL: The structure of the colonic mucosa: The epithelium and subepithelial reticulohistiocytic complex. *Gastroenterology* 49: 496–514, 1965.
60. Kaur P, Potten CS: Effects of puromycin, cycloheximide and noradrenalin on cell migration within the crypts and on the villi of the small intestine. A model to explain cell movement in both regions. Cell Tissue Kinet 19:611–625, 1986.
61. Guth PH, Rosenberg A: In vivo microscopy of gastric microcirculation. *Am J Dig Dis* 17: 391–398, 1972.
62. Guth PH, Smith E: Neural control of gastric mucosal

blood flow in the rat. *Gastroenterology* 69: 935–940, 1975.
63. Gannon B, Browning J, O'Brien P: The microvascular architecture of the glandular mucosa of rat stomach. *J Anat* 135: 667–683, 1982.
64. Nakamura M, Watanabe N, Tsukada N, Oda N, Tsuchiya M: Demonstration of the adrenergic nerves in the rat gastric mucosa — a histofluorescence and electron microscopic study in comparison with the distribution of the cholinergic nerves. *Okajimas Folia Anat Jpn* 59: 65–86, 1982.
65. Nakamura M, Oda M, Yonei Y, Tsukada N, Watanabe N, Komatsu H, Tsuchiya M: Demonstration of the localization of muscarinic acetylcholine receptors in the gastric mucosa — light and electron microscopic autoradiographic studies using ^{3}H–quinuclidinyl benzilate. *Acta Histochem Cytochem* 17: 297–309, 1984.
66. Oda M, Nakamura M, Tsuchiya M: Features of autonomic innervation of the microvascular system in the gastric mucosa. *Int J Micr* 3: 409, 1984.
67. Yarrington JT, Sprinkle DJ, Loudy DE, Diekema KA, McCann PP, Gibson JP: Intestinal changes caused by DL-α-difluoromethylornithine (DFMO), an inhibitor of ornithine decarboxylase. *Exp Mol Pathol* 39: 300–316, 1983.
68. Svanes K, Ito S, Takeuchi K, Silen W: Restitution of the surface epithelium of the in vitro frog gastric mucosa after damage with hyperosmolar sodium chloride: Morphologic and physiologic characteristics. *Gastroenterology* 82: 1409–1426, 1982.
69. Critchlow J, Magee D, Ito S, Takeuchi K, Silen W: Requirements for restitution of the surface epithelium of frog stomach after mucosal injury. *Gastroenterology* 88: 237–249, 1985.
70. Hunt TE: Mitotic activity in the gastric mucosa of the rat after fasting and refeeding. *Anat Rec* 127: 539–550, 1957.
71. Willems G, Vansteenkiste Y, Smets PH: Effects of food ingestion on the cell proliferation kinetics in the canine fundic mucosa. *Gastroenterology* 61: 323–327, 1971.
72. Altmann GG: Influence of starvation and refeeding on mucosal size and epithelial renewal in the rat small intestine. *Am J Anat* 133: 391–400, 1972.
73. Potten CS, Chwalinski S, Swindell R, Palmer M: The spatial organization of the hierarchical proliferative cells of the crypts of the small intestine into clusters of "synchronized" cells. *Cell Tissue Kinet* 15: 351–370, 1982.

Authors' addresses:
Department of Anatomy
Osaka University Medical School
4-3-57 Nakanoshima
Kita-Ku, Osaka 530, Japan

CHAPTER 5

Biology of the duodenal (Brunner's) glands

WILLIAM J. KRAUSE

1. Introduction

The duodenal glands were first observed in 1679 by Wepfer (1) who described them during the autopsy of a beheaded woman. Wepfer described their arrangement in the duodenum and observed that when they were macerated in water, the duodenal glands discharged copious amounts of mucus. Nine years later, Brunner (2,3), who was Wepfer's son-in-law (4), described the duodenal glands in the horse, ox, sheep, deer, dog, and beaver and presented a brief description in man. Brunner considered these glands to be accessory pancreatic tissue and termed them "glandulae duodeni" or "pancreas secundarium." Middeldorpf (5) realized that the secretion of the duodenal glands differed from that of the pancreas and suggested that these glands be designated "the glands of Brunner." Florey and Harding (6,7) found that the region of the duodenum containing the duodenal glands is more resistant to acid erosion than other regions of the small intestine.

2. Distribution of duodenal glands

The duodenal glands have been observed in all mammalian species examined to date but are absent in other vertebrate species. They begin at the gastrointestinal junction and extend for various distances along the proximal small intestine. In most species the glands lie primarily in the submucosa of the duodenum and usually are not found beyond the entrance of the pancreatic duct. In the prototherians, the duck-billed platypus and echidna, Brunner's glands actually lie within the submucosa of the distal stomach and their ducts empty onto the overlying gastric mucosa (8,9). In both species the gastric mucosa is aglandular and lined by stratified squamous epithelium (10). In metatherian (marsupial) and eutherian (placental) mammals, the majority of duodenal glands lie in the submucosa of the proximal small intestine immediately distal to the pyloric sphincter. Here, the duodenal glands form a complex glandular collar that surrounds the beginning of the intestinal tract. In the majority of metatherians, the glands form a narrow band and are arranged in large, complex lobes (11,12,13). The duodenal glands of eutherian species also show their heaviest concentration in the submucosa immediately distal to pyloric sphincter and extend for various distances along the intestinal tract. They become less concentrated distally where they eventually are seen as scattered, smaller, individual glands.

Middeldorpt (5) and Kuczynske (14) divided mammalian species into three categories according to the linear extent of the duodenal glands along the intestinal tract which, in turn, was associated with diet. Herbivores have the longest distribution, omnivores a moderate distribution, and carnivores a short distribution. Although there is a general trend to support such a classification, when species of the same order that vary in dietary habits are compared, this general classification becomes less clear. The duodenal glands of most marsupial species, for example, show similar distributions and glandular organization, despite marked variations in dietary habits (11, 12, 13). Likewise, Forman (15) in examining 27 species of bats, noted little difference in the distribution or composition of duodenal glands, although there were marked differences in diet.

3. Duodenal mucosa

The intestinal mucosa presents numerous spatulate or fingerlike evaginations, villi, which cover the luminal surface of the duodenum. They consist of a central core of lamina propria rich in lymphoid cells and are covered by intestinal epithelium. Numerous simple tubular intestinal glands empty between the bases of the villi, the covering epithelium of which is continuous with that lining the glands. This simple columnar epithelium consists of a heterogeneous population of cells which, depending on the species, includes the intestinal absorptive cells, Paneth cells, caveolated cells, goblet cells, and a wide variety of different types of enteroendocrine cells. Cells of the latter group have been classified ultrastructurally according to the nature of their secretory granules and are immunoreactive for gastrointestinal peptides such as gastrin, secretin, cholecystokinin, enteroglucagon, gastric inhibitory peptide, gastrin releasing polypeptide, motilin, vasoactive intestinal polypeptide, somatostatin, pancreatic polypeptide, and the amine 5H-T in man and other species. Likewise, duodenal goblet cells in several species constitute a heterogeneous population of cells with regard to the type of mucin secreted (Table 5-1). They may contain a neutral mucin, carboxylated or sulfated acidic mucins, or any combination of the three. Generally, in those species that exhibit a heterogeneous population of goblet cells, the more mature superficial cells tend to contain a greater concentration of sulfated mucins. Thus, the intestinal epithelium lining the proximal duodenal mucosa contains a richly diverse population of cell types. All take origin from a narrow proliferative zone near the base of the intestinal glands and differentiate as they migrate up the walls of the intestinal gland and onto the villi to be exfoliated finally into the intestinal lumen. Cells in this region are linked by functional gap junctions which may play a role in the control of cell differentiation and proliferation (16). The life span of cells within the intestinal lining epithelium is four to seven days depending on the cell type. It is onto this heterogeneous epithelial surface that the secretion of the duodenal glands drain.

3.1. Light microscopy of duodenal glands

The majority of duodenal glands are housed within the connective tissue of the submucosa. In occasional species such as the pig, dog (17), moose, mountain goat (18), and man (19), portions of some duodenal glands actually lie within the mucosa.

Secretory units of the duodenal glands generally consist of coiled tubules that show frequent distal branching. Cells comprising the secretory units appear mucous in nature when viewed in routine preparations and show basally positioned oval or round nuclei and a light-staining cytoplasm. They vary in shape from simple cuboidal to simple columnar. In occasional species (primarily ungulates) the secretory units may become dilated and the epithelium assumes a squamous appearance due to the accumulation of secretory material within the lumen (18). Several secretory units eventually are drained by intralobular ducts, the morphology of which is generally similar to the secretory units. Each duodenal gland eventually is drained by a single excretory duct. The height of the epithelium increases in the ducts, particularly near their terminations. The ducts of individual glands pierce the muscularis mucosae to empty into overlying intestinal glands or, in many instances, pass through the mucosa to empty directly into the duodenal lumen between villi (18, 20). In man, the ducts of the duodenal glands drain most often into the midportion of the intestinal glands rather than at their bases (21).

The duodenal glands of several marsupial species are somewhat unique in that they are organized into large, distinct lobes that envelope a central, funnel-shaped depression or sinus lined by intestinal and/or gastric epithelium (12, 13). The duodenal glands of these species are compound tubular-acinar glands, the intralobular ducts of which drain via a single excretory duct into the sinus. The secretory units are lined by large pyramidal cells. The surrounding connective tissue organizes several lobules into the large glandular lobes.

Table 5-1. Mucosubstances in the duodenal glands and duodenal goblet-cells of several mammalian species.

Species	Cells of secretory tubules/acini	Cells of ducts	Duodenal goblet cells
Monotremata			
Duck-billed platypus	Neutral mucin	Neutral mucin	
Echidna	Carboxylated acidic mucin	Carboxylated acidic mucin	Sulfated acidic mucin
Marsupialia			
Opossum	Neutral mucin	Carboxylated acidic mucin	Sulfated acidic mucin
Red kangaroo	Neutral mucin, scattered cells contain carboxylated acidic mucin	Same as secretory units	Carboxylated acidic mucin, occasional cells contain neutral mucin
Eastern native cat	Neutral mucin	Neutral mucin	Sulfated acidic mucin
Marsupial Mouse	Neutral mucin	Neutral mucin	Sulfated acidic mucin, neutral mucin in occasional cells
Short-nosed bandicoot	Neutral mucin	Neutral mucin	Sulfated acidic mucin
Koala	Carboxylated acidic mucin	Carboxylated acidic mucin	Carboxylated acidic mucin
Wombat	Carboxylated acidic mucin	Carboxylated acidic mucin	Sulfated acidic mucin, occasional cells with neutral mucin
Rodentia			
Gerbil	Neutral mucin, occasional, random cells contain sulfated acidic mucin	—	Primarily sialomucin, small amount of sulfated acidic mucin
Hamster	Neutral mucin in deep glands, sulfated acidic mucin in superficial glands	Sulfated acidic mucin	Sulfated acidic mucin
Mouse	Neutral mucin	Sialomucin and sulfated acidic mucin in distal ducts	Primarily sulfated acidic mucin, small amount of sialomucin
Rat	Neutral mucin	Neutral mucin	Both sialomucin and sulfated acidic mucin
Guinea pig	Neutral mucin and abundant sulfated mucin in deeper glands, neutral and sialomucin in superficial glands	—	Sulfated acidic mucin
Lagomorpha			
Rabbit	Neutral mucin in serous acini; neutral, sulfated acidic mucin in superficial mucous tubules; neutral, sialomucin in deep mucous tubules	Neutral, sulfated acidic mucin	Primarily sulfated acidic mucin, a few deeper cells contain sialomucin
Carnivora			
Cat	Neutral mucin	Neutral mucin	Sulfated acidic mucin, deeper cells primarily neutral mucin
Dog	Neutral mucin	Neutral mucin	Both sulfated acidic mucin and sialomucin

Table 5-1. (Continued).

Species	Cells of secretory tubules/acini	Cells of ducts	Duodenal goblet cells
Ferret	Neutral mucin	Neutral mucin	Most cells sialomucin, a few cells contain sulfated acidic mucin
Mongoose	Neutral mucin	Neutral mucin	Sulfated acidic mucin
Edentata			
Armadillo	Neutral mucin	Neutral mucin	Neutral mucin, carboxylated and sulfated acidic mucin
Artiodactyla			
Domestic pig	Acidic mucin	—	—
Cow	Acidic mucin	—	—
Sheep	Acidic mucin	—	—
Mule Deer	Neutral mucin, sulfated acidic mucin	Same as secretory units	Neutral mucin, sulfated acidic mucin
White-tailed deer	Primarily neutral mucin, sulfated acidic mucin in scattered tubules and/or cells	Same as secretory units	Neutral mucin, sulfated acidic mucin
Pronghorn	Primarily neutral mucin, sulfated acidic mucin in scattered tubules and/or cells	More sulfated acidic mucin in cells	Neutral mucin, sulfated acidic mucin
Bison	Primarily neutral mucin, sulfated mucin in peripheral region of glands	Primarily sulfated acidic mucin	Neutral and sulfated acidic mucin
Moose	Primarily neutral mucin, acidic sulfated mucin in scattered glands, tubules or cells	Same as secretory units	Neutral and sulfated acidic mucin
Elk	Primarily neutral mucin, acidic sulfated mucin in scattered tubules and/or cells	Same as secretory units	Neutral and carboxylated acid mucin
Bighorn sheep	Neutral mucin	A few cells in ducts contained carboxylated acidic mucin	Neutral and sulfated acidic mucin
Mountain goat	Primarily neutral mucin, carboxylated acidic mucin in scattered tubules and/or cells	Same as secretory units	Neutral and sulfated acidic mucin
Perissodactyla			
Horse	Carboxylated acidic mucin	—	Sulfated acidic mucin
Sirenia			
West Indian manatee	Primarily neutral mucin, sialomucin, weakly sulfated acidic mucin, primarily in region where secretory units join ductal system	Sialomucin, sulfated acidic mucin	Sialomucin, sulfated acidic mucin

Table 5-1. (Continued).

Species	Cells of secretory tubules/acini	Cells of ducts	Duodenal goblet cells
Chiroptera			
27 species of bat native to North America	Neutral mucin	Neutral mucin	—
Indian fruit bat	Neutral mucin, acidic sialomucin	Same as secretory units	—
Fruit eating bats			
Sturnita lilium	Neutral, acidic mucin	Neutral, acidic mucin	—
Sturnita lulovici	Neutral, acidic mucin	Neutral, acidic mucin	—
Insectivora			
Musk shrew	Neutral mucin	Neutral mucin	Sulfated acidic mucin
Long-tailed shrew	Neutral mucin	Neutral mucin	Sulfated acidic mucin
Pygmy shrew	Neutral mucin	Neutral mucin	Sulfated acidic mucin
Old world water shrew	Neutral mucin	Neutral mucin	Sulfated acidic mucin
Old world mole	Neutral mucin	Neutral mucin	Sulfated acidic mucin
Primates (African & Asian species)			
Sooty mangabey	Primarily neutral mucin, scattered cells contain sulfated acidic mucin	Same as secretory units	Neutral mucin, sulfated acidic and sialomucin
Green monkey	Primarily a neutral mucin, scattered cells contain sulfated acidic mucin	Same as secretory units	Neutral mucin, sulfated acidic and sialomucin
Chimpanzee	Neutral mucin	Neutral mucin	Neutral mucin, sulfated acidic and sialomucin
Western baboon	primarily neutral mucin, scattered cells contain sulfated acidic mucin	Same as secretory units	Neutral mucin, sulfated acidic and sialomucin
Patus monkey	Primarily neutral mucin, scattered cells contain sulfated acidic mucin	Same as secretory units	Neutral mucin, sulfated acidic and sialomucin
Celebese black ape	Neutral mucin	Neutral mucin	Neutral mucin, sulfated acidic and sialomucin
Rhesus macaque	Primarily neutral mucin, scattered cells contain sulfated acidic mucin	Same as secretory units	Neutral mucin, sulfated acidic and sialomucin
Stump-tailed macaque	Primarily neutral mucin, scattered cells contain sulfated acidic mucin	Same as secretory units	Neutral mucin, sulfated acidic and sialomucin
Formosan rock macaque	Primarily neutral mucin, scattered cells contain sulfated acidic mucin	Same as secretory units	Neutral mucin, sulfated acidic and sialomucin

Table 5-1. (Continued).

Species	Cells of secretory tubules/acini	Cells of ducts	Duodenal goblet cells
Primates (South American species)			
Squirrel monkey	Neutral mucin	Neutral mucin	Neutral mucin, sulfated acidic and sialomucin
Night monkey	Primarily neutral mucin, scattered cells contain sulfated acidic mucin	Same as secretory units	Neutral mucin, sulfated acidic and sialomucin
Spider monkey	Neutral mucin	Neutral mucin	Neutral mucin, sulfated acidic and sialomucin
White-fronted capuchin	Primarily neutral mucin, scattered cells contain acidic mucin	Same as secretory units	Neutral mucin, sulfated acidic and sialomucin
Cotton-top marmoset	Neutral mucin	Neutral mucin	Neutral mucin, sulfated acidic and sialomucin
Man	Neutral mucin	Neutral mucin	Sialomucin

* Table 5-1 represents data compiled from references cited in the text.

In all species reported, with the exception of the rabbit, the duodenal glands are composed of a mucous-appearing cell type. In the rabbit, the duodenal glands contain two distinct cell populations, serous and mucous. The mucous cells form secretory tubules, whereas the serous elements generally form discrete acini or demilune-like structures at the ends of the mucous tubules (22). Interestingly, the duodenal glands of another lagomorph, the pika, consist only of the mucous cell type.

On occasion, other cell types derived from the overlying intestinal mucosa are found scattered within the epithelium of the ducts and secretory units of the duodenal glands. Paneth cells have been observed in the rat (23), cat (24), new and old world primates, including man (25), and caveolated cells have been identified in the duodenal glands of the white-tailed deer (26). Several types of enteroendocrine cells have been reported in the duodenal glands of man and other species. Immunoreactivities to some of the enterendocrine cells reported in the duodenal glands are shown in Table 5-2.

3.2. *Duodenal gland kinetics*

By means of flash and cumulative autoradiographic labelling, it has been shown that cell renewal within the duodenal glands of the adult rat occurs in situ without a contribution of cells arising from undifferentiated cells in the intestinal glands (27). The similarity of the ultrastructural characteristics of duodenal gland cells labelled with ^{3}H-thymidine to those that were not labelled suggests that mature epithelial cells within the duodenal glands can divide and supply replacement cells to the glands. The life span of these cells is not fixed and their renewal conforms to a random loss system. The half-life of a duodenal gland cell in the rat is about 40 days (27).

The labelling and/or mitotic index of the duodenal glands increases in response to duodenal ulcer (28). Both the duodenal glands and the intestinal glands contribute cells to the healing process. Similarly, during development of duodenal glands, the mitotic (labelling) index, although high, is insufficient to account for the number of cells needed both for growth and development (29). During their development, the duodenal glands of the mouse receive a major contribution of cells from the overlying intestinal glands.

3.3. *Vasculature of the duodenal glands*

Scanning electron microscopic observations of vascular casts have shown the microvascular bed

Table 5-2. Immunoreactive enteroendocrine cells in the duodenal glands.

Species	Somatostatin	Gastrin	Motilin	CCK	Glucagon	BPP	5-HT
Cow	+	+	+	+			+
Sheep	+	+		+	+		
Horse	+	+		+			
Mink	+	+			+		
Cat	+	+	+				
Vampire bat	+	+					+
Opossum	+	+	+			+	+
Echidna	+		+			+	+

* Data for Table 5-2 were kindly provided by J. Yamada and N. Kitamura of Obihiro University of Agriculature and Veterinary Medicine.

of the duodenal glands in the rat to be clearly separate from that of the duodenal mucosa, although arranged parallel to it (30). Arterial vessels supplying the duodenal glands course along their outer (serosal) surface, providing small vessels to the glandular tissue. Capillaries of the duodenal glands drain into one of several larger veins that lie centrally within the glandular tissue (30).

Arterial vessels supplying overlying duodenal villi pass unbranched on the mucosal side of the submucosal plexus to enter the lamina propria where they branch and provide a subepithelial capillary plexus to each villus. This plexus is continuous with the capillary plexus around adjacent intestinal glands. Venous drainage is by one or two venules within each villus which also receives tributaries from the capillary plexus of the adjacent intestinal glands. These observations give additional support to the concept that the duodenal mucosa has the capacity, due to its vascular supply, to produce adequate bicarbonate for the protection of this region against acid entering from the stomach (31, 32).

3.4. Innervation of duodenal glands

The duodenal glands release secretory product in response to vagal stimulation and are inhibited by the splanchnic nerves (33). Extensive nerve plexuses of cholinergic and adrenergic axons envelop the secretory units of the duodenal glands of the cat and guinea pig (34, 35), and numerous nerves immunoreactive for vasoactive intestinal polypeptide (VIP) are reported around the duodenal glands of the rat (36, 37). These nerve fibers are thought to originate from intramural neurons close to the duodenal glands, since the perikarya of these neurons are also VIP-immunoreactive. Exogenous administration of VIP produces a secretory response similar to that resulting from nerve stimulation, and, in this species, it was concluded that stimulation of the duodenal glands is mediated by VIP-containing nerves (36, 37). VIP- and substance P-immunoreactivities also are found in nerves and neuronal perikarya between the lobules of the human duodenal glands, but very few fibers are seen immediately adjacent to cells of the secretory units (38). J. Yamada and N. Kitamura, using the peroxidase-antiperoxidase immunocytochemical method, have demonstrated VIP-positive nerve elements around the duodenal glands of the echidna, platypus, wombat, koala, rat, guinea pig, cat, dog, pig, sheep, cow, and horse (personal communication).

3.5. Development of duodenal glands

Little work has been done on the development of the duodenal glands and those studies reported deal with species from several orders.

In man, the duodenal glands appear between three to five months gestation arising from intestinal glands (39, 40, 41, 42). The human duodenal glands develop in an anterior-posterior sequence as they do in other species studied. Lomakina (43) observed that in man the duodenal glands are found primarily in the mucosa at four months gestation except near the pylorus where they reach the submucosa. At this time each duodenal gland consists of two or three branching tubules that end in blind enlargements. The duodenal glands increase in size and number from five to seven months, and at eight months are found mainly in the submucosa. Distally they are fewer in number and smaller in size as reported in most other eutherian species.

With the exception of the monotremes, comparative studies have shown that the duodenal glands originate directly from the duodenal epithelium, intestinal glands, or, in some cases, a combination of both sites (44,45). Duodenal glands of the most proximal portion of the duodenum usually form first and often take their origin directly from intestinal epithelium. Those forming more distally and slightly later, generally develop from intestinal glands (44). The developmental origin of these glands explains the considerable variation reported in a number of species with regard to where the excretory duct of an individual duodenal gland terminates. The origin of duodenal glands from the intestinal lining epithelium or developing intestinal glands also explains how other cell types (Paneth, goblet, caveolated, enteroendocrine) come to reside within the ducts or secretory units of the duodenal glands. This concept is supported by the observation that the intestinal glands of the mouse continue to contribute epithelial cells to the developing duodenal glands for up to three weeks into the postnatal period (29). In addition, in one of the monotremes (the echidna), Brunner's glands of the adult and pouch young contain enteroendocrine cells immunoreactive for 5-HT, somatostatin, bovine pancreatic polypeptide, and motilin. These immunoreactive endocrine cells are also found in the simple columnar gastric lining epithelium of the pouch young but disappear when the lining epithelium changes to the stratified squamous of the adult (8). These observations suggest that the enteroendocrine cells in the adult Brunner's glands were derived originally during formation of the glands from the gastric lining epithelium of the pouch young (46).

3.6. *Histochemistry of duodenal glands*

Histochemical studies of the duodenal glands in a wide variety of species have reported them to contain neutral or acidic types of mucin. Acidic mucins are subdivided into sulfated or nonsulfated categories by further histochemical staining. Utilizing this type of histochemistry, the duodenal glands of a number of species have been designated as producing either neutral or acidic mucosubstances or a combination of both types, depending on the species (Table 5-1).

Neutral mucins are reported to be produced by the majority of cells that form the secretory tubules and acini in the duodenal glands. In addition to the neutral mucosubstances produced in the secretory units of the duodenal glands, cells lining the ductal system of some of these species also elaborate small amounts of acidic mucin (Table 5-1). In yet others of these species, scattered cells and/or secretory units or even occasional individual glands may also produce acidic mucins. It should be emphasized, however, that the total secretion from the majority of glands in these species is a neutral mucin, and the acidic mucosubstances form only a minor portion. Acidic mucins are the primary secretory product of the duodenal glands in only a few species (Table 5-1).

Shackleford and Wilborn (47) showed sexual differences in the amount of acidic mucins present in the duodenal glands of the hamster; the duodenal glands of male hamsters contained twice as much acidic mucin as the female. Sexual differences with regard to morphology or histochemical staining have not been reported in any other species examined.

The duodenal glands of the various species studies show marked differences in the histochemical staining properties of mucin despite the similar morphological appearance of constituent cell types. The significance of the wide variation between species and in some cases, within a species, is unclear. Generally, it has been proposed that the type of mucin produced may be related to differences in diet. Such a general correlation is far from satisfactory, since several species that belong to the same order but that have different dietary habits, may show identical histochemical staining (Table 5-1). In the marsupials, for example, the duodenal glands of the herbivorous kangaroo contain neutral mucosubstances as do those of the native cat and marsupial mouse, which are carnivorous and/or insectivorous (12). Similarly, the duodenal glands of 27 species of bat that differ markedly in diet (insect-, fish-, nectar-, blood-, plant-, or animal-feeding species), produce only neutral mucins (15). Nevertheless, it is important to recognize that such differences in mucin production exist in cells of the duodenal glands of different species, which, when stained with routine methods (H & E), appear to represent a uniform cell population.

3.6.1. Glycoproteins of duodenal glands

When glycoprotein (mucin) is secreted from the duodenal glands it forms a continuous thin viscoelastic gel that is adherent to the intestinal epithelium. It is the water-soluble properties of the high molecular weight glycoprotein that are responsible for the gel forming properties of mucin (48). Each glycoprotein subunit has a protein core bristling with short carbohydrate side chains. The carbohydrate component is thought to stabilize the glycoprotein molecule and to protect the protein core from digestion by proteolytic enzymes within the digestive tract. The structure of purified glycoproteins isolated from the duodenal glands of the rat (49, 50) and man (51) have been reported. Glycoprotein of rat and human duodenal gland mucus shows some homology in chemical composition. Both have a relatively high content of protein, contain little sulfate, and lack neuraminic acid. These results confirm the histochemical staining reactions observed in the duodenal glands of both species that were reported to contain neutral mucosubstances (52, 53) and demonstrate the value of such histochemical evaluations.

3.6.2. Immunohistochemistry of duodenal glands

A 53 amino acid peptide, epidermal growth factor (urogasterone), has been identified in the secretory granules of cells comprising the secretory units of the duodenal glands in man (54, 55), rat (56), and mouse (57). Epidermal growth factor exhibits a number of biological effects including inhibition of gastric acid secretion, cytoprotective effects on the duodenal mucosa, and stimulation of cell growth and differentiation (56, 58). Its precise role in the secretion of the duodenal glands is unknown.

Enterokinase, an enzyme in the small intestine that triggers the activation of zymogens in pancreatic juice, was proposed as being produced by Brunner's glands and then subsequently bound in the glycocalyx overlying the brush border of enterocytes (59, 60, 61). With the availability of a specific antiserum to human enterokinase, enterokinase immunoreactivity was observed in the brush border and glycocalyx of the duodenum and proximal jejunum. The human duodenal glands are consistently negative for the presence of enterokinase using this specific immunocytochemical technique (62).

3.6.3. Enzyme histochemistry of duodenal glands

In some mammalian species, the duodenal glands have been associated with certain enzymes. Moe (24) reported strong alkaline phosphatase activity in the duodenal glands of the cat. Other species have been reported as giving a negative alkaline phosphatase reaction with regard to the cytoplasm of cells comprising the duodenal glands (63, 64, 65). Cochrane et al. (64) reported scattered acid phosphatase activity in cells of the duodenal glands of the guinea pig, and acid phosphatase activity has been observed in the marmosct (65) and duckbilled platypus (9) also. Nonspecific esterase is observed in the duodenal glands of the platypus (9), mouse, and rat (66), and lipase has been localized in the serous acini of the rabbit duodenal glands (67).

4. Electron microscopy of duodenal glands

Electron microscopic studies have demonstrated marked species differences in the ultrastructural characteristics of duodenal glands. Although the glands are considered to consist of mucus producing cells, some show ultrastructural features that generally are regarded to be more closely associated with serous cells. In a second group, cells of the duodenal glands are more closely related to typical mucous cells found elsewhere. The duodenal glands of a third group show ultrastructural features intermediate between the former two groups. Although typical organelles associated with mucin synthesis are found in the cells of the duodenal glands in all species, differences primarily are related to the number, location within the cell, and the nature of the secretory granules. The rabbit is the only species examined in which the duodenal glands contain two distinct cell types, mucous and serous.

The ultrastructural features of cells comprising the duodenal glands of the mouse (68), opossum (13), platypus (9), marsupial mouse (12), and three European insectivores (69) more closely resemble those of classical serous cells. In these species, the apical cell membranes of the secretory cells show only a few short microvilli that are irregular in orientation and distribution. The apices of adjacent cells within a secretory tubule are united by typical tripartite junctional complexes. Lateral cell membranes of adjacent secretory cells are in close apposition and show scattered desmosomal attachments. Occasionally, near the apex where secretory cells meet, intercellular secretory canaliculi are found, formed by the plasmalemmae of surrounding cells. Numerous microvilli often project into these narrow intercellular channels. The lateral cell membranes beneath such channels or the apical junctional complexes are relatively smooth but may show some interdigitation with adjacent cell membranes near the base of the epithelium. The basal plasmalemma of all species is relatively smooth and shows few if any infoldings. It rests on a delicate basal lamina.

Nuclei are usually ovoid or spherical and are located in the basal half of the cell. They often show one or two nucleoli and irregular masses of heterochromatin are arranged around the periphery of the nucleus in relation to the nuclear envelope. Nuclear pores within the nuclear envelope are common. The remainder of the chromatin is dispersed in the form of euchromatin and usually comprises most of the nuclear volume.

Mitochondria are moderately numerous, appear ovoid or elongate, and contain numerous cristae. Rough endoplasmic reticulum is located primarily in the basal and perinuclear regions and consists of numerous parallel cisternae. Individual cisternae may partially encircle a mitochondrion or extend into the supranuclear region (Fig. 5-1). Ribosomes occur singly and in small clusters on the external surface of the endoplasmic reticulum. In the supranuclear region, near a well-developed Golgi complex, ribosome-free areas occur on cisternae, and these give rise to numerous small outpocketings and vesicles (Fig. 5-1). The formed smooth-surfaced transport vesicles measure 60 to 90 nm in diameter and link the terminal elements of the rough endoplasmic reticulum to the nearest cisternae of the Golgi complex. The transport vesicles carry formed protein from the rough endoplasmic reticulum to the Golgi complex and contribute membrane to cisternae on the forming face of the Golgi complex.

The Golgi complex in this cell type shows extensive development and is characterized by several large, parallel stacks of smooth-surfaced cisternae. The transport vesicles accumulate along Golgi cisternae at the forming face (Fig. 5-1). The outermost Golgi cisternae of the forming face are extensively fenestrated and may appear as a row of vesicles. The transport vesicles gradually meld with and become incorporated into the Golgi membranes. The formed Golgi cisternae often show fusiform enlargements along their lengths and may exhibit terminal expansions near the edge of the complex. The expanded cisternal segments give rise to condensing vacuoles which eventually mature into secretory granules by addition and concentration of secretory product. It is within the Golgi complex that protein accumulates and the carbohydrate moiety is synthesized and complexed to the protein to form the glycoprotein secreted by cells of the duodenal glands.

Rohr et al. (70) and Schmalbeck and Rohr (71), used ^{3}H-labelled leucine and glucose to demonstrate that protein in the duodenal glands of the mouse is synthesized in the rough endoplasmic reticulum and then transported, via transport vesicles, to the Golgi complex where the synthesized carbohydrate is coupled to the protein core of the glycoprotein.

In other species — the cat (24), rat (23), echidna (8), wombat, short-nosed bandicoot, and grey kangaroo (12) — cells of the duodenal glands appear to be more mucous than serous with regard to their internal fine structure. Numerous free ribosomes are present within the basal cytoplasm as are some profiles of rough endoplasmic reticulum. Several Golgi complexes are found within the supranuclear cytoplasm, but their development and concentration is not as marked as in the duodenal glands of species in the former group. In the latter group, the secretory

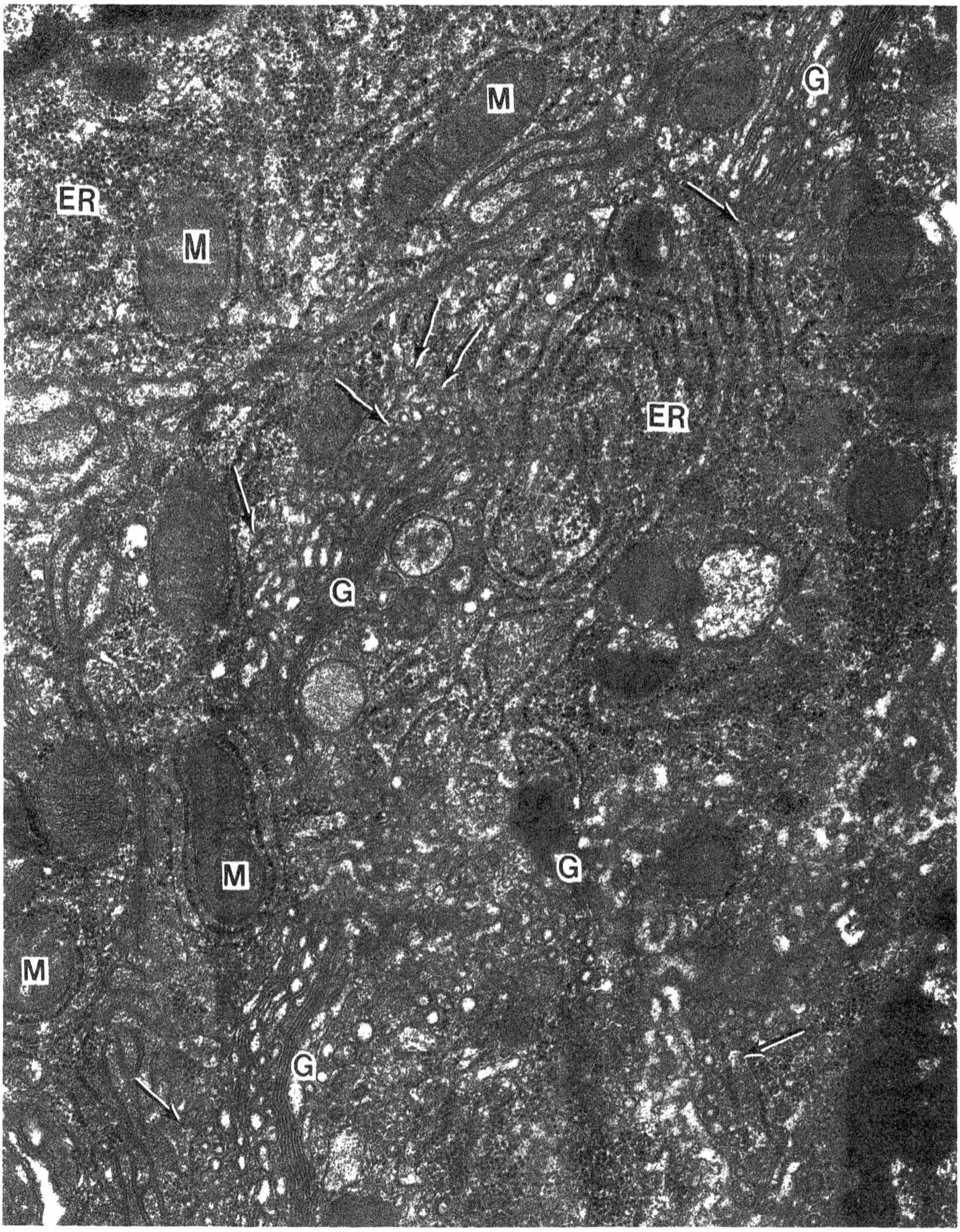

Figure 5-1. Extensive networks of granular endoplasmic reticulum (ER) and well-developed Golgi complexes (G) with forming secretory granules occupy the supranuclear region of three adjacent cells taken from a duodenal gland of a white-footed mouse (*Peromyscus leucopus*). Note the numerous transport vesicles (arrows) originating from smooth membrane regions of the granular endoplasmic reticulum and their association with the Golgi membranes. The relationship of mitochondria (M) enveloped by cisternae of rough endoplasmic reticulum also is clearly demonstrated. X 25,000.

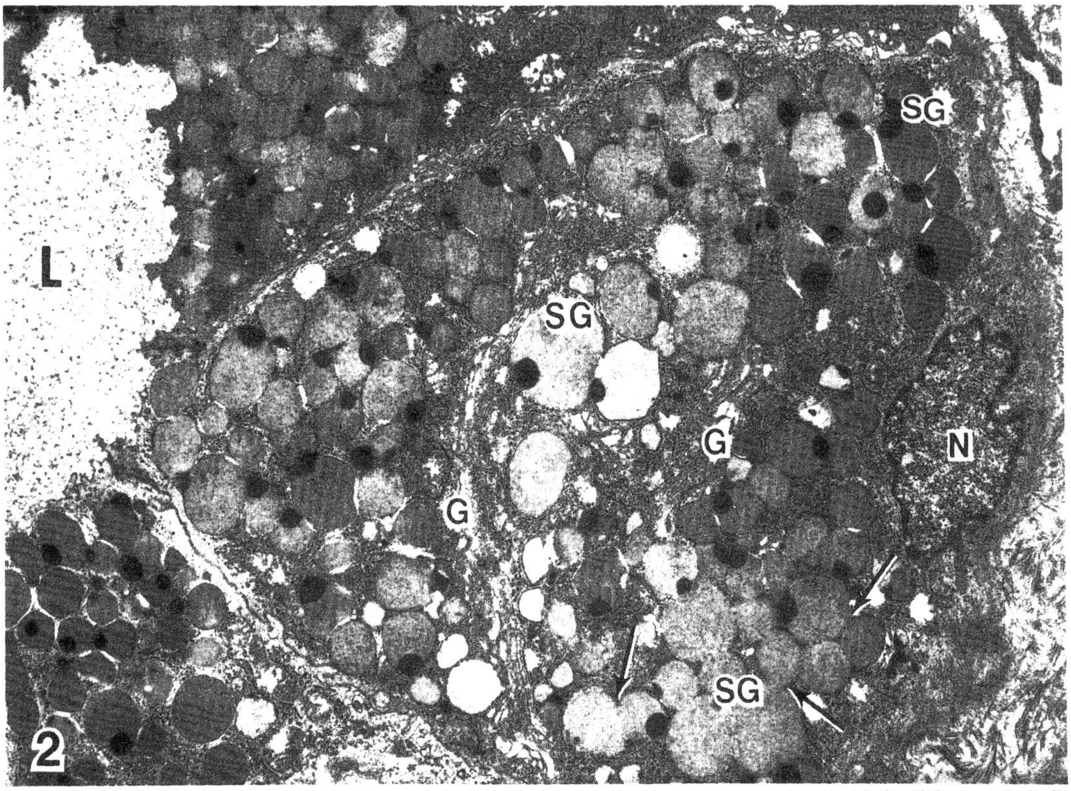
L
SG
SG
G
N
G
SG
2

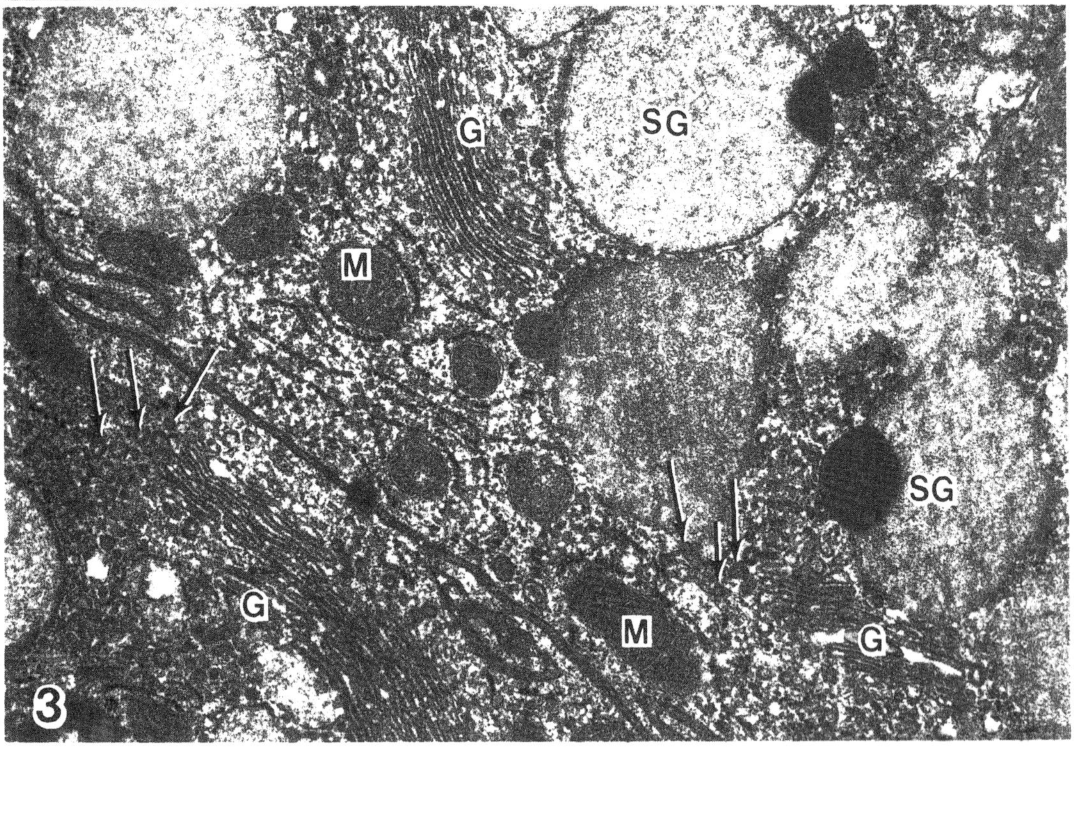
G
SG
M
SG
G
M
G
3

granules tend to fuse into irregular complexes and generally are electron-lucid.

Studies using ultrastructural histochemistry have demonstrated carbohydrate in Golgi cisternae, condensing vacuoles, and secretory granules in duodenal gland cells of the rat (72). The pathway of glycoprotein synthesis by the rat duodenal glands appears identical to that reported in the mouse. The incorporation patterns of radioactive precursors in the duodenal glands is similar to other mucus-secreting cells such as goblet cells (73, 74).

Sugai and Oosaki (75) have demonstrated carbonic anhydrase activity in cells of the secretory units and ducts of the rat duodenal glands at the ultrastructural level. Their electron microscopic observations indicate the activity to be located primarily in microvilli at the cell apex and in those projecting into intercellular canaliculi. Regions of interdigitation of lateral cell membranes as well as the basal cytoplasm also shows some activity. Intestinal epithelial cells covering the duodenal villi also gave a strong reaction for carbonic anhydrase activity.

In constrast to the former two groups of animals, cells comprising the duodenal glands of the guinea pig (64), man (76), and marsupial cat (12) appear more mucous in nature. In these species, the nuclei and organelles of the cells of the duodenal glands are similar to those seen in the other two groups but are confined primarily to the basal and lateral regions of the cell. The most prominent feature of the secretory cells of the duodenal glands in this particular group are the secretory droplets. These are large and concentrated apically, but generally are so numerous that they fill the cytoplasm of the cell (Fig. 5-2). The secretory droplets are membrane-bound although the limiting membrane is often indistinct (Figs. 5-3, 5-4). The secretory droplets tend to fuse into irregularly-shaped complexes, are electron lucid and often show a fine, fibrillar content (Figs. 5-3, 5-4). In man the secretary granules often show a region of increased electron density which gives them a mottled appearance (Figs. 5-2, 5-3, 5-4, 5-5). Similar mottled appearing secretory granules have been reported in the duodenal glands of bandicoot (12) and manatee (77), as well as in mucous cells of the secretory tubules in the human cardiac and pyloric glands (78, 79).

Secretory granules of all species examined to date are released from duodenal gland cells by exocytosis. The limiting membranes of the secretory granules tend to fuse with the apical plasmalemma and release their contents extracellularly (Fig. 5-5). The membranes usually become incorporated into the cell membrane but, on occasion, may be released intact with a secretory granule (12). Depletion of secretory granules after stimulation of the duodenal glands does not change the pathway or kinetics of radioactive precursors into the organelles involved in the synthetic process (80).

5. Control of duodenal gland secretion

Secretion by the duodenal glands can be stimulated or inhibited by neural, hormonal, and/or luminal factors. Two extensive reviews (81, 82) have addressed the question as to what factor or factors influence duodenal gland secretion, and some of the conclusions are summarized in Table 5-3.

Although several factors have been attributed to influencing duodenal gland secretion directly, these are complicated by duodenal motility, spontaneous secretion, and secretory response to feeding (82). In addition, the secretory response itself appears complex as shown by the duodenal glands in response to acidification of the duodenum. In this case, the secretory response that occurs is diphasic. The initial response is a short-lasting, motor-dependent, secretory response that results

←

Figure 5-2. A cell taken from a secretory tubule of a human duodenal gland is filled with secretory granules (SG) compressing the nucleus (N) towards the base. Well-developed Golgi complexes (G) are present and the secretory granules show a tendency to coalesce into irregularly shaped complexes (arrows). The lumen (L) of the secretory tubule is shown at the left. X 7,300.

Figure 5-3. The lateral, supranuclear cytoplasm of two adjacent cells in a secretory tubule from a human duodenal gland exhibits transport vesicles (arrows) originating from smooth membrane areas of the granular endoplasmic reticulum, well-developed Golgi complexes (G), mitochondria (M) enveloped by profiles of granular endoplasmic reticulum, and membrane-bound secretory granules (SG). Note the region of increased electron density within individual secretory granules. X 25,000.

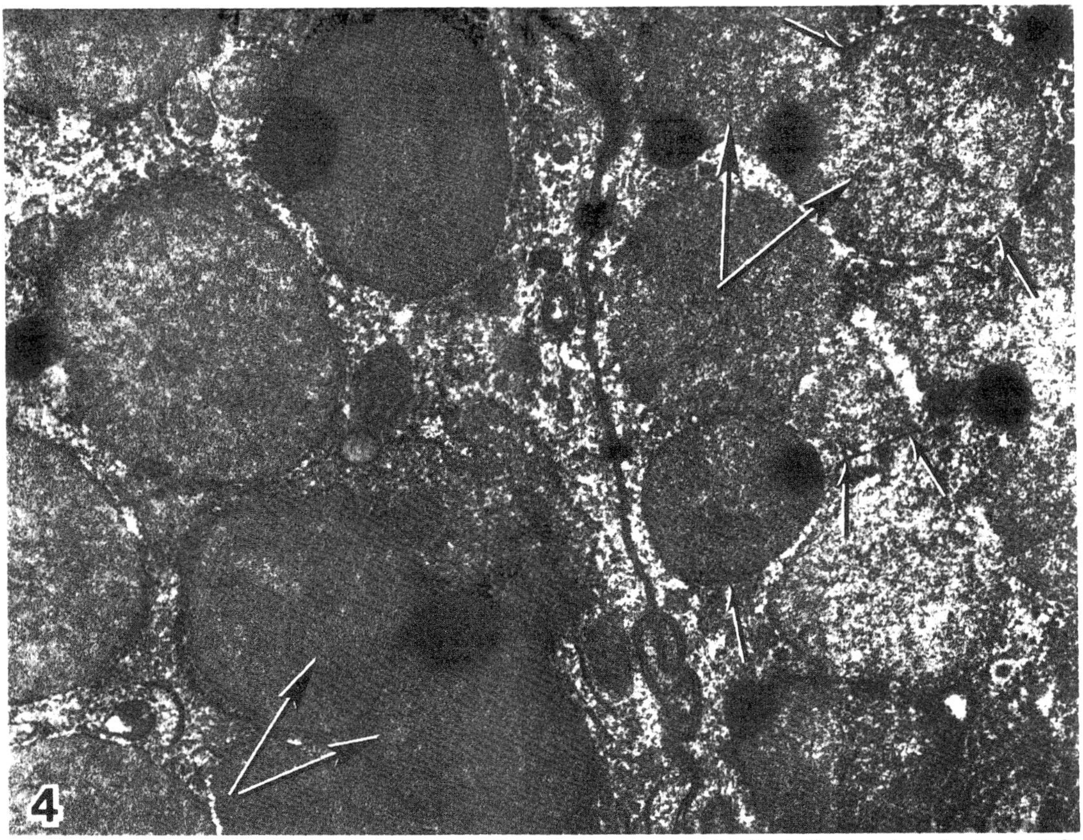
4

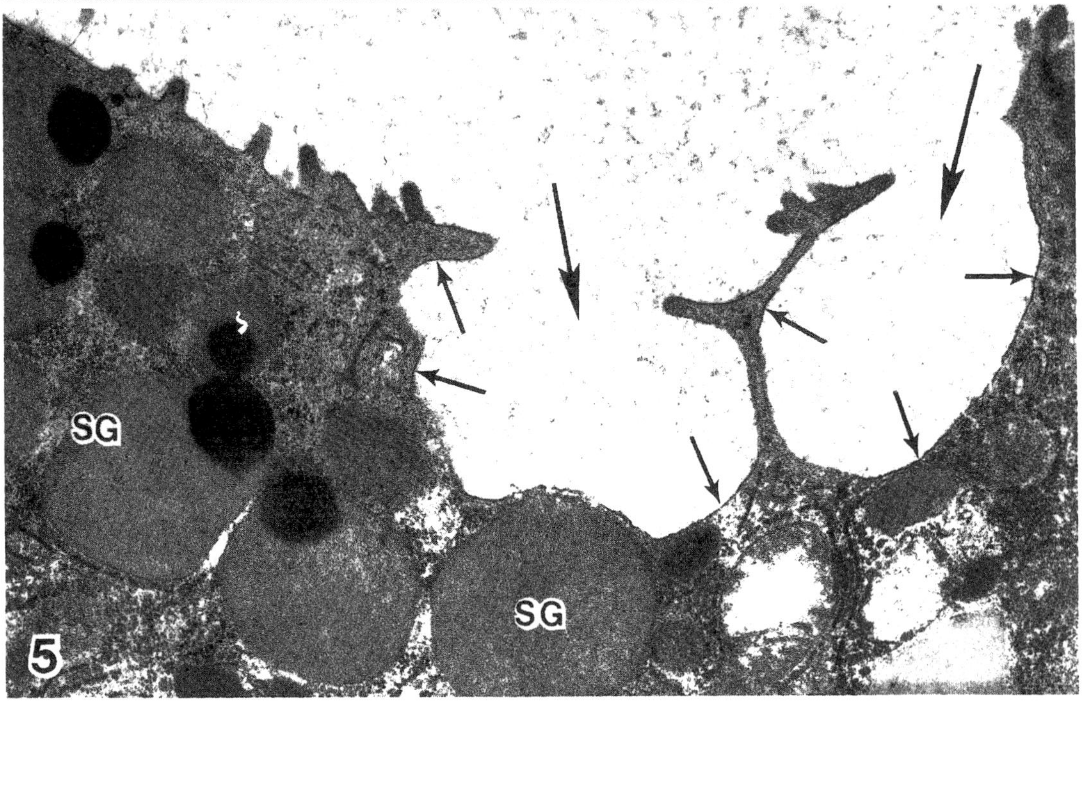
SG
SG
5

Table 5-3. Factors that stimulate(+) or inhibit(−) duodenal gland secretion.

Neural	Luminal	Hormonal
vagus(+)	acid (HCL, H_2SO_4, butyric)(+)	cholecystokinin(+)
sympathics(−)	mechanical distention(+)	secretin(+)
VIP containing neurons(+)	feeding(+)	gastrin(+)
	intestinal motility(+)	glucagon(?)
		somatostatin(−)
		histamine(+)
		serotonin(+)

from the expulsion of mucus already present in the lumina of ducts and secretory units. A second, delayed, long-lasting secretory period occurs and is thought to be due to the synthesis and release of new glycoprotein (81, 82). VIP, acetylcholine, and secretin have been reported to stimulate the secretion of glycoprotein, epidermal growth factor, and bicarbonate from the duodenal glands. Glucagon has no effect on duodenal gland secretion, and the effects of VIP-stimulated secretion are inhibited by somatostatin (37, 56, 83). Precisely what major factor or factors control duodenal gland secretion and how they are interrelated physiologically remains controversial. Further studies using modern technologies are needed.

6. Function of the duodenal glands

The early literature concerning duodenal gland function has been reviewed thoroughly by Grossman (84) and Cooke (85). The secretion of the duodenal glands contains only small amounts of digestive enzymes, is highly viscous due to the glycoprotein content, and is less alkaline than pancreatic secretion. The secretion has little digestive capacity and cannot, by itself, neutralize the entering acid pepsin from the stomach to any great extent. The primary function of the duodenal glands, acting through the glycoproteins they produce, is thought to be the protection of the proximal duodenal mucosa from acidic and/or abrasive gastric contents (32, 84, 86, 87). The region of the duodenum containing the duodenal glands is more resistant to damage from either acid or acid pepsin than other regions of the small intestine (84, 88) and has the capacity to dispose of acid entering the lumen (84, 89, 90, 91). Recent studies have shown that bicarbonate secretion originating from surface (intestinal) epithelium in response to intraluminal acid is important in protection of the duodenal mucosa (31, 32). This response has been demonstrated in mammalian and amphibian species (which lack duodenal glands), and stimulation is thought to be mediated through mucosal prostaglandins, hormonal agents, and perhaps neural mechanisms as well (31, 32). This concept is supported, in part, by experiments using a variety of ulcerogenic agents such as cinchophen (92), 3-4-tolenediamine (93), and cysteamine (94, 95). Cysteamine, in addition to inhibiting duodenal gland synthesis and secretion, impairs the ability of the duodenal mucosa to respond to intraluminal acid (96).

Glycoproteins secreted by the duodenal glands form a tightly adherent gel that covers the mucosal surface. This barrier has short-term resistance to bile, pepsin, hypertonic NaCl, acid, and ethanol (97). However, it is readily permeable to

←

Figure 5-4. The secretory granules show a tendency to coalesce into irregularly shaped complexes (large arrows). The membranes limiting the secretory granules may continue to partially limit the secretory complex and fragments may be trapped within the secretory mass (small arrows). A tight junction is shown between cell apices at the top-center of the micrograph. Human duodenal gland. X 25,000.

Figure 5-5. The irregularly shaped secretory masses or individual granules may be discharged into the lumen of the secretory tubule (large arrows). A membrane (small arrows) continues to line the concavities formed following the discharge of secretory material. Human duodenal gland. X 22,400.

small molecules and ions and will not prevent these agents from reaching the underlying epithelial cells (48). Although the layer mucus allows penetration of several substances to the underlying mucosa it remains intact, thereby minimizing damage and enhancing repair. The mucous layer provides a zone of stability on the epithelial surface and is thought to delay the diffusion of acid toward the mucosa, allowing enough time to be neutralized by the bicarbonate originating from the surface epithelium.

7. Concluding remarks

It is of interest that the duodenal glands are present in all mammals but absent in other vertebrate species. The gastric mucosa of birds, reptiles, amphibians, and fishes also elaborate acid pepsin, and in these species bicarabonate production by the duodenal mucosa is sufficient for protection against gastric contents entering the small intestine. Furthermore, if the duodenal glands of the rat are surgically removed, lesions fail to develop in the mucosa up to ten days postoperatively, even following starvation and/or administration of gastrin (W.J. Krause, unpublished observations). The comparative observations together with the ablation study leave some doubt concerning the accepted interpretation of duodenal gland function. Isolation and characterization of glycoproteins produced by duodenal glands in the rat and man have shown earlier histochemical tests to be highly accurate. Previous ultrastructural studies have dealt with characterizing duodenal gland cells of various species and determining the intracellular pathway of glycoprotein synthesis. Further ultrastructural studies that use modern immunocytochemical and histochemical methods will play a pivotal role in localizing duodenal gland glycoproteins in the intestinal tract and in determining their precise influence on the intestinal epithelium in the intestinal glands and on villi.

References

1. Wepfer JJ: Cicutae aquaticae historia et noxae. Basileae, 1679.
2. Brunner JC: De glandulis in intestino duodeno hominis detectis. Dissertation, Heidelberg, 1688.
3. Brunner JC: Glandulae duodeni seu pancreas secondarium in intestino duodeno hominis primum abhinc in ahis quoque animalibus detectum. Heidelberg: University of Heidelberg, 1715.
4. Dobson J: Anatomical Eponyms, 2nd ed. London: E & S Livingstone, 1962.
5. Middeldorpf AT: Disquisitio de Glandulis Brunnianis. Vratislavia, 1846.
6. Florey HW, Harding HE: The function of Brunner's glands and the pyloric end of the stomach. *J Pathol Bacteriol* 37: 431–453, 1933.
7. Florey HW, Harding HE: Further observations on the secretion of Brunner's glands. *J Pathol Bacteriol* 39: 255–276, 1934.
8. Krause WJ: Brunner's glands of the echidna. *Anat Rec* 167: 473–488, 1970.
9. Krause WJ: Brunner's glands of the duckbilled platypus (*Ornithorhynchus anatinus*). *Am J Anat* 132: 147–166, 1971.
10. Krause WJ, Leeson CR: The gastric mucosa of two monotremes: The duckbilled platypus and echidna. *J Morphol* 142: 285–300, 1974.
11. Krause WJ: The distribution of Brunner's glands in 55 marsupial species native to the Australian region. *Acta Anat* 82: 17–33, 1972.
12. Krause WJ: Morphological and histochemical features of the duodenal glands in six marsupial species. *J Morphol* 140: 321–342, 1973.
13. Krause WJ, Leeson CR: Studies of Brunner's glands in the opossum. I. Adult morphology. *Am J Anat* 126: 255–274, 1969.
14. Kuczynske A: Beitrag zur histologie der brunnerschen drusen. *Intern Monatschr Anat Phys*, 7: 419–446, 1890.
15. Forman GL: Histochemical differences in gastric mucus of bats. *J Mammal* 52: 191–193, 1971.
16. Bjerknes M, Cheng H, Erlandsen S: Functional gap junctions in mouse small intestinal crypts. *Anat Rec* 212: 364–367, 1985.
17. Carleton A: The distribution of Brunner's glands in the duodenum of mammals. *Proc Zool Soc Lond* 1: 385–390, 1935.
18. Krause WJ: Morphological and histochemical observations on the duodenal glands of eight wild ungulate species native to North America. *Am J Anat* 162: 167–181, 1981.
19. Landboe-Christensen E: The duodenal glands of Brunner in man, their distribution and quantity. *Acta Pathol Microbiol Scand* (suppl) 52: 1–267, 1944.
20. Krause WJ: A brief note on the morphological and histochemical features of the duodenal glands in 5 species of insectivore. *Anat Anz* 147: 118–124, 1980.
21. Treasure T: The ducts of Brunner's glands. *J Anat* 127: 299–304, 1978.
22. Leeson CR, Leeson TS: The fine structure of the Brunner's glands in the rabbit. *Anat Rec* 159: 409–

420, 1967.
23. Leeson CR, Leeson TS: The fine structure of the Brunner's glands in the rat. *Anat Rec* 156: 253–267, 1966.
24. Moe H: The ultrastructure of Brunner's glands of the cat. *J Ultrastruct Res* 4: 58–72, 1960.
25. Krause WJ: The duodenal glands of fourteen primate species. *Acta Anat* 93: 580–589, 1975.
26. Krause WJ: Caveolated cells observed in the duodenal glands of the white-tailed deer. *Experientia* 36: 341–342, 1980.
27. Niki H: Studies on architectures and cellular kinetics of duodenal mucosa. *J Kyoto Prefect Univ Med* 89: 717–734, 1980.
28. Fuse Y: Kinetic studies on the Brunner's glands in healing process of duodenal ulcer. *J Kyoto Prefect Univ Med* 93: 711–729, 1984.
29. Obuoforibo AA, Martin BF: Postnatal growth of Brunner's glands in the mouse. *J Anat* 124: 779–790, 1977.
30. Browning J, Gannon B: The microvascular architecture of rat proximal duodenum, with particular reference to Brunner's glands. *Biomed Res* 5: 245–258, 1984.
31. Flemstrom G, Garner A: Some characteristics of duodenal epithelium. *Ciba Found Symp* 109: 94–108, 1984.
32. Flemstrom G, Turnberg LA: Gastroduodenal defense mechanisms. *Clin Gastroenterol* 13: 327–354, 1984.
33. Wright RD, Jennings MA, Florey HW, Lium R: The influence of nerves and drugs on secretion by the small intestine and an investigation of the enzymes in intestinal juice. *Q J Exp Physiol* 30: 70–120, 1940.
34. Kyosolo K: Innervation of the duodenal submucosal Brunner's glands. *Acta Physiol Scand* 101: 498–500, 1977.
35. Stach VW, Hung N: Light microscopic and histochemical results on the innervation of Brunner's glands in the duodenum of laboratory animals. *Anat Anz* 144: 25–35, 1978.
36. Kirkegaard P, Lundberg JM, Poulsen SS, Olsen PS, Fahrenkrug J, Hokfelt T, Christiansen J: Vasoactive intestinal polypeptidergic nerves and Brunner's gland secretion in the rat. *Gastroenterology* 81: 872–878, 1981.
37. Kirkegaard P, Olsen PS, Nexo E, Holst JJ, Poulsen SS: Effect of vasoactive intestinal polypeptide and somatostatin on secretion of epidermal growth factor and bicarbonate from Brunner's glands. *Gut* 25: 1225–1229, 1984.
38. Ferri G-L, Botti P, Biliotti G, Rebecchi L, Bloom SR, Tonelli L, Labo G, Polak JM: VIP, substance P-, and metenkephalin-immunoreactive innervation of the human gastroduodenal mucosa and Brunner's glands. *Gut* 25: 948–952, 1984.
39. Brand E: Beitrage zur entwickelung der magen und darmwand. *Verhandl d Phys -Med Gesellsch in Wurzb* 9: 243–256, 1877.
40. Heisina VI: Histogenesis of the duodenum in man. *Arkh Anat Gistol I Embriol* 34: 100–102, 1957.
41. Johnson FP: The development of the mucous membrane of the esophagus, stomach and small intestine in the human embryo. *Am J Anat* 10: 521–561, 1910.
42. Patzelt V: Die feinere ausbildung des menschlichen darmes von der funften woche bis zur beburt. *Z mikrosk Anat Forsch* 27: 269–518, 1931.
43. Lomakina LI: Age-conditioned characteristics of the human duodenal glands. *Trudy III Nauchnol Konferentsii po Voprosam Vozrastnoi Morfologii, Fizologii Ibigrimi* (Moskva) p 569–574, 1959.
44. Krause WJ, Leeson CR: The origin, development and differentiation of Brunners's glands in the rat. *J Anat* 101: 309–320, 1967.
45. Krause WJ, Leeson CR: Studies of Brunner's glands in the opossum. II. Postnatal development. *Am J Anat* 126: 275–290, 1969.
46. Krause WJ, Yamada J: A brief report on enteroendocrine cells observed in the gastrointestinal tract of the suckling echidna, *Tachyglossus aculeatus*. *Z Mikrosk-Anat Forsch* 100: 464–468, 1986.
47. Schackleford JM, Wilborn WH: Sexual differences among duodenal glands of hamster. *Am J Anat* 152: 441–446, 1978.
48. Bell AE, Sellers LA, Allen A, Cunliffe WJ, Morris ER, Ross-Murphy SB: Properties of gastric and duodenal mucus: Effect of proteolysis disulfide reduction, bile, acid, ethanol, and hypertonicity on mucus gel structure. *Gastroenterology* 88: 269–280, 1985.
49. Smits HL, Van Kerkhof PJM, Kramer MF: Composition and structure of rat Brunner's glands and gastric mucus glycoprotein. *Adv Exp Med Biol* 144: 159–161, 1982.
50. Smits HL, Van Kerkhof PJM, Kramer MF: Isolation and partial characterization of rat duodenal-gland (Brunner's-gland) mucus glycoprotein. *Biochem J* 203: 779–786, 1982.
51. Smits HL, Kramer MF: Human duodenal gland (Brunner's gland) mucus glycoprotein analysis. *Arch Biochem Biophys* 228: 64–70, 1984.
52. Sheahan DG, Jervis JR: Comparative histochemistry of gastrointestinal mucosubstances. *Am J Anat* 146: 103–132, 1976.
53. Ganter P, Marche C: Histochimie des mucines gastrointestinales de l'homme. *Ann Anat Path* 15: 321–346, 1970.
54. Elder JB, Williams G, Lacy E, Gregory H: Cellular localization of human urogasterone/epidermal growth factor. *Nature* 271: 466–467, 1978.
55. Heitz PH, Kasper M, VanNoorden S, Polak JM, Gregory H, Pearse AGE: Immunohistochemical localization of urogastrone to human duodenal and submandibular glands. *Gut* 19: 408–413, 1978.
56. Kirkegaard P, Olsen PS, Poulsen SS, Nexo E: Exocrine secretion of epidermal growth factor from Brunner's glands. Stimulation by VIP and acetylcholine. *Reg Pept* 7: 367–372, 1983.
57. Van Noorden S, Heitz P, Kasper M, Pearse AGE: Mouse epidermal growth factor: Light and electron microscopical localization by immunocytochemical staining. *Histochemistry* 52: 329–340, 1977.
58. Hollenberg MD: Epidermal growth factor-urogasterone a polypeptide acquiring hormonal status. *Vitam Horm* 37: 69–110, 1979.
59. Eggermont E, Molla AM, Rutgeerts L, Tytgat G: The source of human enterokinase. *Lancet* 2: 369, 1971.
60. Rutgeerts L, Tytgat G, Eggermont E: Localization of enterokinase in man. *Gastroenterology* 63: 381–382, 1972.
61. Woodley JF, Keane R: Enterokinase in normal intestinal biopsies and those from patients with untreated coeliac

disease. *Gut* 13: 900–902, 1972.
62. Hermon-Taylor J, Perrin J, Grant DAW, Appleyard A, Bubel M, Magee AI: Immunofluorescent localization of enterokinase in human small intestine. *Gut* 18: 259–265, 1977.
63. Kar AB, Banerjee S, Ghosh A: Studies in the distribution and concentration of alkaline phosphatase in different tissues of alloxan-diabetic rats. *Anat Anz* 98: 336–343, 1952.
64. Cochrane W, Davies DV, Palfrey AJ, Stockwell RA: The histochemistry and electron microscopy of Brunner's glands in the guinea pig. *J Anat* 98: 1–10, 1964.
65. Miraglia T, Ledoux L, Castelo Branco AL: Histological and histochemical data on the intestinal tract of the marmoset, *Callithrix jacchus*. *Acta Anat* 68: 459–472, 1967.
66. Richterich R: Uber die lokalisation einiger esterassen in verschiedenen organen der albinomaus. *Acta Anat* 14: 342–353, 1952.
67. Martin BF: Serous cells in Brunner's glands of the rabbit. *Nature* 174: 1195–1196, 1954.
68. Friend D: The fine structure of Brunner's glands in the mouse. *J Cell Biol* 25: 563–576, 1965.
69. Weise VB: Zur feinstruktur der brunnerschen drusen, der panethschen zellen and der becherzellen dreier mitteleuropaischer insektivorenarten (Erinaceus europaeus, Talpa europa, Sorex araneus). *Zoo Jb Anat* 94: 129–142, 1975.
70. Rohr HP, Schmalbeck J, Felhege A: Electronenmikroskopish-autoradiographische unterschuhungen uber die eiweisse-synthese in der brunnerschen druse der maus. *Z Zellforsch* 80: 183–204, 1967.
71. Schmalbeck J, Rohr H: die mukopolysaccharid-synthese in ihrer beziehung zur eiweiss-synthese in der brunnerschen druse der maus. *Z Zellforsch* 80: 329–344, 1967.
72. Smits HL, Kramer MF: Glycoprotein synthesis in the mucous cells of the vascularly perfused rat stomach. III. Mucous cells of the antrum and the duodenal glands. *Am J Anat* 161: 365–374, 1981.
73. Peterson M, Leblond CP: Synthesis of complex carbohydrates in the Golgi region as shown by radioautography after injection of labeled glucose. *J Cell Biol* 21: 143–148, 1964.
74. Neutra M, Leblond CP: Synthesis of the carbohydrate of mucus in the Golgi complex as shown by electron microscopic radioautography of goblet cells from rats injected with glucose H^3. *J Cell Biol* 30: 119–136, 1966.
75. Sugai N, Oosaki T: Cytochemistry for carbonic anhydrase activity in rat Brunner's gland. *Fukushima J Med Sci* 29: 63–75, 1983.
76. Leeson TS, Leeson CR: The fine structure of the Brunner's glands in man. *J Anat* 103: 263–276, 1968.
77. Reynolds JE, Krause WJ: A note on the duodenum of the West Indian manatee, *Trichechus manatus*, with emphasis on the duodenal glands. *Acta Anat* 114: 33–40, 1982.
78. Krause WJ, Ivey KJ, Baskin WN, MacKercher PA: Ultrastructure of the human pyloric glands with emphasis on the mucous cell component. Part I. *Acta Anat* 99: 1–10, 1977.
79. Krause WJ, Ivey KJ, Baskin WN, MacKercher PA: Morphological observations on the normal human cardiac glands. *Anat Rec* 192: 59–72, 1978.
80. Scott CA, Flickinger CJ: Secretory process in Brunner's glands during recovery from stimulation with a single dose of pilocarpine. *Anat Rec* 206: 267–282, 1983.
81. Lang IM, Tansy, MF: Neural and hormonal control of Brunner's gland secretions. *Life Sci* 30: 409–417, 1982.
82. Lang IM, Tansy MF: Brunner's glands. *Int Rev Physiol* 28: 85–102, 1983.
83. Kirkegaard P, Olsen PS, Poulsen SS, Holst JJ, Schaffalitzky de Muckadell OB, Christiansen J: Effect of secretin and glucagon on Brunner's gland secretion in the rat. *Gut* 25: 264–268, 1984.
84. Grossman MI: The glands of Brunner. *Physiol Rev* 38: 675–690, 1958.
85. Cooke AR: The glands of Brunner. In: Handbook of Physiology, Section 6, Alimentary canal, 2: Secretion. CF Code (ed) Wash DC: Physiol Soc p 1087–1095, 1967.
86. Florey HW, Wright RD, Jennings MA: The secretions of the intestine. *Physiol Rev* 21: 36–69, 1941.
87. Florey HW: Mucin and the protection of the body. *Proc Sco Lond* (Biol) 143: 147–158, 1955.
88. Florey HW, Jennings MA, Jennings DA, O'Connor RC: The reactions of the intestine of the pig to gastric juice. *J Pathol Bacteriol* 49: 105–123, 1939.
89. Griffith CA, Harkins HN: The role of Brunner's glands in the intrinsic resistance of the duodenum to acid-peptic digestion. *Ann Surg* 143: 160–172, 1956.
90. Harmon JW, Woods M, Gurll NJ: Different mechanisms of hydrogen ion removal in stomach and duodenum. *Am J Physiol* 235: E692–E698, 1978.
91. Himal HS, Werther HW, Chapman ML, Janowitz HO, Rudick J: Acid absorption in the canine duodenum. *Ann Surg* 185: 481–487, 1977.
92. Hartiala K, Ivy AC, Grossman MI: Effect of feeding cinchophen on secretion of juice by duodenal glands in dog. *Am J Physiol* 162: 110–114. 1950.
93. Perkins WE, Green TJ: Effect of 3, 4-toluenediamine on output from in situ rat Brunner's gland pouches. *Proc Soc Exp Biol Med* 149: 991–994, 1975.
94. Kirkegaard P, Poulsen SS, Halse C, Loud FB, Olsen K, Christiansen J: The effect of cysteamine on the Brunner gland secretion in the rat. *Scand J Gastroent* 16: 93–96, 1981.
95. Poulsen SS, Kirkegaard P, Olsen PS, Christiansen J: Brunner's glands of the rat during cysteamine ulceration. *Scand J Gastroent* 16: 459–464, 1981.
96. Briden S, Flemstrom G, Kivilaakso E: Cysteamine and propionitrile inhibit the rise of duodenal mucosal alkaline secretion in response to luminal acid in rats. *Gastroenterology* 88: 295–302, 1985.
97. McQueen S, Hutton D, Allen A, Garner A: Gastric and duodenal surface mucus gel thickness in rat: Effects of prostaglandins and damaging agents. *Am J Physiol* 245: G388–G393, 1983.

Author's Address:
Department of Anatomy
School of Medicine
University of Missouri-Columbia
Columbia, Missouri 65212, USA

CHAPTER 6

Jejunum and villi: Structural basis of intestinal absorption

TORAO YAMAMOTO

1. Introduction

The jejunum is the main portion of the intestine involved in the digestion and absorption of nutrients in the digestive canal. According to an old concept, the food materials were believed to be first broken down into micromolecules, such as amino acids or monosaccharides, by digestive enzymes in the lumen of the digestive canal and then absorbed through the epithelial lining into the blood capillaries or lymphatics. However, since it has been revealed that terminal digestive enzymes such as disaccharidases and dipeptidases are concentrated in the brush border membrane of absorptive cells (1,2), this concept had to be revised.

There have been many attempts to clarify the mechanism of intestinal absorption morphologically, biochemically, and physiologically. Electron microscopic studies have revealed the ultrastructure of intestinal absorptive cells and contributed a great deal to our better understanding the mechanism of intestinal absorption. With the availability of recent advanced techniques for electron microscopy, the mechanism of intestinal absorption has been considerably revealed.

The present review deals with the structural basis of intestinal absorption in the jejunum.

2. Morphology of the jejunum and intestinal villi

2.1. Specialization of the mucosal surface

In order to make its absorptive function more efficient, the jejunum has the structural specialization that increases the surface area of the mucosa, thus vastly increasing its absorptive surface. First of all, the plicae circulares or Kerckring's plicae, which are the folds of mucous membrane having cores of submucosa, are found throughout the small intestine but are most well-developed in the jejunum, and they serve to increase the absorptive surface. Secondly, the luminal surface over and between the folds is provided with numerous tiny tongue-like to finger-shaped projections of the mucous membrane (Fig. 6-1), which are 0.5 to 1.0 mm in height and are called the intestinal villi. At the core of these villi is the lamina propria which houses the absorptive microvasculatures. It is apparent that these villi do remarkably increase the surface area and, correspondingly, the absorptive area of the jejunal mucosa. At the bottom of the intestinal villi, there are tubular invaginations of the epithelium between villi which are called the intestinal crypts (crypts of Lieberkühn). Dome-like structures which represent the surface of lymphatic nodules are sometimes encountered among the villi.

2.2. Structure of the villi

The intestinal villi consist of two layers, an epithelial lining and a supporting core, the lamina propria. The epithelium is composed of a simple columnar variety of cells, and most of the cells are of the absorptive type, with moderate numbers of goblet cells. A few endocrine cells are scattered among the absorptive cells. The core of the villi, the lamina propria, consists of thin, loose connective tissue which has an abundant number of cells of the immune system and contains a number of blood capillaries and central lacteal. Thus, the food materials absorbed across the epithelial lining can easily gain entrance to the blood capillaries and to the central lacteal (Fig. 6-2).

2.3. General ultrastructure of the intestinal absorptive cells

To date, the fine structure of intestinal absorptive cells from various vertebrates has been described in detail by many authors. The luminal surface of absorptive cells is covered with a thick homogeneous layer, the "striated border" or "brush border" seen in light micrographs (Fig. 6-2). This layer is tightly packed with cytoplasmic microprojections, now called microvilli, of about 1 μm in length and 0.1 μm in width (Fig. 6-3). These microvilli, together with the plicae circulares and intestinal villi, tremendously increase the absorptive surface. The plasma membrane of microvilli shows a clear trilaminar structure and is associated with fine filaments radiating toward the intestinal lumen (Fig. 6-4). These filamentous structures may represent the chains of acid mucopolysaccharides constituting the glycocalyx covering the surface of the microvilli (3). The cytoplasm of the microvilli lacks cell organelles but contains a bundle of longitudinal fine filaments which extend from the apex of the microvilli to the terminal web beneath the microvilli (Fig. 6-4). These filaments have been verified to be actin in nature and are thought to play a role in the movement of microvilli, where they seem to be linked with myosin filaments in the terminal web (4,5).

At the level below the microvilli, the lateral surface is attached tightly to the neighboring cells by the junctional complex. This is composed of the tight junctions (zonula occludens), the intermediate junctions (zonula adherens), and desmosomes (macula adherens) (Fig. 6-5). Apart from these points of attachment, there are interdigitations of the lateral cell membranes between adjacent cells at the levels above and below the nucleus (Fig. 6-3). The basal surface is relatively

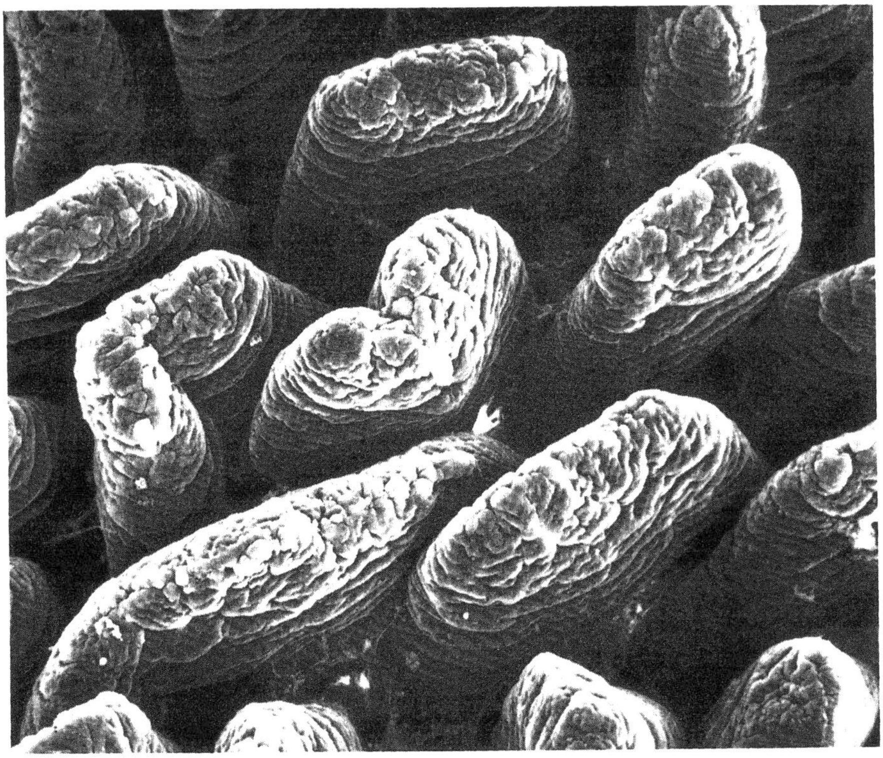

Figure 6-1. A scanning electron micrograph of the intestinal villi of the rat jejunum. X 180.

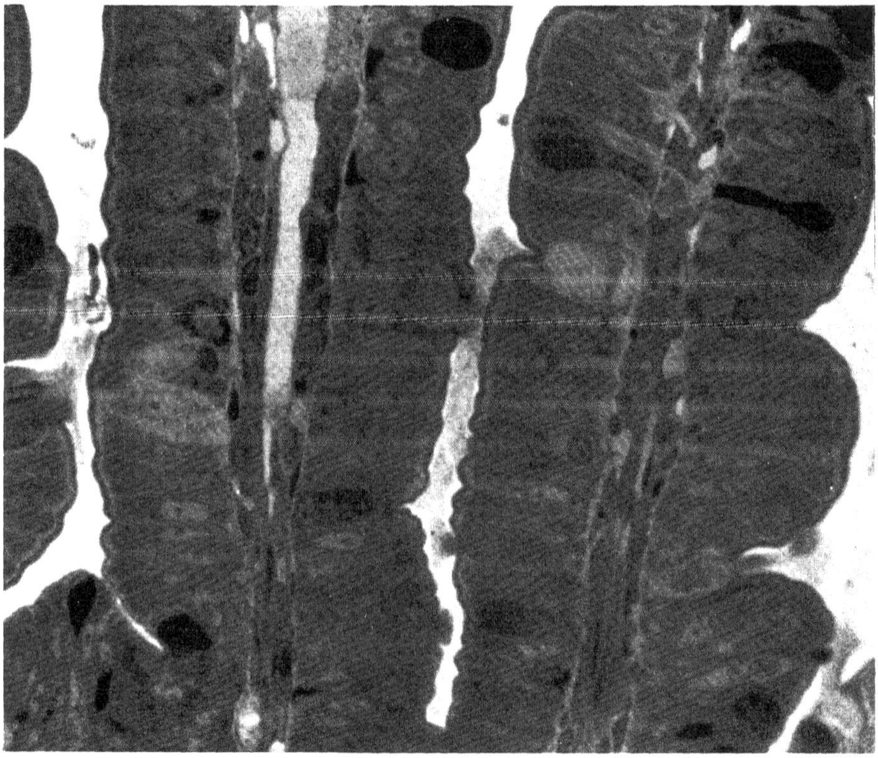

Figure 6-2. A light micrograph of the intestinal villi of the rat jejunum. The luminal surface of the epithelial lining is covered with a thick homogeneous layer, the brush border. The core of villi contains a central lacteal and blood capillaries. X 750.

Figure 6-3. An electron micrograph of the epithelium of rat jejunum at low magnification. X 3,700.

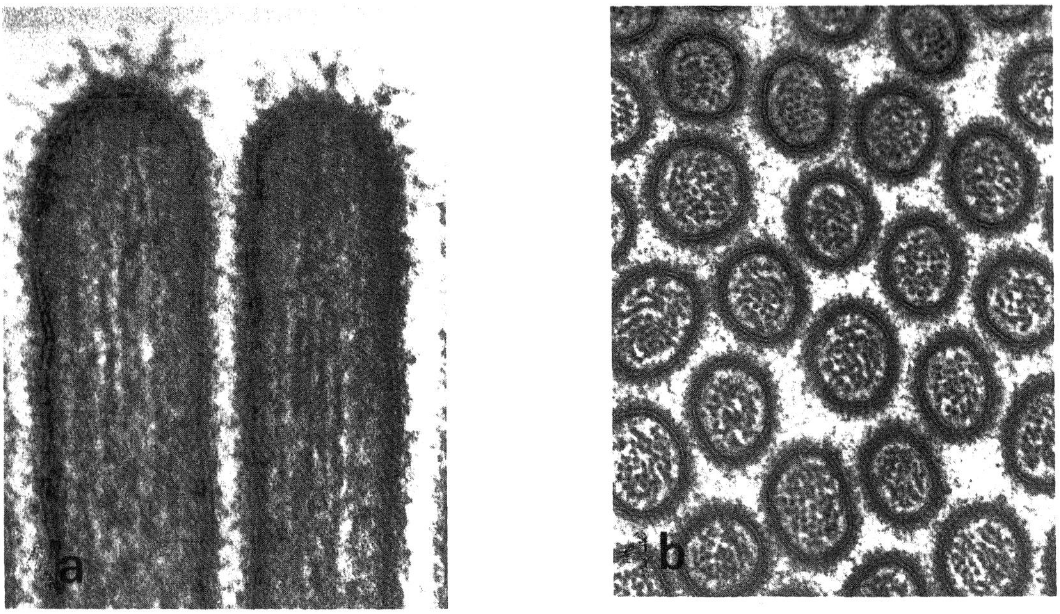

Figure 6-4. (a) Longitudinal section of microvilli. The limiting membranes show trilaminar structures, from which filamentous structures of the glycocalyx are seen radiating toward the lumen. Longitudinal actin filaments can be seen in the core. X 187,000. (b) Cross section of microvilli. X 94,000.

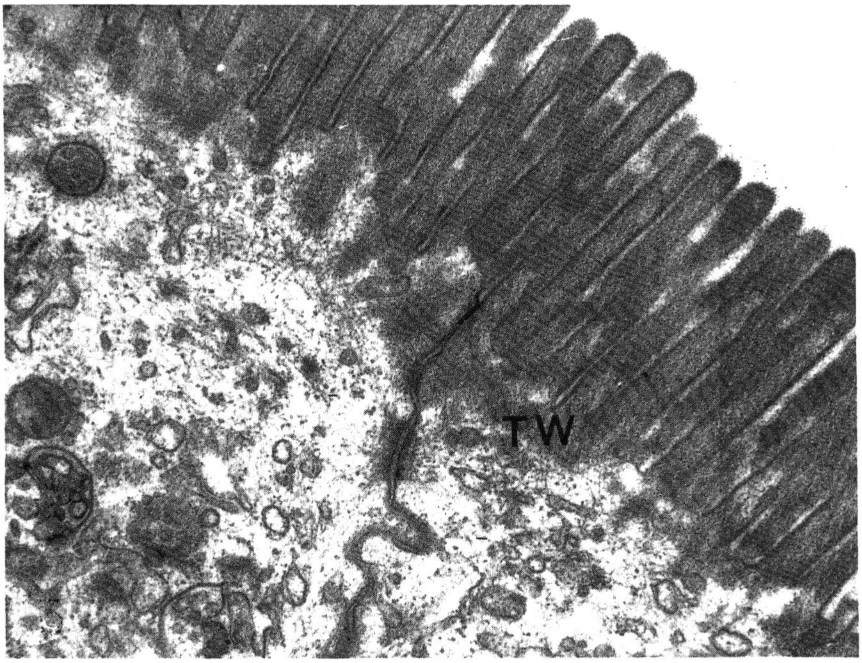

Figure 6-5. Junctional complex between two adjoining absorptive cells. TW, terminal web. X 36,000.

smooth and rests on a thin basal lamina (Fig. 6-3).

The apical cytoplasm above the nucleus, which is somewhat elongated at the basal third of the cell body, contains cell organelles such as the endoplasmic reticulum, mitochondria, the Golgi apparatus, and free ribosomes. The basal cytoplasm below the nucleus is provided with a small amount of granular endoplasmic reticulum and relatively abundant mitochondria (Fig. 6-3).

3. Structural basis of intestinal absorption

Both digested and undigested nutrients in the intestinal lumen must be transported across the epithelial lining for absorption. There may be two possible routes for this transport: one way is via the absorptive cells, and the other is the passageway through the intercellular junctions.

3.1. Absorption via the absorptive cell (Fig. 6-6)

3.1.1. Absorption from microvilli

The microvilli covering the luminal surface constitute the first site of entrance of food materials into the cell for absorption. Before entering the microvilli, food materials must encounter the glycocalyx which is bound firmly with the microvillous membrane itself and represents an integral part of the external leaflet of the membrane (3). Since the glycocalyx in vertebrates contains such acidic groups as sialic acid, uronic acid, and sulfates, it can be easily demonstrated cytochemically by using ruthenium red, which is positively charged (Fig. 6-7) (6). Although the real functional implication of the glycocalyx in intestinal absorption is not clear, a variety of roles have been suggested. By virtue of the chemical nature of existing groups, the glycocalyx can bind certain ions, molecules, or particles and, in addition, can trap some substances such as digestive enzymes, thus providing the site for luminal digestion in its meshwork of polysaccharide chains. Therefore, the glycocalyx serves as a filter system for the materials absorbed (7).

It is of particular interest to know how the materials can pass through the microvillous membrane. From experimental studies on the thermodynamic aspects of membrane components, the membrane model, namely the "fluid mosaic membrane model" has been proposed and is now widely accepted (8). According to this model, the matrix of the membrane consists of the lipid bilayer, in which globular proteins are embedded like "icebergs" in a "sea of lipids." Some protein molecules penetrate across the bilayer of lipid and others are partially embedded in half of the lipid bilayer. Since these membrane proteins are amphipathic, as are phospholipids, they are thought to be partly embedded in and partly protruding from the membrane. Recent freeze-fracture replica techniques have made it possible to visualize the structural organization of the membrane, particularly the distribution of membrane protein particles (9–12). The P-F face of the microvillous membrane, the fractured face of the cytoplasmic leaflet of the membrane, shows numerous intramembranous particles representing membrane proteins on the smooth background, whereas the E-F face, the fractured face of the external leaflet of the membrane, exhibits a smooth surface studded with a few particles (Fig. 6-8). Electron micrographs of these freeze-fracture replicas at higher magnification demonstrate some intramembranous particles which are irregular in outline and appear to be an assemblage of three or more subunit particles, forming a central pit. This structural feature of intramembranous particles in replicas resembles that of the particles of the gap junction (13–15). Since it is known that the gap junction is a site of ionic coupling between adjoining cells and that a central pit of intramembranous particles implies the presence of a transport channel across the cell membrane (17, 18), the similarity in structural appearance of the microvillous membrane particles to the gap junction particles may suggest the presence of transmembrane channels formed by protein subunits in the microvilli. This interpretation may well be consistent with the hypothesis proposed by Singer that the water-filled channels for transport of hydrophilic molecules through the membrane may be formed by the assembling of four or more protein subunits (19).

In an earlier theory on the mechanism of transmembrane transport, either free or embedded carrier proteins, localized within the membrane, were thought to bind to the transported ions or hydrophilic molecules and move or rotate with them across the membrane from the outside to the inside of the cell or vice versa (20). This theory was applied to the interpretation of sugar transport across the microvillous membrane (21). If this type of carrier-mediated transport takes place in reality, such carriers, which possess hydrophilic side groups, would require a large input of energy to diffuse through the hydrophobic intrior of the membrane (19). Considering carrier-mediated transport on the basis of the new membrane model, however, it is possible that the protein subunits, constituting a trans-

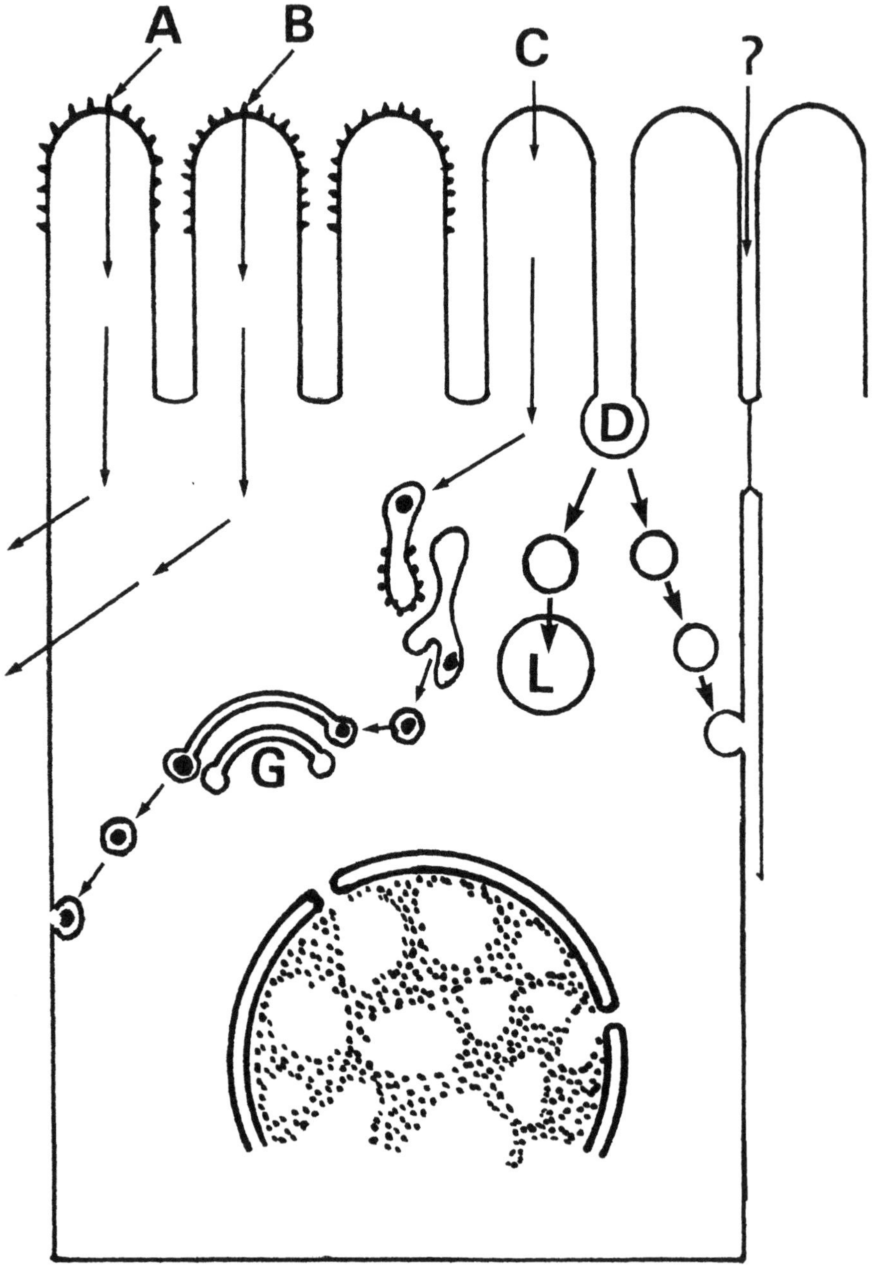

Figure 6-6. A schematic drawing of absorption pathways through the intestinal absorptive cell. A dot on the microvillous membrane represents terminal digestive enzymes. A: Pathway for disaccharides. B: Pathway for protein (oligopeptides). C: Pathway for lipids. D: Endocytic absorption pathways. L: Lysosome, G: Golgi apparatus See the text for explanation.

membrane channel, play a role as a carrier and are involved in the carrier-mediated transport through the transmembrane channels. Recently, convincing data supporting this presumption have been accumulating biochemically (22).

The localization of disaccharidases such as maltase, sucrase, and lactase in microvillous membrane was first described using biochemical

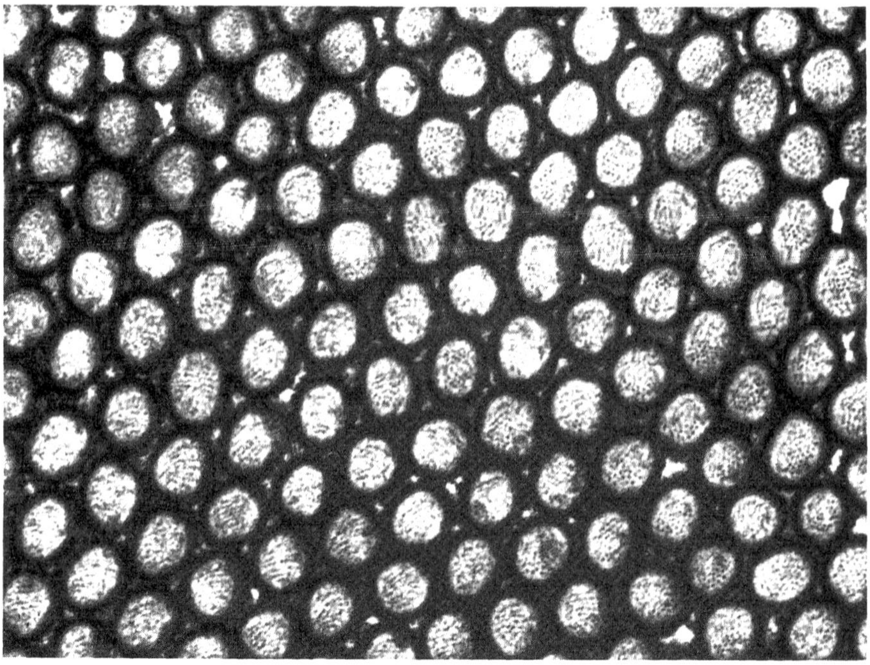

Figure 6-7. Cross sections of microvilli treated with ruthenium red. Glycocalyx is remarkably enhanced in electron density. X 52,000.

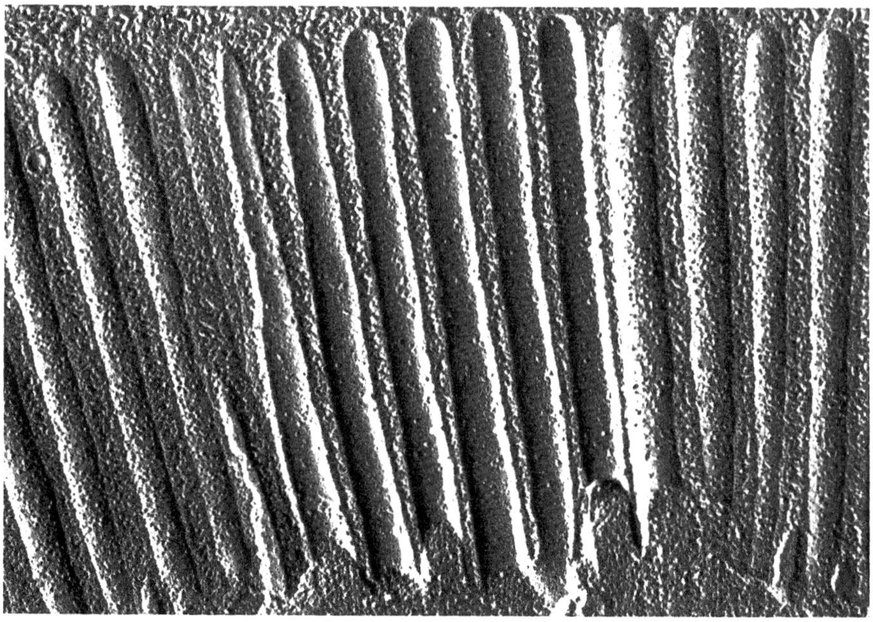

Figure 6-8. A freeze-fracture image of the microvillous membranes of an absorptive cell. X 46,000.

methods by Miller and Crane (1), and then was later verified morphologically (23–26). Later still, oligopeptidase was demonstated to reside in the microvillous membrane (2). These digestive enzymes are known to constitute membrane proteins of the microvilli, their active sites protruding into the intestinal lumen from the membrane surface (25, 27). The granular appearance of the external surface of microvilli, which is demonstrated by a rapid-freeze-etching technique or scanning electron microscopy (Fig. 6-9), may represent the protruding portions of these enzymes. Some enzymes are also known to exist in the microvillous membranes; one example is the complex of sucrase and isomaltase (27, 28). Further, a disaccharidase-related sugar transport system has been reported to exist in the membrane of microvilli (1, 29–33). Thus, the possibility may be suggested here that some of these terminal digestive enzymes, such as disaccharidase, assemble into an intramembranous particle together with some other protein subunits to form a transmembrane channel in the microvilli. If this is the case, the close coupling of terminal digestion with the transport of sugar across the membrane could be clearly understood.

Morphological information on the transmembrane transport of hydrophobic substances has not been available at present, thought there have been many reports dealing with the intestinal absorption of neutral fat (34–36). In rats subjected to oral administration of corn oil, fine lipid droplets in discrete form can be found initially within the cavities of the agranular endoplasmic reticulum in the apical cytoplasm without any structural changes of the microvillous membranes (Fig. 6-10). Subsequently, these droplets appear in the granular endoplasmic reticulum and cavities of the Golgi lamellae and vacuoles. The endocytic uptake of lipid droplets, as described by Palay and Karlin (35), has not been observed in our experiments under normal conditions. Our results confirm the observations on the mechanism of fat absorption described by Cardell et al. (36). Thus, fats are first digested into fatty acids and monoglycerides by lipase in the intestinal lumen, and then these products are transported by simple diffusion through the lipid bilayer of the microvillous membrane, after producing micelles with bile salts. After entering the microvillous cytoplasm, they are again combined with triglyceride synthetase localized in the membrane of the endoplasmic reticulum, to produce chylomicrons together with phospholipids and proteins (36).

3.1.2. Absorption by endocytosis

Endocytosis is a common cellular process involving the cell membrane by which macromolecules are taken up into the cells. It is known that the intestinal absorptive cells in newborn rats (37, 38) and in some stomachless fish (39) are particularly characterized by endocytic absorption. In mammals, active endocytosis can be observed at the bottom of the clefts between microvilli of the absorptive cells throughout the small intestine. This is seen during the first three to four weeks after birth, in part relating to the uptake of antibodies from maternal milk (37, 38, 40). Recently, it has been revealed in newborn rats that selective uptake of antibody takes place only in the proximal small intestine by endocytosis, whereas the distal small intestine is involved in nonselective absorption of macromolecules (38, 41). This antibody transport is reported to be mediated by antibody receptors localized in the cell membrane of absorptive cells, and these receptors are present only in the proximal intestine, being absent in the distal intestine (41). Antibodies from homologous species, after binding with their receptors, are trapped into endocytic invaginations of the intermicrovillous membrane and then transported into the intercellular spaces by means of cytoplasmic vesicles. While being transported, there is no disintegration by lysosomal enzymes, probably because of the antibodies' protective binding to the receptors (38, 40, 42).

In the distal intestine, however, maternal milk is nonselectively taken up into endocytic invaginations of intermicrovillous membranes, and then, via tubular vesicles formed by pinching off from invaginations, transported into phagosomes for intracellular digestion by lysosomal enzymes (38). In the ileal absorptive cells of suckling rats, the luminal surface of endocytic invaginations is known to be studded with latticed particles which contain N-acetylglucosamidase (43). This enzyme is presumed to be engaged in excising sugar moieties in the maternal milk (44). Our freeze-fracture deep-etch studies have revealed that

Figure 6-9. A high resolution scanning electron micrograph of the tips of microvilli. Note granular appearance of their surfaces. X 80,000.

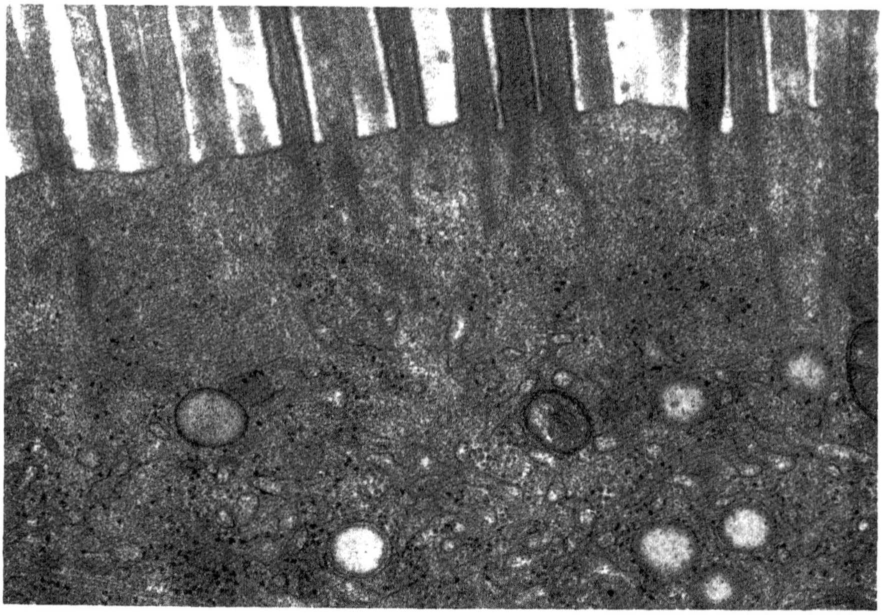

Figure 6-10. Apical cytoplasm of the jejunal absorptive cell from the rat at 30 minutes after oral administration of corn oil. Lipid droplets are seen confined to the endoplasmic reticulum. X 40,000.

particles of these enzymes are present not only on the luminal surface of endocytic membranes, but are also observed on almost the entire apical membrane between microvilli and frequently on the microvilli (45). These particles seemed to be partly integrated in the outer half of the membrane layer. Therefore, the possibility that these latticed particles were transferred to the apical surface along the outer half of the membrane layer and released or secreted into the intestinal lumen (45) was suggested. In suckling rats, the endocytic absorption in the distal small intestine is apparently linked with extracellular sugar digestion.

Similar endocytic adsorption can be observed in the absorptive cells of goldfish hindgut for their entire life span (39, 46). These absorptive cells are characterized by the presence of invaginations of intermicrovillous membranes and an abundance of cytoplasmic tubules, vesicles, and vacuoles of varying size, in the apical cytoplasm (39, 46). In our studies, it became apparent that all these apical membrane components were linked to each other in terms of structure and function (47–50). These membrane invaginations release vesicles from their bases by a pinching-off mechanism, and then the vesicles fuse to each other to form apical vacuoles which are free of acid phosphatase activity and are here referred to as the endosomes. The cytoplasmic tubules which are not open to the cell surfaces are usually located close to the endocytic invaginations (Fig. 6-11) and seem to be formed by budding-off from the endosomes (49, 50). Our studies have suggested that all the apical membrane structures mentioned above are involved in a membrane recycling system (46, 48, 49, 50). In addition to endosomes, there are lysosomal vacuoles which are characterized by the presence of acid phosphatase activity. In the deeper cytoplasm, fusion between endosomes and lysosomes was sometimes observed. Thus, it is suggested that materials internalized by endocytosis are transported through the endosome–lysosome system to a giant food vacuole for intracellular digestion (50).

In the adult small intestine, it has been controversial whether macromolecules such as intact proteins could be taken up by endocytosis. Some authors believe that endocytic absorption does not take place in the intestinal absorptive cells (37, 42, 51, 52), but some others report that it does (53–60). However, it should be noted that endocytic uptake in the adult intestine has been observed in the absorptive cells of ligated loops of the small intestine in vivo and of cultured intestinal tissues, but has not been observed in physiological conditions. We have examined endocytic absorption in the jejunum of rats after oral or intestinal administration of horseradish peroxidase (HRP) or ferritin. In these studies, HRP reaction products or ferritin particles were absorbed only onto the glycocalyx covering microvilli, and any evidence indicating active endocytic adsorption could not be obtained under physiological conditions (Fig. 6-12). Therefore, endocytic absorption takes place very infrequently in the adult mammalian intestines, though their absorptive cells are potentially able to perform endocytosis, which may appear, in reality, under abnormal conditions (61). In the adult intestine affected with malabsorption syndrome, endocytic absorption is enhanced in the absorptive cells, in which numerous short pits of intermicrovillous membranes, increased small vesicles, and lysosomes are observed characteristically in the apical cytoplasm (Fig. 6-13). Though all these membrane components are apparently linked to each other, such cytoplasmic tubules and endosomes as observed in the goldfish hindgut cannot be encountered in the absorptive cells of the adult mammalian intestine. This means that such a membrane recycling mechanism like the one suggested in the goldfish does not work in endocytic absorption of the adult mammalian intestine.

3.2. Absorption through the intercellular junction

The intestinal epithelial cells are attached to neighboring cells by the junctional complex (Fig. 6-5). Among these attachment devices, the tight junction is thought to play a role in tightly sealing the intercellular spaces. Whether the intercellular junctions are permeable to the intestinal contents has been controversial. Our studies have revealed that the tight junction in the jejunal epithelium is permeable to neither horseradish peroxidase nor ferritin injected into the intestinal lumen (Fig. 6-11) (61). Furthermore, the tight junction fixed in glutaraldehyde-containing tannic acid is not

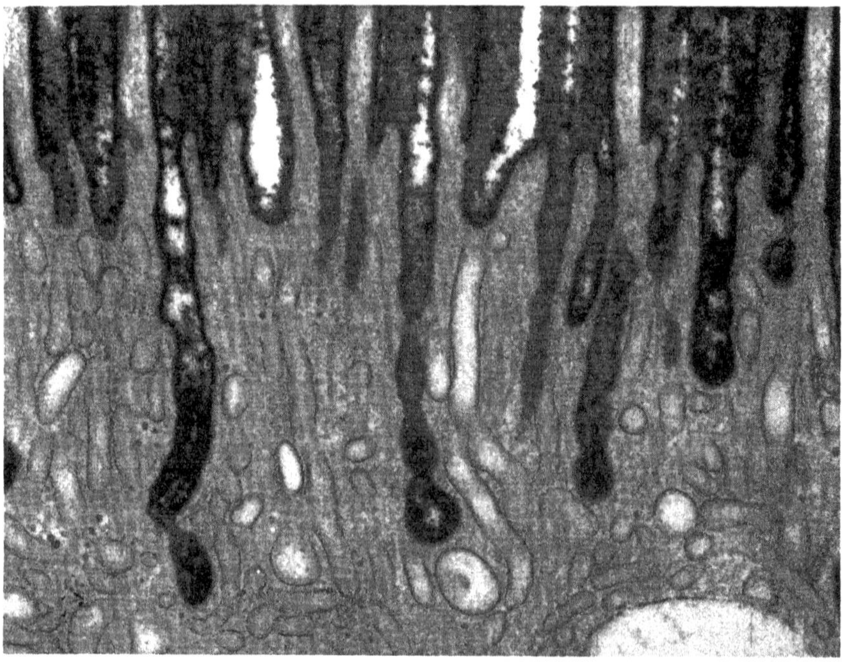

Figure 6-11. An electron micrograph showing the apical cytoplasm of absorptive cell in the goldfish hindgut treated with ruthenium red. Cytoplasmic tubules can be clearly distinguished from the endocytic invagination labelled with ruthenium red. X 43,000.

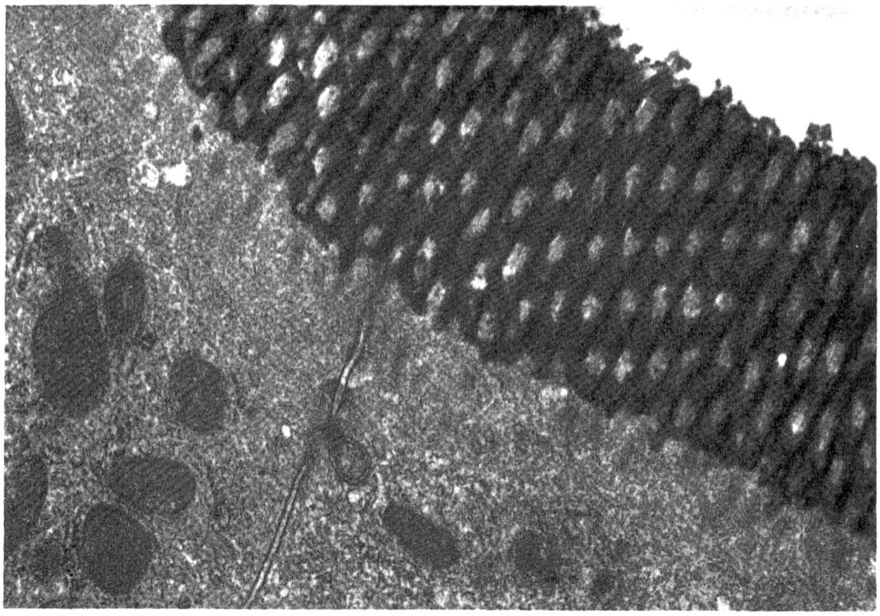

Figure 6-12. Intercellular junction between two adjacent absorptive cells of the rat after the injection of HRP into the jejunum. HRP reaction product is only adsorbed onto the surface of microvilli, but is not observed in the intercellular space beyond the tight junction. X 24,000.

labelled with electron-dense deposits after osmification, suggesting here that this attachment is also not permeable to tannic acid. From these observations, it may be concluded that the tight junction in the rat jejunum prevents the passage of such small molecules as tannic acid and of macromolecules as well. There is, however, a report that the tight junction in the intestinal epithelium is permeable to ionic lanthanum (62).

Freeze-fracture replica studies have demon-

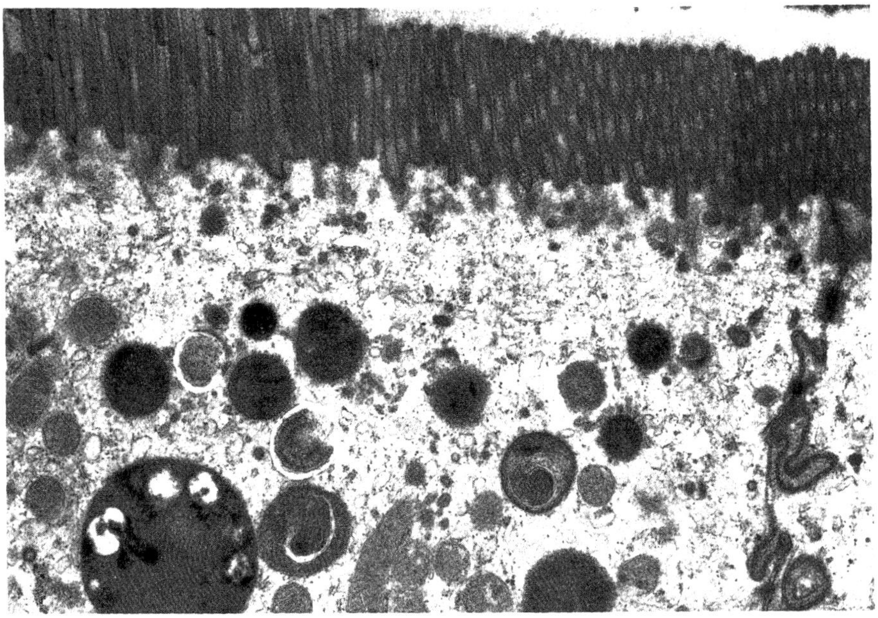

Figure 6-13. Apical cytoplasm of absorptive cells in the jejunal biopsy from a 7-year-old patient affected with intractable diarrhea and malabsorption syndrome. Note the many intermicrovillous pits and a variety of lysosomes. X 14,000.

Figure 6-14. A freeze-fracture image of the junctional complex in the absorptive cell of the rat jejunum. X 63,000.

strated that the tight junction in the rat jejunum exhibits a continuous beltlike meshwork, consisting of strands on the P-F face (Fig. 6-14) (16, 63). These strands are variable in number from place to place, being reduced to two or three at some locations and discontinuous at others. The structural organization of strands in the tight junction of the rat jejunum seems to resemble quite well that described in the rabbit gallbladder as a leaky tight junction (63). Judging from the observations mentioned above, the tight junction in the rat jejunal epithelium is probably permeable in some places, but not in all, to micromolecules such as ions or water, depending on the degree of junctional organization. It seems, therefore, that the intercellular passageway is not a main route for intestinal absorption, and materials to be transported through the intercellular junction are apparently very limited in quality and quantity.

References

1. Miller D, Crane RK: The digestive function of the epithelium of the small intestine. II. Localization of disaccharide hydrolysis in the isolated brush border portion of intestinal epithelial cells. *Biochim Biophys Acta* 52: 293–298, 1961.
2. Peters TJ: The subcellular localization of di- and tripeptide hydrolase activity in guinea pig small intestine. *Biochem J* 120: 195–203, 1970.
3. Ito S: The enteric surface coat on cat intestinal microvilli. *J Cell Biol* 27: 475–491, 1965.
4. Mooseker MS, Tilney LG: Organization of an actin filament-membrane attachment in the microvilli of intestinal epithelial cells. *J Cell Biol* 67: 725–743, 1975.
5. Mooseker MS, Pollard TD, Fujiwara K: Characterization and localization of myosin in the brush border of intestinal epithelial cells. *J Cell Biol* 79: 444–453, 1978.
6. Luft JH: Electron microscopy of cell extraneous coats as revealed by ruthenium red staining. *J Cell Biol* 23: 54A, 1964.
7. Bennett HS: Morphological aspects of extracellular polysaccharides. *J Histochem Cytochem* 11: 14–23, 1963.
8. Singer SJ, Nicolson GL: The fluid mosaic model of the structure of cell membranes. *Science* 175: 720–731, 1972.
9. Branton D: Fracture faces of frozen membranes. *Proc Nat Acad Sci USA* 55: 1048–1056, 1966.
10. Pinto da Silva P, Branton D: Membrane splitting in freeze-etching. *J Cell Biol* 45: 598–605, 1970.
11. Goodenough UW, Staehelin LA: Structural differentiation of stacked and unstacked chloroplast membranes: Freeze-etch electron microscopy of wild-type and mutant strains of *chlamydomonas*. *J. Cell Biol* 48: 594–619, 1971.
12. Staehelin LA, Chlapowski FJ, Bonneville MA: Lumenal plasma membrane of the urinary bladder. I. Three-dimensional reconstruction from freeze-etch images. *J Cell Biol* 53: 73–91, 1972.
13. Kreutzinger GO: Freeze-etching of intercellular junctions of mouse liver. *Proc 26th Electron Microscopy Soc Amer*, Baton-Rouge: Claitor's Publ p 234–235, 1968.
14. Chalcroft JP, Bullivant S: An interpretation of liver cell membrane and junction structure based on observation of freeze-fracture replicas of both sides of the fracture. *J Cell Biol* 47: 49–60, 1970.
15. McNutt NS, Weinstein RS: The ultrastructure of the nexus: A correlated thin-section and freeze-cleave study. *J Cell Biol* 47: 666–688, 1970.
16. Friend, DS, Gilula NB: Variations in tight and gap junctions in mammalian tissues. *J Cell Biol* 53: 758–776, 1972.
17. Gilula NB, Reeves OR, Steinbach A: Metabolic coupling, ionic coupling and cell contacts. *Nature* (London) 235: 262–265, 1972.
18. Peracchia C: Low resistance junctions in crayfish. II. Structural details and further evidence for intercellular channels by freeze-fracture and negative staining. *J Cell Biol* 57: 66–76, 1973.
19. Singer SJ: Architecture and topography of biologic membranes. In: Cell Membranes: Biochemistry, Cell Biology and Pathology. G Weissman and R Claiborne (eds), New York: HP Publ, p 35–44, 1975.
20. Rosenberg T, Wilbrandt W: Carrier transport uphill. I. General. *J Theor Biol* 5: 288–305, 1963.
21. Crane RK: Na-dependent transport in the intestine and other animal tissues. *Fed Proc* 24: 1000–1006, 1965.
22. Muecker M, Caruso C, Baldwin SA, Panico M, Blench I, Morris HR, Allard WJ, Lienhard GE, Lodish HF: Sequence and structure of a human glucose transporter. *Science* 229: 941–945, 1985.
23. Dahlqvist A, Brun A: A method for the histochemical demonstration of disaccharidase activities: Application of invertase and trehalase in some animal tissues. *J Histochem Cytochem* 10: 294–302, 1962.
24. Doell RG, Rosen G, Kretchmer N: Immunological studies of intestinal disaccharidases during normal and precocious development. *Proc Nat Acad Sci USA* 54: 1268–1273, 1965.
25. Oda T, Seki S: Molecular basis of structure and function of the plasma membrane of the microvilli of intestinal epithelial cells. In: Electron Microscopy, Vol II Biology R Ueda (ed), Tokyo: Maruzen Co, p 387, 1966.
26. Benson RL, Sacktor B, Greenwalt JW: Studies on the ultrastructural localization of intestinal disaccharidases. *J Cell Biol* 48: 711–716, 1971.
27. Nishi Y, Takesue Y: Localization of intestinal sucrase-isomaltase complex on the microvillous membrane by electron microscopy using non-labeled antibodies. *J Cell Biol* 79: 516–525, 1978.

28. Braun HA, Cogoli A, Semenza G: Dissociation of small-intestinal sucrase-isomaltase complex into enzymatically active subunits. *Europ J Biochem* 52: 475–480, 1975.
29. Ugolev AM: Membrane (contact) digestion. *Physiol Rev* 45: 555–595, 1965.
30. Parson DS, Prichard LJ: Relationships between disaccharide hydrolysis and sugar transport in amphibian small intestine. *J Physiol* 212: 299–319, 1971.
31. Storelli C, Vogeli H, Semenza G: Reconstruction of a sucrase-mediated sugar transport system in lipid membranes. *FEBS Letters* 24: 287–292, 1972.
32. Malathi P, Ramaswamy K, Caspary WF, Crane RK: Studies on the transport of glucose from disaccharidases by hamster small intestine in vivo. I. Evidence for a disaccharidase-related transport system. *Biochim Biophys Acta* 307: 613–626, 1973.
33. Semenza G: Small intestinal disaccharidases: Their properties and role as sugar translocators across natural and artificial membranes. In: The Enzymes of Biological Membranes, Vol 3. Martonosi (ed), New York: Plenum Press p 349–382, 1976.
34. Weiss J: The role of the Golgi complex in fat absorption as studied with the electron microscope with observations on the cytology of duodenal absorptive cells. *J Exp Med* 102: 775–782, 1955.
35. Palay SL, Karlin LJ: An electron microscopic study of the intestinal villus. II. The pathway of fat absorption. *J Biophys Biochem Cytol* 5: 373–384, 1959.
36. Cardell RR, Badenhausen C, Porter KR: Intestinal triglyceride absorption in the rat: An electron microscopical study. *J Cell Biol* 34: 123–155, 1967.
37. Clark SL: The ingestion of proteins and colloidal materials by columnar absorptive cells of the small intestine in suckling rats and mice. *J Biophys Biochem Cytol* 5: 41–50, 1959.
38. Rodewald R: Selective antibody transport in the proximal small intestine of neonatal rat. *J Cell Biol* 45: 635–640, 1970.
39. Yamamoto T: An electron microscope study of the columnar epithelial cell in the intestine of fresh water teleosts; goldfish (*Carassius auratus*) and rainbow trout (*Salmo irideus*). *Z Zellforsch* 72: 66–87, 1966.
40. Kraehenbuhl JP and Campiche MA: Early stages of intestinal absorption of specific antibodies in the newborn. An ultrastructural, cytochemical and immunological study in the pig, rat and rabbit. *J Cell Biol* 42: 345–365, 1969.
41. Rodewald R: Distribution of immunoglobulin G receptor in the small intestine of the young rat. *J Cell Biol* 85: 18–32, 1980.
42. Rodewald R: Intestinal transport of antibodies in the newborn rat. *J Cell Biol* 58: 189–211, 1973.
43. Jakoi ER, Zampighi G, Robertson JD: Regular structures in unit membranes. II. Morphological and biochemical characterization of two water-soluble membrane proteins isolated from the suckling rat ileum. *J Cell Biol* 70: 97–111, 1976.
44. Robertson JD, Knutton S, Limbrick AR, Jakoi ER, Zampighi G: Regular structures in unit membranes. III. Further observations on the particulate component of the suckling rat ileum endocytic membrane complex. *J Cell Biol* 70: 112–122, 1976.
45. Shibata Y, Arima T, Arima T, Yamamoto T: Regular structures on the microvillar surface membrane of ileal epithelial cells in suckling rat intestine. *J Ultrastruct Res* 85: 70–81, 1983.
46. Yamamoto T: Absorption across the plasma membrane of the intestinal absorptive cells. *Acta Histochem Cytochem* 5: 266–268, 1972.
47. Iida H, Yamamoto T: Morphological studies of the goldfish hindgut mucosa in organ culture. *Cell Tiss Res* 238: 523–528, 1984.
48. Yamamoto T, Arima T, Shibata Y, Iida H. Hirakawa Y: Freeze-fracture study of membrane systems in absorptive cells of the hindgut of goldfish (*Carassius auratus*). *Proc of the 3rd Asia-Pacific Conference on Electron Microscopy* (Singapore) p 390–390, 1984.
49. Iida H, Yamamoto T: Intracellular transport of horseradish peroxidase in the absorptive cells of goldfish hindgut in vitro, with special reference to the cytoplasmic tubules. *Cell Tiss Res* 240: 553–560, 1985.
50. Iida H, Shibata Y, Yamamoto T: The endosome-lysosome system in the absorptive cells of goldfish hindgut. *Cell Tiss Res* 243: 449–452, 1986.
51. Graney DO: The uptake of ferritin by ileal absorptive cells in suckling rats. An electron microscope study. *Am J Anat* 123: 227–254, 1968.
52. Cooper M, Teichberg S, Lifshitz F: Alterations in rat jejunal permeability to a macromolecular tracer during a hyperosmotic load. *Lab Invest* 38: 447–454, 1978.
53. Bockman DE, Winborn WB: Light and electron microscopy of intestinal ferritin absorption. Observations in sensitized and nonsensitized hamsters (*Mescoricetus auratus*). *Anat Rec* 155: 603–622, 1966.
54. Cornell R, Walker WA, Isselbacher KJ: Small intestinal absorption of horseradish peroxidase: A cytochemical study. *Lab Invest* 25: 42–48, 1971.
55. Warshaw AL, Walker WA, Cornell R, Isselbacher KJ: Small intestinal permeability to macromolecules: Transmission of horseradish peroxidase into mesenteric lymph and portal blood. *Lab Invest* 25: 675–684, 1971.
56. Walker WA, Cornell R, Davenport LM, Isselbacher KJ: Macromolecular absorption: Mechanism of horseradish peroxidase uptake and transport in adult and neonatal rat intestine. *J Cell Biol* 54: 195–205, 1972.
57. Worthington BS, Syrotuk J: Intestinal permeability to large particles in normal and protein deficient adult rats. *J Nutr* 106: 20–32, 1976.
58. Warshaw AL, Walker WA, Isselbacher KJ: Protein uptake by the intestine: Evidence for absorption of intact macromolecules. *Gastroenterology* 66: 987–992, 1974.
59. Williams EW, Hemmings WA: Intestinal uptake and transport of proteins in the adult rat. *Proc Roy Soc London* B 203: 177–189, 1978.
60. Blok J, Mulder-Stapel AA, Ginsel LA, Daems WTH: Endocytosis in absorptive cells of cultured human small intestinal tissue: Horseradish peroxidase, lactoperoxidase and ferritin as markers. *Cell Tiss Res* 216: 1–13, 1981.
61. Yamamoto T: Ultrastructural basis of intestinal absorption. *Arch Histol Japon* 45: 1–22, 1982.
62. Machen TE, Erlij D, Wooding FBP: Permeable junc-

tional complexes: The movement of lanthanum across rabbit gall bladder and intestine. *J Cell Biol* 54: 302–312, 1972.

63. Claude P, Goodenough DA: Fracture faces of zonulae occludentes from "tight" and "leaky" epithelia. *J Cell Biol* 58: 390–400, 1973.

Author's address:
Department of Anatomy
Faculty of Medicine
Kyushu University
Fukuoka 812, Japan

CHAPTER 7

Ultrastructure of the normal and diseased appendix and cecum

CARL J. PFEIFFER & BRIAN R. MACPHERSON

1. Introduction

The ultrastructure of the gut has received widespread attention over the years and has been summarized in two previous atlases on this subject (1,2). Reports on small intestinal fine structure have been most numerous, while ultrastructural findings of the adjacent cecum and appendix have been perhaps the least reported. This paradoxical situation, in view of the clinical attention given to the appendix, is likely to be reversed as the functional significance and immunological role of the appendix becomes further elucidated. The present chapter briefly describes a current view of the ultrastructure of the appendix and cecum in both normal and pathological situations from a comparative perspective of human and lower animal species.

2. Functional-anatomic relationship of the appendix/cecum

2.1. *The appendix: Human and other primates*

In the human adult, the appendix averages around 7 cm in length (5–20 cm range) and arises from the cecum at a point typically 2.5 cm from the ileocecal valve, though four types of cecal-appendiceal junctions have been described (3). In children and fetuses, the appendix arises generally from the apical end of a funnel-shaped cecum, and in adults the appendix most commonly opens toward the ileocecal valve due to sacculation of the lateral aspect of the cecum, displacing the appendicular opening away from the apex of the cecum.

The ultrastructure of the appendix is undoubtedly related to its function, but in humans the function of the appendix remains unknown, or at least in dispute. In herbivorous animals the appendix is well developed and is a site for intraluminal digestion of cellulose and is thus a reservoir of bacterial flora. For many years the human vermiform appendix has been considered to be a vestigial organ since it is poorly developed and its surgical extirpation seems to be without adverse effects. However, recent evidence suggests that it may be far from being a vestigial organ. The evidence for this latter notion is as or follows: (a) histologic recognition of copious amounts of lymphoid tissue suggests that the appendix is a significant source of lymphocytes, some of which become IgA-secreting plasma cells, therefore making the appendix an important component of the gut-associated lymphoepithelial tissue or GALT, and (b) the morphologic evidence in primates that, although absent in prosimians and New World monkeys, the vermiform appendix began to develop progressively in selected Old World monkeys, becoming fully developed in the anthropoid apes (4). In the orangutan there is no obvious junction between the cecum and appendix, and all of the appendix and the proximal cecum are covered by an external muscle layer prior to division of the external muscle into taenia coli at the level of the ileocecal junction. A similar anatomic situation is found in human infants. The adult human and the gorilla have a similar anatomic arrangement in which three fully formed taeniae coli fuse at the base of the appendix. Rare bizarre anomalies in location of the human vermiform appendix have also been noted, including placement of the appendix within the posterior wall of the cecum, beneath the serosa (5).

2.2. The appendix: Non-primates

In domestic mammals, the vermiform appendix is generally absent. In the rabbit, however, it is highly developed and serves, along with the sacculus rotundus (the enlargement at the distal aspect of the terminal ileum) and Peyer's patches of the ileum, as a central lymphoid organ for humoral immunity. Neonatal surgical excision of these organ sites leads to a specific antibody deficiency syndrome in which cell-mediated immune reactions remain intact. These organ sites are now considered homologous to the avian bursa of Fabricius (6).

The bursa of Fabricius is present in most young birds as an endodermal cloacal derivative that opens into the cloacal orifice just beyond the openings of the urogenital ducts. It functions as a lymphoid organ during early life but degenerates as the animal approaches sexual maturity. In this respect it resembles the activity of the human thymus gland, another lymphoid organ.

Since its discovery, the bursa of Fabricius has been attributed a significant function in the development of the immune system of the fowl, similar to that attributed to the human thymus (6). Bockman and Cooper (7) compared the micropinocytotic capabilities of epithelial cells associated with lymphoid follicles in the bursa of Fabricius in the chicken, in the appendix of the rabbit, and in the Peyer's patch of the mouse. Epithelial cells capable of transporting ferritin and India ink tracers from the lumen were demonstrated in all these areas. Epithelial cells not associated with the lymphoid follicles in the bursa and appendix did not exhibit this pinocytotic activity. By monitoring these areas through development, they concluded that follicle-associated epithelium having pinocytotic activity does not induce initial migration of lymphatic stem cells into these areas along the gastrointestinal tract, but that the transepithelial flow may provide a significant stimulus for attraction, proliferation, and egression of lymphocytes to these areas.

The lymphoid follicles, which are very numerous in the appendix, are characterized by an apical dome, a germinal center, and a large basal nodule, as well as lateral extending thymus-dependent areas. Both columnar absorptive cells and specialized follicle associated epithelial cells (FAE) comprise the dome epithelium, and scanning electron microscopy frequently reveals clustering of luminal bacteria to the surface of the FAE (8). Normal development of lymphoid follicles of the appendix requires the exposure of the appendix surface to intestinal contents, and the FAE are efficient at transporting luminal material to the underlying lymphoid tissue. Macromolecules may also be transported from the lymphoid follicle into the lumen through the epithelium. Macrophages, shown by transmission electron microscopy to be numerous within the dome epithelium, can process and present (possibly along with the FAE) antigens to reactive lymphocytes. Such antigens appear to stimulate proliferation of lymphocytes within the lymphoid follicles. Most of the appendiceal lymphocytes are transported through the lymphatic route, ultimately to the gut lamina propria where they differentiate into IgA-producing plasma cells. The IgA subsequently become secretory IgA. Scanning electron microscopy of the luminal surface of the appendiceal epithelium reveals openings to the crypts of Lieberkühn, swellings due to presence of domes, and microvilli present on the surface of columnar absorptive cells of the dome epithelium. FAE cell surfaces can also be observed scattered amongst the absorptive cells, and they extend microvilli toward the lumen.

2.3. The cecum: Human and other primates

In the human, the cecum, the initial part of the large intestine, is the portion below the horizontal line across the bowel at the level of the ileocecal valve (in the standing individual). In lower primates the proportional length of the cecum, compared to the rest of the colon, decreases as the species becomes elevated in the primate evolutionary scale, and in others, particularly herbivorous species, the cecum is quite large (9).

Functionally, Hill and Rewell (10) considered the primate cecum to exhibit no correlation between its morphology and nutritional habits.

2.4. The cecum: Non-primate species

The cecum displays a multiplicity of form among the Rodentia and even between closely related species (11–13). While controversial, the significance of nutrition on the determination of intestinal development (14, 15) has been downplayed recently in favor of phylogeny (12, 13, 16). The concept that all herbivores possess a voluminous cecum while that of carnivores is rudimentary, or absent, cannot be strictly upheld. The giant panda (*Ailuropoda melanoleuca*) is a strict herbivore but does not possess a cecum (17), while the dog, a pure carnivore, has a cecal component to its gastrointestinal tract. There is a tendency, however, for most herbivores to possess a large well-developed cecum.

The function of the cecum is variable among different animals. It may be considered to: (a) provide space for retention of partially digested foodstuffs and some degree of digestion of carbohydrate, (b) increase mucosal area for absorption, (c) develop special localized areas of lymphoid tissue (i.e., rabbit appendix; (18, 19), and (d) possess secretory functions. Transmission electron microscopy reveals the epithelial sheet of the cecum to be composed of surface epithelial cells and goblet cells. The cecum, often in conjunction with the initial part of the ascending colon, has been documented as a region of efficient absorption and/or storage of water and electrolytes (20, 21), volatile fatty acids (22–26), amino acids (27, 28), carbohydrates (29, 30), and limited amounts of triglycerides (31). Absorption of the breakdown products of microbial fermentation also occurs in the cecum (22, 23, 29, 32, 33). The most important of these are the short-chain fatty acids that are the major anions in the feces of some species including goat, pony, rat, and rabbit [34].

To date very few morphological studies, particularly at the ultrastructural level, have been devoted to the cecum. As important processes of digestion occur here, morphological information is indispensible. Structural differentiation revealed by these studies may well aid in a more complete integration of physiological data on digestive processes. When the relationship between dietary habit and structure of the cecum are taken into account a comparative functional-anatomic approach will evolve and will supply additional information for assessing similarities and differences in function and structure among different orders or even within closely related species. Many questions arise. How does the cecum of a carnivore differ from that of a herbivore or omnivore, especially when the functional aspect and vestigiality are considered? Why has the cecum been retained throughout phylogeny? Is the carnivore cecum vestigial?

3. Ultrastructural aspects

3.1. Ultrastructure of normal human appendix

The ultrastructure of the normal human appendix has been studied in several laboratories, usually in relationship to its role as a source of lymphocytes or in relation to its fetal development. Auböck reported early findings (35), and Gorgollon (36) more recently studied normal appendices from children by combined light and electron microscopy. Two main types of lymphocytes were observed by electron microscopy, light and dark cells. Some intermediate forms were observed as well. Light lymphocytes were observed to be more abundant in the epithelium and within parafollicular postcapillary venules. These cells formed clusters between epithelial cells and appeared to mature into plasma cells in the subepithelial region. Few intraepithelial lymphocytes were found in crypt epithelium, but a marked concentration was seen in the luminal surface epithelium. Fewer goblet cells were evident in the appendiceal epithelium than were found in the similar-appearing large intestinal epithelium. The columnar cells of the outer epithelial surface displayed short, irregularly-arranged microvilli and numerous mitochondria. Acidophilic leukocytes, mast cells, and some degenerating intraepithelial leukocytes were observed. Lymphocytes were not observed to pass through the apex of the epithelial cells, though they were in contact with the free surface (36). The germinal

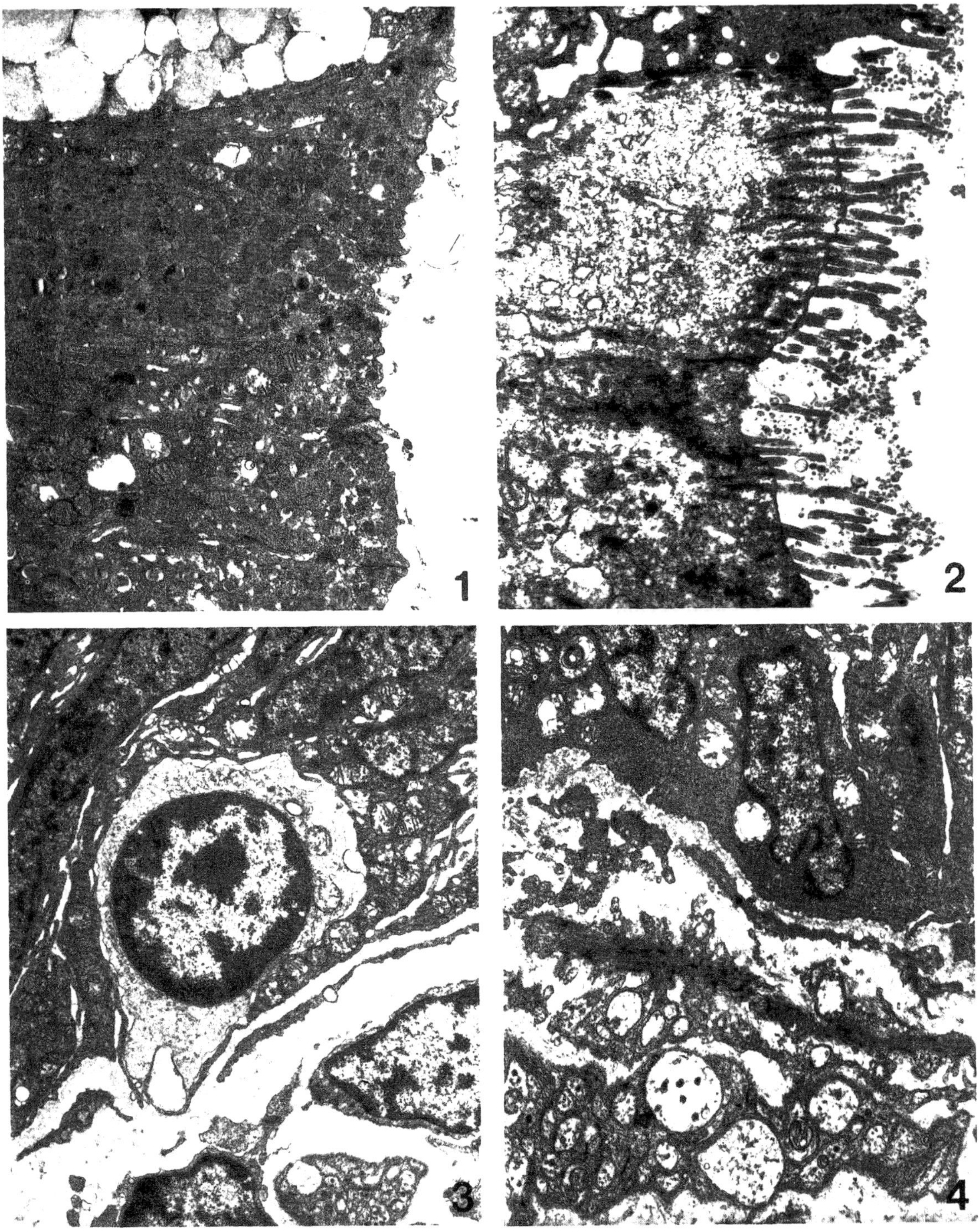
1
2
3
4

centers or lymphoid follicles of the normal appendix contained large blast cells showing numerous polyribosomes and dendritic reticular cells. The association of light versus dark lymphocytes with T versus B lymphocytes remains uncertain. Clusters of lymphocytes, such as seen by Gorgollon in the appendix of children, have also been reported by one of the present authors within the mucosal epithelium of other species in other regions of the gastrointestinal tract (1).

The fine structure of the human appendix has been studied by Bockman and Cooper (37) from the earliest fetal stages of lymphoid development to definitive development such as that found in children up to eight years of age. In later life, the appendix becomes less active with respect to its lymphoid function and structure. Follicular accumulations of lymphocytes were seen initially in 11.5 cm (crown-rump length) fetuses in the mesenchyme subjacent to the epithelium which contained mainly goblet cells on the surface and in the crypts. They were found in aggregates. In older fetuses lymphocytes were closely associated with the surface epithelium. In 17.5 cm fetuses the follicle-associated epithelium (FAE) of the developing appendix was ultrastructurally very similar to the avian bursae, the rabbit appendix, and Peyer's patch FAE of mouse, monkey, and human tissue. Irregular microvilli and numerous vescicles were found in the FAE of appendices of five- to eight-year-old children. Haar (38) reported ultrastructural data of stratified columnar cells in the developing human appendix of fetuses at 7.5 weeks gestation. The epithelium had transformed to single columnar type by 13.5 weeks gestation, and goblet cells and argentafin cells were observed. By the same age, free lamina proprial leukocytes as well as migrating intraepithelial lymphocytes were observed. At 18 weeks gestation, electron microscopy revealed a secretory status of goblet cells and single lymphocytes in the lamina propria and epithelium.

The neuroendocrine cells in the normal human appendix have received considerable attention. These neurosecretory cells have been given various names, including enterochromaffin or argentaffin cells (39,40), argyrophilic cells, APUD cells, and EC cells, depending upon their staining characteristics, trends in the literature, etc. They are centered mostly at the base of the crypts in the appendix and in close proximity to the submucosal nerve plexus. It should be noted, from the standpoint of light microscopic findings, that the autonomic nerve supply of the appendix is not typical of the remainder of the bowel. In the appendix there are relatively few submucosal ganglion cells, and the intramuscular ganglion cells are scattered randomly rather than in a plexiform pattern as is found elsewhere in the intestines. This atypical condition also prevails in the cecum. Ultrastructurally, there is a close association between the extraepithelial enterochromaffin cell-nerve fiber complex. This close association supports the concept of neuroendocrine programmed ectoblasts, a common origin for nerves and the neuroendocrine cells (41,42).

The neuroendocrine cells contain at least two types of granules: small, irregularly shaped dense granules (0.1–0.4 μm) like those observed in enterochromaffin cells elsewhere in the gut (1), and round, small granules (0.1–0.2 μm) of variable electron density (36). It has recently been demonstrated (43) that the neurosecretory cells in the ganglionated neural plexuses in the human appendix have the same pleomorphic neurosecretory granules as found within the en-

←

Figure 7-1. Apical aspect of surface epithelial cells of appendix of human with appendicitis. A portion of a goblet cell with mucin granules can be seen at the top. Note abnormal flattening of microvilli at luminal surface (right) and degenerating mitochondria. X 16,800.

Figure 7-2. Human appendicitis. The apical aspects of these epithelial cells from this patient show normal microvilli surrounded by viral-like particles in the lumen and a cell degenerating in situ (top). X 25,000.

Figure 7-3. Human appendicitis. Basal portion of glandular mucosa. Note the lymphocyte (center clear cell) migrating from the lamina propria through basement membrane, between epithelial cells which contain degenerating mitochondria. A nerve bundle can be seen at the lower right. X 10,500.

Figure 7-4. Human appendicitis. Region basal to glandular mocosa. Note degenerating mitochondria within epithelial cells (top) and large nerve bundle in submucosa at bottom. Unmyelinated neuronal processes in close association with muscle cells of muscularis mucosae can be seen in the center. X 13,300.

terochromaffin cells, and each contains serotonin. This was demonstrated by electron immunocytochemical methods. The fine structure of these neuroendocrine cells of the appendix has been described in detail (44), and currently four types of neurosecretory cells have been classified.

3.2. Ultrastructure of the human appendix during pathologic states

Pathological changes in the human appendix can be rather diverse, and many such conditions have not yet been evaluated at the ultrastructural level. An example of the diversity of appendiceal abnormalities can be observed from the review by Dynock (45) of South Australia of 1,000 consecutive, simple appendectomies. Of these appendices, 14% were subsequently diagnosed as normal; 561 (56%) exhibited acute appendicitis (acute inflammation predominantly in mucosa, submucosa, and/or inner muscular wall); fecoliths causing luminal dilatation, thinning of the appendiceal wall, obstruction, etc. were seen in 183 (18%), of which 88 (9%) were concurrent with acute appendicitis; and pinworm (*Enterobius vermicularis*) infections were seen in 30 cases (3%), 4 of which were concurrent with acute appendicitis. Other appendiceal pathology included melanosis (2.2%), lymphoid hyperplasia (2.2%), carcinoid tumors (1.4%), mucocele (0.7%), mesothelial inclusion cysts (0.7%), diverticulosis (0.7%), pseudodecidual changes (0.4%), metaplasia (0.3%), local visceral arteritis (0.3%), endometriosis (0.3%), primary carcinoma (0.1%), and secondary carcinoma situated in the appendix (0.1%). The following paragraphs outline some of the pathologic conditions of the human appendix which have had ultrastructural analysis or which require much study in the future. Figures 7-1 to 7-4 illustrate some of the degenerative and inflammatory fine structural changes we have observed in acute appendicitis of unknown origin in man.

3.3. Batten-Spielmeyer-Vogt disease

Juvenile amaurotic idiocy, currently referred to as Batten-Spielmeyer-Vogt disease (Batten's disease), is a distinct malady characterized by lipid-accumulation in a variety of tissues including blood vessel walls and thyroid and by intracellular granular substance accumulations in liver, myocardium, spleen, lymph nodes, kidneys, etc. The ultrastructural study of the appendix of a five-year-old child with Batten's disease has aided the early diagnosis of this illness (46). Earlier investigations by light microscopy have characterized the cellular inclusions as ceroid-lipofusoin material or granules. Electron microscopy of the appendix revealed intracellular accumulations of curviform inclusion bodies in smooth muscle cells, nerve cells, and Schwann cells of the mysenteric plexus of Auerbach. These inclusions did not resemble typical lipofusion pigment but did resemble the lipid bodies described in this disease by other workers, as found in cerebral biopsies of neuronal, glial, and vascular tissue. Further studies on the cytoplasmic inclusions in the vermiform appendix by Rapola and Haltia (47) on ten patients with Batten's neuronal ceroid-lipofuscinosis were undertaken. Granular bodies, present in smooth muscle cells of the appendix wall, were usually located near the nuclear poles and were also found in large macrophages of the mucosal lymphoid tissue. The curvilinear fine structural profile of the appendiceal cytosome in smooth muscle cytoplasm appeared identical to that reported by earlier workers, as described above. Thus, ultrastructural examination of tissue from a simple appendectomy, instead of from a hazardous brain biopsy, can provide helpful diagnostic or confirmatory information for patients with suspected Batten's disease.

3.4. Neurogenic appendicopathy

The diathesis of neurogenic appendicopathy has been described at the ultrastructural level by Auböck and Ratzenhofer (41), Ratzenhofer et al. (48), and Höfler (49) and has been classified into three types. These include: (a) mucosal neurogenic appendicopathy, the most frequently observed variant; (b) central or axial neuroma, which is found in cases in which the apex of the appendix is obliterated; and (c) the neuromuscular proliferation type, in which marked proliferation of smooth muscle fibers and axons near the mucosa or lamina propria occurs, resulting in formation of neuromuscular tangles. This latter form is uncommon. Höfler and coworkers (50) studied the neuroendocrine system in 17 appendices from patients with mucosal neurogenic appendicopathy, central neuromas, or with normal appendices. They localized by immunocytochemical methods epithelial endocrine cells containing substance P, somatostatin, enteroglucagon, 5-hydroxytryptamine, and vasoactive intestinal polypeptide. A large number of proliferating nerves were visualized in neurogenic appendicopathy independent of neurotransmitters. The number of endocrine cells in the mucosa of the appendix was greatly increased in neurogenic appendicopathy for unknown reasons. These cells contained 5-hydroxytryptamine, somatostatin, and possibly substance P. Whether the hyperplasia of stromal endocrine cells was a cause or effect of neurogenic appendicopathy is open to speculation.

3.5. *Malacoplakia of the appendix*

Rodriques and associates (51) described a case history of malacoplakia involving the appendix of an 11-year-old boy with a history of severe diarrhea and reported electron microscopic findings. The lesions described contained a proliferation of histiocytes with typical cytoplasmic inclusions known as Michaelis-Gutmann bodies, giant bodies which showed progressive stages of crystallization of the matrix. Further study is needed in order to clarify these findings, and x-ray probe analysis of the cytoplasmic inclusion bodies would be informative.

3.6. *Electron microscopy of appendiceal cancers*

Adenocarcinoma of the human appendix is rare and as its fine structure has not yet been characterized, it will be important in the future to compare its cytomorphology with that of typical adenocarcinoma of adjacent areas. The well-developed immune system within the appendix may be responsible for enhanced immunosurveillance within this organ, providing greater resistance to chemical carcinogenesis.

Villous adenomas of the vermiform appendix are also rare tumors. Villous adenomas are considered benign but premalignant lesions. Only 44 cases have been reported through 1982 (52), and the ultrastructure of such lesions has not been described up to 1987.

Carcinoid is the commonest tumor of the appendix and, indeed, about half of all gastroenterological carcinoids originate from the appendix. Since carcinoid is a malignant tumor arising from endocrine glands, early detection is important and this can be aided by electron microscopic studies of biopsies, as well as by endocrine and light microscopic studies. The staining patterns and cell morphology of carcinoids of the appendix have been described elsewhere (53), and the electron microscope has been used to confirm the presence of neurosecretory granules within the endocrine cells from which carcinoids have their origin. The genesis of carcinoid tumors remains uncertain. They appear to arise from subepithelial neurosecretory cells, which may be a component of the subepithelial nervus plexus derived from the neuroectoderm. The subepithelial neurosecretory cells contain a neuron-specific enolase, which provides evidence of neuroectodermal origin. Enterochromaffin cells, of which many types have been described in the gut by ultrastructural morphology and which were at one time considered as progenitors of carcinoid tumors because of common peptide products, are most likely of endodermal origin. According to Rode et al. (54), the fine structure and histochemistry, as well as arrangement in tissue of nests and rosettes of carcinoid tumor cells are identical to that of subepithelial neurosecretory cells. The neurosecretory cells observed by Rode and his associates in the lamina propria in 69 of 100 human appendices examined appear to be comparable to the neuroendocrine cells in the appendix described by Höfler et al. (50) which contain 5-hydroxytryptamine, substance P, somatostatin, enteroglucagon, and vasoactive intestinal polypeptide immunoreactivity as mentioned above.

A special type of carcinoid of the appendix, termed goblet cell carcinoid, has been described by Subbuswamy et al. (55). On the basis of 12 tumors of the appendix which showed histologic features different from adenocarcinoma or from the typical carcinoid of the appendix, a tumor type was identified which was composed mainly of goblet cells. This tumor had a low grade of malignancy on the basis of histologic criteria and patient prognosis. The lesion frequently contained Paneth and argentaffin cells. Paneth cells are not normally observed in the appendix. It has not yet been characterized by electron microscopy and offers fertile ground for future ultrastructural study.

3.7. *Viral appendicitis*

It has been established that appendicitis may, upon rare occasion, be induced by viral infection (56). However, it is frequently difficult to prove such an association, though viruses have been isolated from mesenteric lymph nodes of patients with adenovirus-induced ileocecal intussusception, and nuclear inclusion bodies have been observed in epithelial cells of appendix mucosa during appendicitis of children. Surface epithelium budlike proliferations of the appendiceal mucosa have been observed by light microscopy. This is another clinical condition in which future electron microscopic studies are needed in order to seek morphologic evidence of viral intrusion.

3.8. *Ultrastructural pathology in human cecum*

Numerous studies have been reported on ultrastructural pathology of the human large intestine, but they generally have not aided diagnostic capabilities, and few have focused on the cecal portion of the large bowel. A notable exception is the report by Mora-Galindo et al. (57) which depicts by scanning electron microscopy the disruption of the cecal surface epithelium by *Entamoeba histolytica*. Localized zones of epithelial desquamation occurred after 30 minutes incubation with this agent, allowing invasion of the microbe.

3.9. *Cecal ultrastructure in the rat*

Ono (58) examined the surface of the principal cells of the cecum during postnatal development in the rat and found that from day 1–14 these cells were characterized by apical tubulo-vacuolar systems and large supranuclear vacuoles similar to those observed in the absorptive cells of the neonatal ileum. By day 16 the special membrane system had disappeared. This system was not observed in distal colonic segments and was hypothesized to aid the cecum and ascending colon in the absorption of protein macromolecules during early postnatal development. The ultrastructural appearance of the cecal epithelial cells does not differ from that of the remaining colonic segments in the adult rat.

Ono (59) also noted significant changes in cecal morphology during the postnatal period in rats. On the day of birth, numerous cecal villi of varying size and shape were present on the cecal mucosa. After five days of age, the villi decreased rapidly in length and number, with a constriction appearing at their base. Between five and nine days of age, the villi separate and disappear from the mucosa. A similar phenomenon was observed in the postnatal development of bovine ceca by Wille (60) who attributed a limited, but unknown, function to these villi, related solely to the fetal state. The cecal surface of the normal adult rat is illustrated in Figure 7-5.

Portella-Gomes et al. (61) reported numerous enterochromaffin cells in the rat cecum. The granules found within the basal region of these cells were reduced in size when compared to those of the small intestine but were larger than those of the colon, suggesting a gradient in granule size throughout the gastrointestinal tract.

The cecum of germ-free rats has long been noted to be enlarged as much as five times that of conventional rats (62). Gustafsson and Maunsbach (63) reported that the cecal epithelial cells in this animal were taller and had larger nuclei and longer microvilli than conventional rats. There was larger variability in the arrangement of the crypts, and the lamina propria was devoid of plasma cells but rich in mast cells. Only one type of elongated bacteria was observed within the crypts, but bacteria were present in large number. The alteration in crypt morphology was attributed, in part, to simple mechanical distension of the cecal wall in the germ-free rat. The consistently enlarged cecum of the germ-free animals was paralleled by a change in the crypts, their epithelium, and the bacterial flora in the cecal lumen. The endosymbiotic relationship that exists between gut flora and the cecal crypts was altered by their change in morphology, suggesting that these crypts are a crucial factor in the normal symbiotic relationship. Numerous protozoan parasites can be found within the cecal lumen of normal rats (64). Figures 7-6 and 7-7 illustrate that protozoa within the normal rat cecum may lie close to the microvilli of the surface epithelium and may even invade the apical aspect of goblet cells without inducing adverse morphologic damage.

Pfeiffer et al. (1) divide the rat cecum into a large basal portion, the cecum proper, and a small apex that contains large amounts of lymphoid tissue. This latter region corresponds to the appendix. Mucosal glands are less numerous in this region while numerous goblet cells fill the epithelial sheet. The principal cells contain numerous mitochondria in the apical cytoplasm, have poorly developed microvilli, and basally-located nuclei. The muscularis externa of the rat cecum is ultrastructurally similar to that of other regions of the rat intestinal tract (Figure 7-8).

3.10. Cecum in the ground squirrel (gopher)

The Richardson's ground squirrel, a species commonly referred to as a gopher, is currently being used for digestive tract studies. As revealed in Figure 7-9, its normal cecal epithelium contains relatively few goblet cells, and elaborate interdigitations are apparent between adjacent epithelial cells.

3.11. Cecum in the cat

Snipes (65) stated that while macroscopically undifferentiated, the cat cecum is characterized microscopically by an abundance of goblet cells and enterocytes, rich in organelles, that suggest high functional activity. The internal surface of the cecum is characterized by slightly raised interconnected rims of tissue. A large number of goblet cells are found on these rims and the central orifices represent the openings to the crypt which are closely arranged, straight, single tubes. Mu and Lingham (66) used histological evidence of lymphoid aggregates to determine the presence of a rudimentary appendix in the cat. This appendiceal region took the appearance of a conical projection from the blind end of the cecum. They felt its composition and placement correlated well with Arey's (67) embryological derivation of the appendix as the lower blind end of the cecum that in other mammals fails to enlarge as rapidly as the cecum and retains its narrow shape, thus appearing as a diverticulum.

One of the often overlooked functions of the cecum is that of secretion, a feature which may be the most important function of the cat cecum. The abundance of goblet cells indicates that this cecal secretion is rich in mucus. This very active process, as well as the ultrastructural evidence of organelles characteristic of cells performing functionally active roles, makes it difficult to conclude that the cecum of the cat is a rudimentary organ (65).

3.12. Cecum in the mouse, hamster, and vole

Wesley and Jensen (68) reported crystal-containing cells which can be found with regularity in the cecal epithelium of the mouse. They contained aggregated rhomboidal crystals — a membrane-bound, semi-crystalline granular inclusion with a 60 Å lamellar periodicity. Actinomycin D treatment increased the number of crystal-containing cells as well as the number of crystals within each cell. The increase in cell number was shown to arise from migratory cells passing into the cecal epithelium from the underlying lamina propria. These precursor cells were considered to be an intermediate cell type with a morphology similar to that of granulocytes that most likely arose from mast cells. These crystal-containing cells lacked both desmosomes and microvilli. Hendrikson (69) did not mention these crystal-containing cells but noted the ultrastructure of the lining epithelium and goblet cells were constant throughout the crypts. He also observed a close relationship of bacteria with the crypts and an increased number of apical granules in the cells thought to be a component of the secretory immune system of the gastrointestinal tract. These granules were more numerous in the intracryptal cells than in those lining the luminal surface and were thought to be incorporated into the glycocalyx.

Snipe (70) divided the cecum of the hamster into an ampulla ceci and a corpus ceci. The former contains the ileocecal orifice and a circumventing fold. The mucosa contains wide-open crypts that give a ring-walled and looped configuration on SEM examination. A similar mucosal appearance was described in the vole (71) and rabbit (19) cecum. This mucosal appearance had previously been referred to as "rolling hills" by Pfeiffer et al. (1) and Toner et al. (2). The

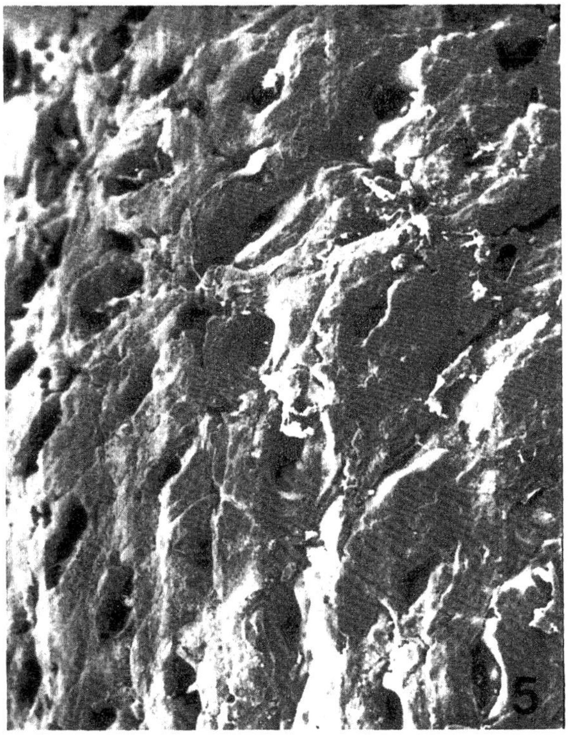

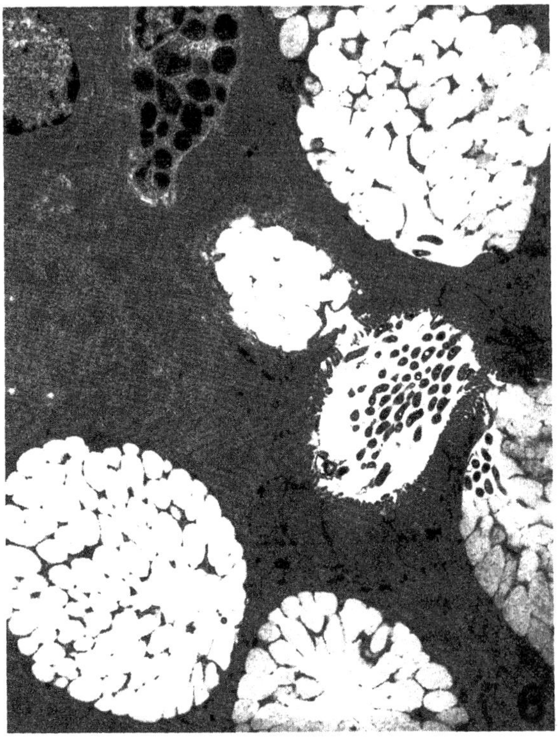

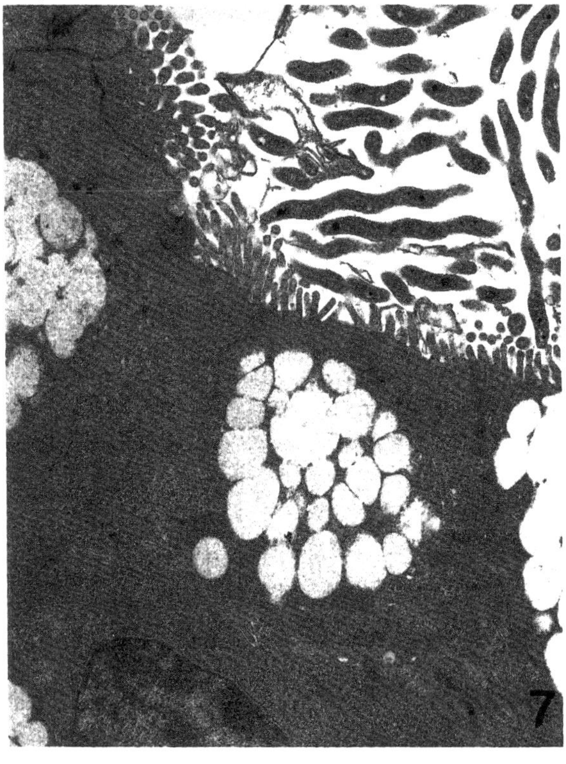

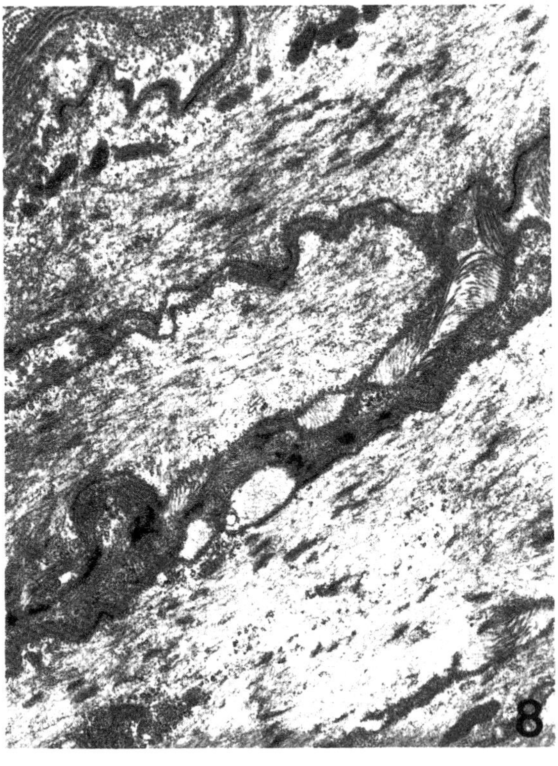

Figure 7-5. Luminal surface of cecum of normal rat. Note openings to crypts of Lieberkühn. X 340.

Figure 7-6. Cross section through crypt of Leiberkühn in normal rat cecum showing mucous granules in goblet cells and presence of protozoa in crypt lumen, even extending into apical aspect of goblet cell (center, right). X 4,775.

Figure 7-7. Higher magnification of normal rat cecum illustrating close proximity of protozoa to microvilli of epithelial cells. These commensal protozoa morphologically resemble *Trypanosoma lewisi*, more commonly found in rat blood. X 11,200.

Figure 7-8. Fine structure of smooth muscle cells in muscularis externa of normal rat cecum. Note mitochondria (center, top), collagen fibrils (left, top), glycogen deposit (left, bottom), and myofilaments within the cells (diffuse). X 11,400.

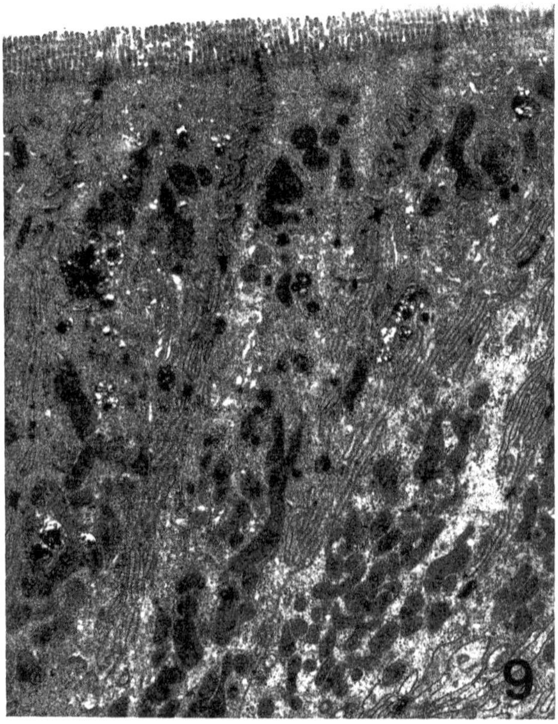

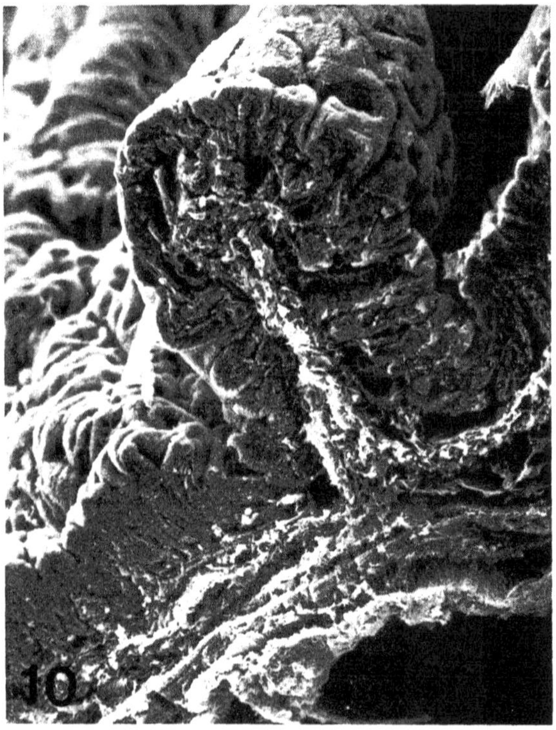

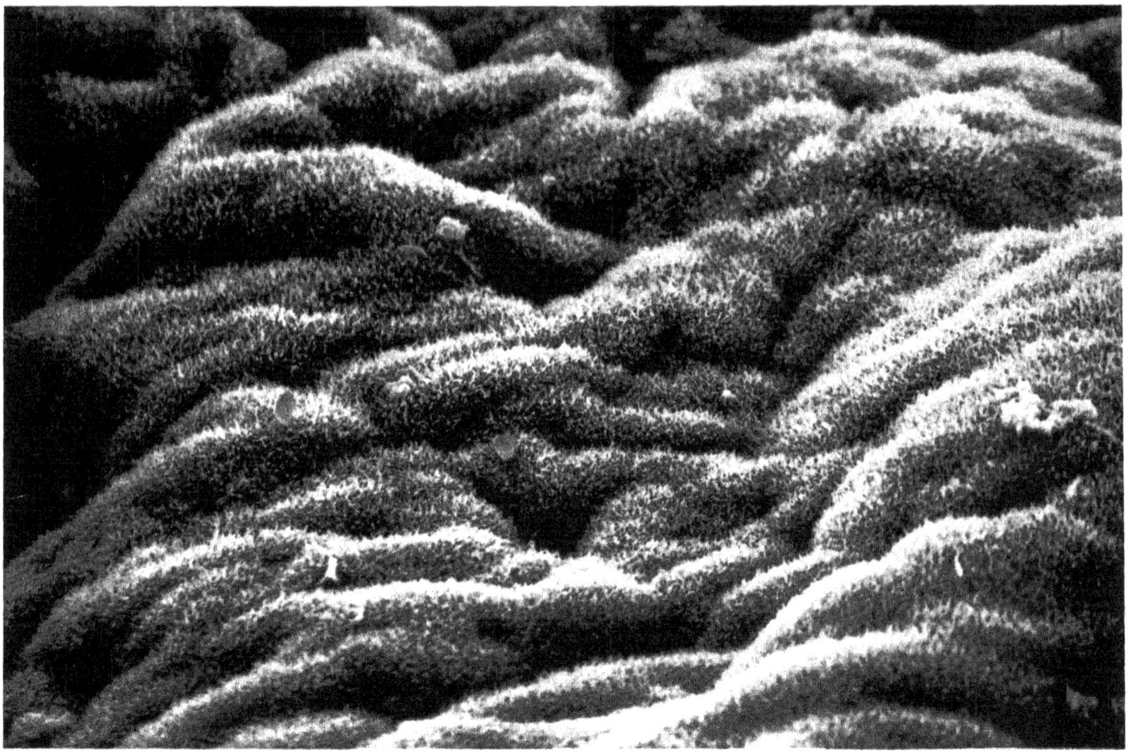

Figure 7-9. The columnar epithelial cells lining the normal cecum of Richardson's ground squirrel contain numerous mitochondria and are separated by elaborate, interdigitatng membranous folds. The surface is lined by microvilli. X 7,225.
Figure 7-10. A cross section through the entire thickness of the cecum of the normal guinea pig reveals clearly by scanning electron microscope the mucosa, submucosa, and muscularis externa. X 9,375.
Figure 7-11. Low power surface scanning of the guinea pig cecum shows extensive surface area for transport processes due to microvilli. X 4,685.

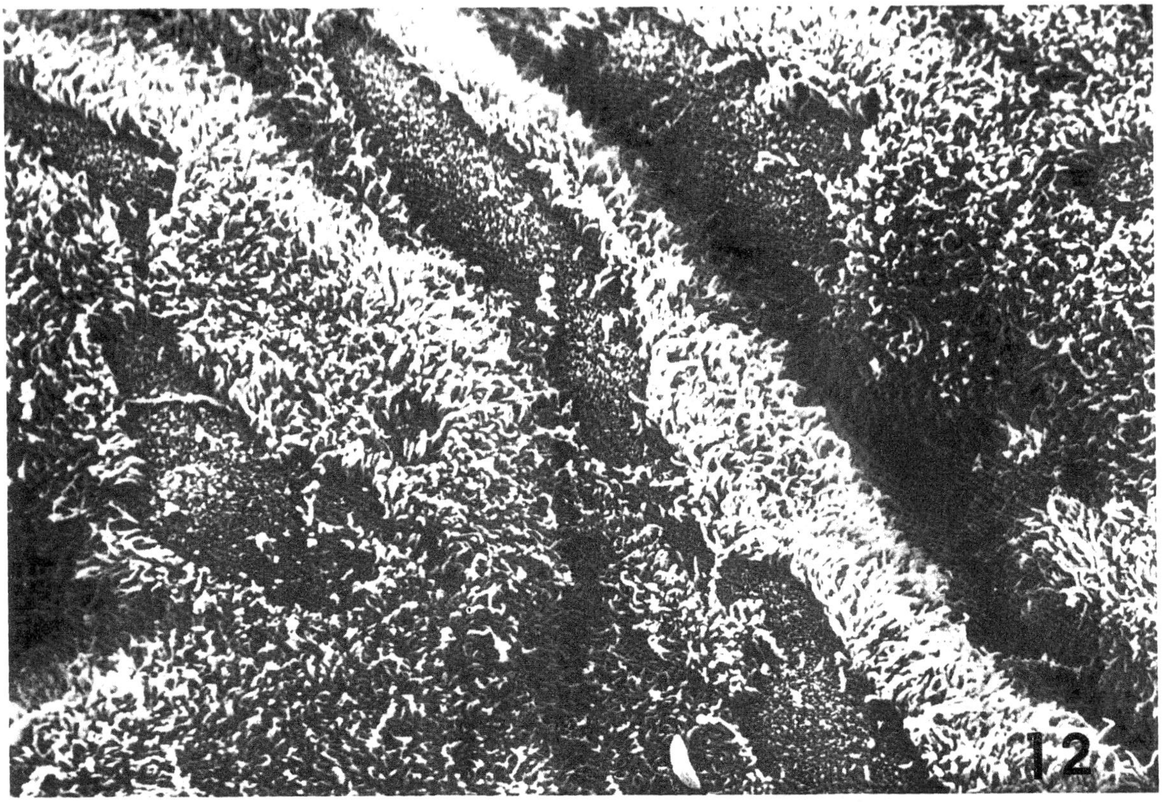
12

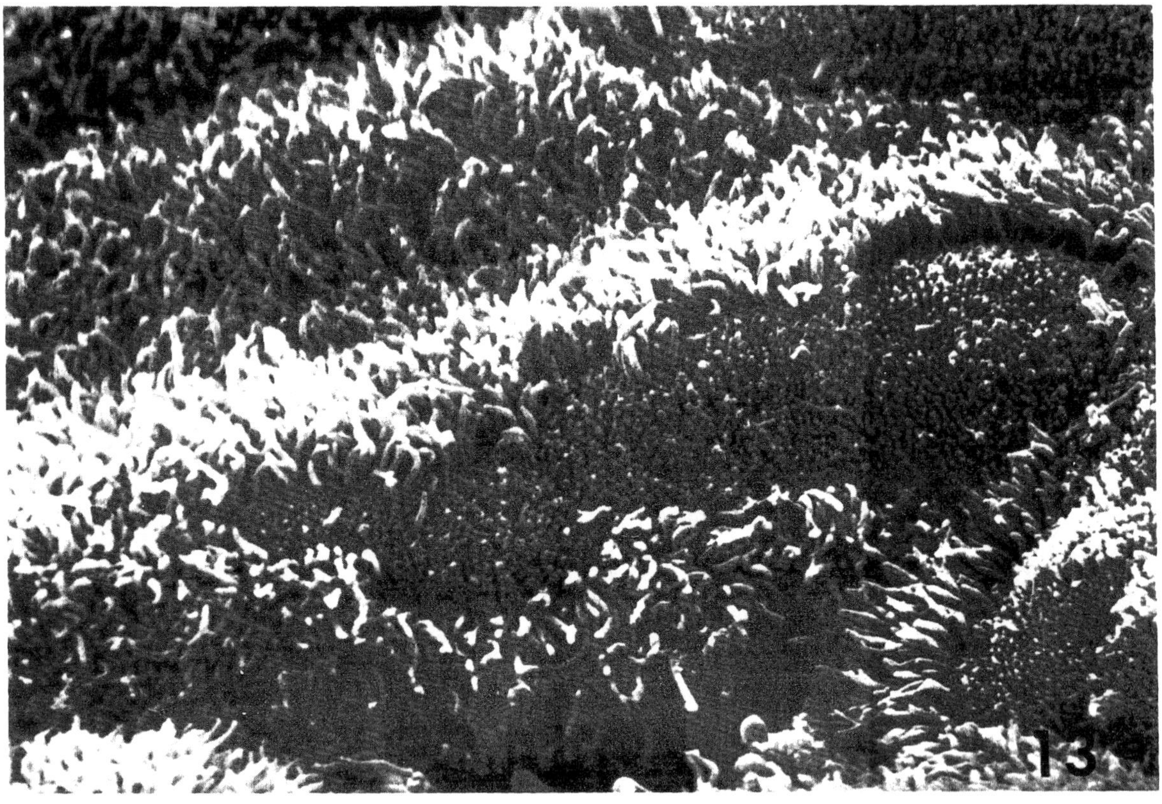
13

mitochondrial profiles within the cells were ovoid to elongated and lamellar cristae were present. Those of the vole (71) also had tubular cristae in addition to the lamellar configuration. The lining epithelium was similar in both regions. Lymphoid follicles were present beneath the lamina propria in the apex of the corpus ceci, a feature reminiscent of the appendix.

3.13. Cecum in the guinea pig

Snipes (72) reported a highly-folded cecal mucosa in the guinea pig consisting of irregular mounds and crevasses. TEM revealed typical absorptive enterocytes, the apical surface of which had a close intermicrovillar relationship with bacteria. Taenia were present on the external aspect of the organ but not along the remainder of the colon. These structural features are related to the fact that considerable fermentation occurs in this organ, particularly the breakdown of cellulose, a process requiring a significant population of bacteria. Bacteria adherent to the epithelial wall have been demonstrated to be common in ruminants (73). The actual mode of adherence may involve an overlying mass of polysaccharide glycocalyx not often visualized on transmission EM due to the processing. Adherence by pili of bacteria to the epithelium has not yet been demonstrated. The luminal mucin layer provides a favorable habitat for microorganisms (74). Snipes (72) considered the amount of bacteria observed in the guinea pig to be excessive, compared with other rodent species. Despite the overall size of the organ, however, the morphology and ultrastructure of the cecal wall of the guinea pig is remarkably similar. The fine structure of the normal adult guinea pig cecum is illustrated in Figures 7-10–7-13.

3.14. Cecal ultrastructure in domestic animals

Wille (75) examined the cecal mucosa of the pig, several ruminants, and the horse and noted marked species differences. Irregular-shaped openings of the cecal crypts were features of the horse and goat, while that of the pig, ox, and sheep were a uniform mosaic. These differences were attributed to different functional states with only a slight possibility of true species-bound differences (75). The apical region of the columnar cells in the equine cecum were observed to contain numerous secretory granules whose electron density and histochemistry suggested a mucoid nature. Typical goblet cells were not observed in this species (76).

Witlock et al. (77) examined the mucosal pattern of the chicken cecum with the SEM. They found four types of epithelial surfaces. The villous configuration in the neck region of the cecal pouch resembled that normally found in the intestine of the chicken and other animals. Three of the four types of mucosal architecture resembled pathologic changes of intestinal mucosa induced by malabsorptive disease states in other animals. The majority of these changes involved reduction of mucosal villi and the opening of crypt orifices directly onto the luminal surface, a transformation from small to large intestinal morphology. The stages observed, if considered individually, could be linked to that observed in areas of the large intestine involved in disease. Chicken ceca infected with *Eimeria tenella* presented with lesions of the neck, middleposition, and distal end of the cecal pouch. The cecal neck was not involved; these organisms preferentially affected the dilated portions of the organ. The lesions extended into the lamina propria, revealing masses of connective tissue and cellular debris (77).

3.15. The cecal tonsil in the chicken

The cecal tonsil develops in the initial 4 to 18 mm of each cecal pouch. The basic unit of the structure was identified by Glick et al. (78), using scanning electron microscopy. as a central crypt with several primary branches, dense lymphoid

←

Figure 7-12. Higher magnification of normal guinea pig cecum, showing diverse lengths of microvilli on different epithelial cells. X 12,700.

Figure 7-13. At higher magnification by scanning electron microscopy, two types of microvilli can be discerned on the luminal surface of the normal guinea pig cecum. X 41,100.

tissue, and germinal centers. The structural design of the cecal tonsil corresponds to the mammalian palatine and lingual tonsils. The numerous branching crypts make it dissimilar to the structural organization of the appendix. Transmission EM of the proximal region of the cecum, in the general area of the cecal tonsil, revealed bacterial attachment to the villi. The surface morphology of the cecal tonsil did not differ markedly from the surrounding proximal cecum. Witlock et al. (77), however, claimed that the villi of the proximal region may be shorter and broader than that of the tonsil, and thick sections revealed more germinal centers in the tonsil region. Epithelial cell-associated bacteria have been described in the chicken (79, 80) associated with the luminal aspect of the crop, ileum, and cecum. Segmented flagellated forms were more numerous in litter-reared animals, but it was not possible to differentiate the tonsil region from that of the surrounding proximal cecum by the presence of bacteria (81). The bacteria exhibited end-to-end, as well as end-on, attachment to the cecal tonsil, occasionally appearing almost embedded in the epithelial cells. While firmly attached, they did not penetrate the cells and their presence is known to be influenced by the environment within the cecum (79). Do these organisms in turn exhibit an influence on cellular development of the cecal tonsil?

3.16. Disease-associated changes in the animal cecum

Marcus and Watt were the first to show that seaweed extract (carrageenan) caused an ulcerative colitis-like disease reaction in the cecum and large bowel of the guinea pig (82) and rabbit (83). In addition to cecal lesions, this agent also caused hyperplastic mucosal changes in the rabbit cecum (84). Electron microscopy revealed that damage to cecal epithelium occurred only after the macrophages of the lamina propria ingested the carrageenan and incorporated it into their lysosomes (85). Other workers observed cecal changes as early as 24 hours following ingestion, such as small, rounded foci of degeneration and inflammation in the surface epithelium. These foci later formed superficial ulcerations that eventually grew into the linear lesions characteristic of this disease. This initial ulcerative process was associated with epithelium overlying the large submucosal vessels. Nonulcerated parts of the cecum were not altered until after 7 to 14 days of administration, when accumulation of macrophages under these areas was evident (86). A similar phenomenon was observed in electron microscopic studies of experimental *Entamoeba histolytica* infection in the guinea pig (87). Olsen and Poulsen (86) offered no explanation for this relationship of the early focal lesions with the submucosal vessels in the guinea pig but stated that the subsequent ulceration associated with the macrophage accumulation was attributable to a foreign body reaction.

Onderonk et al. (88) used the carrageenan model to test the effect of metronidazole as a protective agent in ulcerative colitis. They found that coliform concentrations increased significantly during the initial stages of ulceration, Pretreatment with this antianaerobic microbial agent prevented carrageenan ulceration in the majority of animals where other antimicrobials directed against coliforms failed. A subsequent study (89) evaluated the bacteriologic aspect of this experimental model and suggested that the bacteria were an important factor in the development of these lesions. These workers isolated a cytotoxic substance in the stools and sera of both ulcerative colitis patients and these guinea pigs that were not present in individuals with other inflammatory bowel diseases. They felt that this might provide a link between the experimental model and the human disease. In view of the clinical importance of this phenomenon, further ultrastructural investigations are warranted.

4. Concluding remarks

The morphology of the appendix and cecum is closely related to its functional role, though the significant immunologic and other functions of the primate appendix and the transport and related functions of the cecum are not completely resolved. Indeed, investigators are still debating the functional significance and phylogenetic/evolutionary relationships of the appendix and cecum to their gross anatomic structure. Accordingly, detailed studies of the ultrastructural morphology can only help to clarify these interrelationships. Such investigations must be done by both transmission and scanning electron microscopy because only the former can provide extensive detailed information on cellular morphology, and only the latter can provide an overview of surface perspectives. Other adjunctive electron microscopic techniques (freeze-fracturing, electron-probe x-ray microanalysis, immunocytochemical procedures, etc.) can also be expected to clarify the ultrastructural morphology of the cecum and appendix, particularly in the future, as these techniques have not yet been widely used for these tissues. In addition to further exploitation of these techniques which are already available for future studies of the phylogenetic/evolutionary relationships of cecal and appendiceal structures, another general approach must be followed in the future. This is simply the methodical ultrastructural analysis of normal cecum, and appendix when present, of multiple species with different dietary and phylogenetic backgrounds. Ideally, this analysis should be done for fetal, newly born, and adult representatives, since these tissues have been documented to undergo significant structural and functional change with age. Also with respect to the ultrastructure of the normal cecum and appendix, interrelationships of the morphology to the microbial populations present should be studied, since such microflora are well known to alter morphology of the regions of the gut and since the cecum and appendix are significant reservoirs of microorganisms.

There is much opportunity for future elucidation of disease processes in the appendix and cecum by electron microscopic investigation. In most lower animal species, cecal disease is not common but some examples of its occurrence have been reported. In the horse, which is highly susceptible to intestinal colic and in which the distal digestive tract is the major site for digestion and absorption, ultrastructural studies of the cecum are greatly needed. In humans, the appendix is a frequent site for pathologic processes, yet it has been studied little by ultrastructural analyses. Preliminary electron microscopic data have been reported on human neurogenic appendicopathy, malacoplakia of the appendix, appendiceal cancers, and viral appendicitis, though these entities warrant further ultrastructural study. One of the significant electron microscopic advances made in this field has been the facilitation of clinical diagnosis of Batten-Spielmeyer-Vogt disease by EM analysis of appendectomy tissue.

References

1. Pfeiffer CJ, Rowden G, Weibel J: Gastrointestinal Ultrastructure. Stuttgart: Georg Thieme Publishers, p 1–294, 1975.
2. Toner PG, Carr KE, Wyburn GM: The Digestive System — An Ultrastructural Atlas and Review. London: Butterworths, p 1–303, 1971.
3. Morson BC, Dawson IMP: Gastrointestinal Pathology. Oxford: Blackwell Scientific Publications, p 391–393, 1972.
4. Scott GBD: The primate cecum and appendix vermiformis: A comparative study. *J Anat* 131: 549–563, 1980.
5. Abramson DJ: Vermiform appendix located within the cecal wall. *Dis Colon Rectum* 26: 386–389, 1983.
6. Perey DY, Cooper MD, Good RA: The mammalian homologue of the avian bursa of Fabricius. *Surgery* 64: 614–621, 1968.
7. Bockman DE, Cooper MD: Pinocytosis by epithelium with lymphoid follicles in the bursa of Fabricius, appendix and Peyer's patches. An electron microscopic study. *Am J Anat* 136: 1973.
8. Bockman DE: Functional histology of the appendix. *Arch Histol Jpn* 46: 271–292, 1983.
9. Simic V, Ilic A: Characteristiques et differences essentielles comparatives, morphologiques et topographiques, du caecum de l'homme et des mammiferes domestiques. *Acta Anat* 94: 299–310, 1976.
10. Hill WCO, Rewell RE: The caecum of primates — its appendages, mesenteria, and blood supply. *Trans Zool Soc Lond* 2: 199–256, 1948.
11. Mitchell C: On the intestinal tract of mammals. *Trans Zool Soc Lond* 17: 437–536, 1905.
12. Gorgas M: Vergleichend-anatomische untersuchungen am magen-darm-kanal der *Sciuromorpha*, *Hystricomorpha* and *Caviomorpha* (*Rodentia*). *Z Wiss Zool* 175: 237–404, 1967.
13. Behmann H: Vergleichend- und funktionell-anatomische untersuchungen am caecum and colon myomorpher nagetiere. *Z Wiss Zool* 186: 173–294, 1973.
14. Vorontsov NN: The ways of food specialization and evolution of the alimentary system in *Muroidea*. *Symp Ther Internat Symp Meth, Mammal Invest Brno* p 360–377, 1960.
15. Vorontsov NN: Die ungleichmassigkeit der tempi der umgestaltung der organe und der prinzips der kompensation der funktionen. *Zool J* 9: 1289–1305, 1963.
16. Janis C: The evolutionary strategy of the *Equidae* and

the origins of rumen and cecal digestion. *Evolution* 30: 757–774, 1976.
17. Davis DD: The giant panda. A morphological study of evolutionary mechanisms. *Fieldiana Zool Mem* 3: 1–337, 1964.
18. Barg M, Draper LR: Development of the rabbit appendix. I. Electron microscopic observations. *Acta Anat* 98: 106–115, 1977.
19. Snipes RL: Anatomy of the rabbit cecum. *Anat Embryol* 155: 57–80, 1978.
20. Loeschke K: Adaption der elektrolyt- und wasserresorption. *Fortschr Med* 93: 1613–1615, 1975.
21. Snipes RL, Clauss W, Weber A, Hornicke H: Structural and functional differences in various divisions of the rabbit colon. *Cell Tissue Res* 225: 331–346, 1982.
22. Leng E: Absorption of inorganic ions and volatile fatty acids in the rabbit caecum. *Br J Nutr* 40: 509–519, 1978.
23. Murray RM, Marsh H, Heinsohn GE, Spain AV: The role of the midgut caecum and large intestine in the digestion of sea grasses by the dudong (Mammalia: Sirenia). *Comp Biochem Physiol* 56A: 7–10, 1977.
24. Mottaz P, Worbe JF: Transfert des acides gras volatile dans la paroi du caecum isole de rat. *Comptes Red Seances Soc Biol* 171: 375–380, 1977.
25. Parker D: The measurement of production rates of volatile fatty acids in the caecum of the concious rabbit. *Br J Nutr* 36: 61–70, 1976.
26. Ly J: Cecal function in the pig. VFA content and utilization by the caecal wall. *Cuban J Agr Sci* 8: 247–254, 1974.
27. Olszewski A: Interrelationships between amino acids absorbed from the isolated pig caecum in situ. *Acta Physiol Pol* 29: 168–176, 1978.
28. Olszewski A, Buraczewski S: Absorption of amino acids in isolated pig caecum in situ. *Acta Physiol Pol* 29: 67–77, 1978.
29. Rerat A: Digestion and absorption of carbohydrates and nitrogenous matters in the hindgut of the omnivorous nonruminant animal. *J Anim Sci* 46: 1808–1837, 1978.
30. Bond JH, Levitt MD: Fate of soluble carbohydrate in the colon of rats and man. *J Clin Invest* 57: 1158–1164, 1976.
31. Snipes RL: Limited fat absorption in the intestine of mice. A morphological study. *Acta Anat* 99: 435–439, 1977.
32. McKenzie RA: The caecum of the koala, *Phascolarctos cinereus*: Light, scanning and transmission electron microscopic observations on its epithelium and flora. *Aust J Zool* 26: 249–256, 1978.
33. Bayley HS: Comparative physiology of the hindgut and its nutritional significance. *J Anim Sci* 46: 1800–1803, 1978.
34. Kerlin P, Phillips S: Absorption and secretion of electrolytes by the human colon. In: Colon Structure and Function. L Bustos Fernandez (ed), New York, London: Plenum Med Book Co, p 17–37, 1983.
35. Auböck L: Die elektronenikroskopie der histologisch "normalen" appendices. *Oster Chirurgenkongress Kongressband* 10: 317–322, 1969.
36. Gorgollon P: The normal human appendix: A light and electron microscopic study. *J Anat* 126: 87–101, 1978.
37. Bockman DE, Cooper MD: Early lymphoepithelial relationships in human appendix: a combined light- and electron-microscopic study. *Gastroenterology* 68: 1160–1168, 1975.
38. Haar JL: Epithelium and associated lymphocytes of developing human fetal appendix. *Biol Neonate* 31: 94–102, 1977.
39. Kazzaz BA: Argentaffin and argyrophil cells in the appendix. *J Pathol* 104: 206–209, 1971.
40. Helmstader V: Immunohistologische und elektronenmikroskopische untersuchungen an endokrinen zellen desgastroenteropankreatischen systems. Habilitationsschrift, Universitat, Heidelberg, 1979.
41. Auböck L, Ratzenhofer M: "Extraepithelial enterochromaffin cell-nerve fiber complexes" in the normal human appendix, and in neurogenic appendicopathy. *J Pathol* 136: 217–226, 1982.
42. Millikin PD: Extraepithelial enterochromaffin cells and Schwann cells in the human appendix. *Arch Pathol Lab Med* 107: 189–204, 1983.
43. Rode J, Dhillon AP, Papadaki L: Serotonin-immunoreactive cells in the lamina propria plexus of the appendix. *Human Pathol* 14: 464–469, 1983.
44. Papadaki L, Rode J, Dhillon AP, and Dische FE: Fine structure of a neuroendocrine complex in the mucosa of the appendix. *Gastroenterology* 84: 490–497, 1983.
45. Dymock RB: Pathological changes in the appendix. A review of 1000 cases. *Pathology* 9: 331–339, 1877.
46. Van Haelst UJGM, Gabreels FJM: The electron microscopic study of the appendix as early diagnostic means in Batten-Spielmeyer-Vogt disease. *Acta neuropath* (Berl) 21: 169–175, 1972.
47. Rapola J, Haltia M: Cytoplasmic inclusions in the vermiform appendix and skeletal muscle in two types of so-called neuronal ceroid-lipofuscinosis. *Brain* 96: 833–840, 1973.
48. Ratzenhofer M, Aubock L, Becker H: Elektronen- und fluoreszenmikroskopishe untersuchungen der appendicitite neurogene. *Verh Dtsch Ges Pathol* 53: 218–224, 1969.
49. Höfler H: "Appendicite neurogene" — Eine zu gestellte diagnose. *Langenbecks Arch Chir* 351: 171–178, 1980.
50. Höfler H, Kasper M, Heitz PH U: The neuroendocrine system of normal human appendix, ileum and colon, and in neurogenic appendicopathy. Virchows Arch (*Pathol Anat*) 399: 127–140, 1983.
51. Rodriques MAM, Matter NJ, Campos JCP, Mitidieri HAC, Camargo JLV: Malacoplakia of the ascending colon and appendix, in childhood. *Arq Gastroent* (S Paulo) 18: 25–29, 1981.
52. Lall KS, and Mavrelis WP: Villous adenoma of the appendix. Report of a case and review of the literature. *Dis Colon Rectum* 25: 716–718, 1982.
53. Jones RA, Dawson IMP: Morphology and staining patterns of endocrine cell tumors in the gut, pancreas and bronchus and their possible significance. *Histopathology* 1: 137–150, 1977.
54. Rode J, Dhillon AP, Papadaki L, Griffiths D: Neurosecretory cells of the lamina propria of the appendix and their possible relationship to carcinoids. *Histopathology* 6: 69–79, 1982.
55. Subbswamy SG, Gibbs NM, Ross CF, and Morson BC: Goblet cell carcinoid of the appendix. *Cancer* 34: 338–

344, 1974.

56. Reif RM: Viral appendicitis. *Human Pathol* 12: 193–195, 1981.
57. Mora-Galindo J, Martinez-Palomo A, Gonzalez-Robles A: Interaccion entre entamoeba histolytica y epitelio cecal del cobayo. Estudio cuantitativo. *Arch Invest Med* (Mex) 13 (suppl) 3: 233–243, 1982.
58. Ono K: Ultrastructure of the surface principal cells of the large intestine in postnatal developing rats. *Anat Embryol* 149: 155–171, 1976.
59. Ono K: Changes of the cecal villi during postnatal development in rats. *Cell Tissue Res* 208: 253–259, 1980.
60. Wille K-H: Raster-elektronenmikroskopische beobachtungen an der blinddarm-schleimhaut des rinderfetus. *Verh Anat Ges* 69S: 891–898, 1975.
61. Portela-Gomes GM, Grimelius L, Bergstrom R: Enterochromaffin (Argentaffin) cells of the rat gastrointestinal tract. An ultrastructural study. *Acta Path Microbiol Immunol Scand* (Sect A) 92: 83–89, 1984.
62. Pleasants JP: Rearing germ-free caesarean-born rats, mice and rabbits through weaning. *Ann N Y Acad Sci* 78: 116–126, 1959.
63. Gustafsson BE, Maunsbach AB: Ultrastructure of the enlarged cecum in germ-free rats. *Z Zellforsch* 120: 555–578, 1971.
64. Wenrich DH: Protozoan parasites of the rat. In: The rat in laboratory investigation, 2nd ed. EJ Farris, JQ Griffith, Jr (eds), Philadelphia: JB Lippincott Co, p 486–501, 1949.
65. Snipes RL: Anatomy of the cecum of the cat. *Anat Embryol* 170: 177–185, 1984.
66. Mu, MM, Lingham R: The rudimentary appendix of the cat (*Felis domesticus*). *Acta Anat* 87: 119–123, 1974.
67. Arey LB: Developmental anatomy. A textbook and laboratory manual of embryology, 7th ed. Philadelphia, London: WB Saunders, p 249–255, 1966.
68. Wesley RKA, Jensen TE: The origin of crystal-containing cells in the cecal epithelium of mice. *J Ultrastr Res* 27: 306–318, 1969.
69. Henrikson RC: Ultrastructural aspects of mouse cecal epithelium. *Z Zellforsch Mikrosk Anat* 140: 445–449, 1973.
70. Snipes RL: Anatomy of the cecum of the dwarf hamster (*Phodopus sungorus*). *Anat Embryol* 157: 329–346, 1979.
71. Snipes RL: Anatomy of the cecum of the vole, *Microtus agrestis*. *Anat Embryol* 157: 181–203, 1979.
72. Snipes RL: Anatomy of the guinea pig cecum. *Anat Embryol* 165: 97–111, 1982.
73. Cheng KJ, Costerton JW: Adherent rumen bacteria — their role in the digestion of plant material, urea and epithelial cells. In: Digestion Physiology and Metabolism in Ruminants. Lancaster, England: MTP Press Ltd, 1980.
74. Sakata T, von Engelhardt W: Luminal mucin in the large intestine of mice, rats and guinea pigs. *Cell Tissue Res* 215: 629–635, 1981.
75. Wille K-H: Uber die schleimhautoberflache des blinddarmes einiger haussauger eine raster-elektronmikroskopische untersuchung. *Zbl Vet Med* 4: 265–273, 1975.
76. Kanakoudis GG: Secretory granules in the columnar cells of the cecum and the great colon of the horse. *Anat Histol embryol* 2: 295–299, 1973.
77. Witlock DR, Lushbaugh WB, Danforth HD, Ruff MD: Scanning electron microscopy of the cecal mucosa in *Eimeria tenella*-infected and uninfected chickens. *Avian Dis* 19: 293–304, 1975.
78. Glick B, Holbrook KA, Olah I, Perkins WD, Stinson R: An electron and light microscope study of the caecal tonsil: The basic unit of the caecal tonsil. *Develop Comp Immunol* 5: 95–104, 1981.
79. Fuller R: Ecological studies on the lactobacillus flora associated with the crop epithelium of the fowl. *J Appl Bact* 36: 131–139, 1973.
80. Fuller R, Turvey A: Bacteria associated with the intestinal wall of the fowl (*Gallus domesticus*). *J Appl Bact* 34: 617–622, 1971.
81. Glick B, Holbrook KA, Olah I, Perkins WD, Stinson R: A scanning electron microscope study of the caecal tonsil: The identification of a bacterial attachment to the villi of the caecal tonsil and the possible presence of lymphatics in the caecal tonsil. *Poultry Sci* 57: 1408–1416, 1978.
82. Watt J, Marcus R: Ulcerative colitis in the guinea pig caused by seaweed extract. *J Pharm Pharmacol* 21S: 187–188, 1969.
83. Watt J, Marcus R: Ulcerative colitis in rabbits fed degraded carrageenan. *J Pharm Pharmacol* 22: 130–131, 1970.
84. Watt J, Marcus R: Hyperplastic mucosal changes in the rabbit colon produced by degraded carrageenin. *Gastroenterology* 59; 760–768, 1970.
85. Abraham R, Goldberg L, Coulston F: Uptake and storage of degraded carrageenan in lysosomes of reticuloendothelial cells of the rhesus monkey, *Macaca mulattae*. *Exp Mol Pathol* 17: 77–93, 1972.
86. Olsen PS, Poulsen SS: Stereomicroscopic and histologic changes in the colon of guinea pigs fed degraded carrageenan. *Acta Path Microbiol Scand* (Sect A) 88: 135–141, 1980.
87. Takeuchi A, Jervis HR, Phillips BP: Electron microscope studies of experimental *Entamoeba histolytica* infection in the guinea pig. III. histolysis of the cecum. *Virchows Arch B (Cell Path)* 245: 263–277, 1977.
88. Onderonk AB, Hermos, JA, Dzink JK, Bartlett JG: Protective effect of metronidazole in experimental colitis. *Gastroenterology* 74: 521–526, 1978.
89. Onderonk AB, Bartlett JG: Bacteriological studies of experimental ulcerative colitis. *Am J Clin Nutr* 32: 258–265, 1979.

Authors' addresses:
Carl J. Pfeiffer
Department of Vet. Biosciences
Virginia-Maryland Regional College of Veterinary Medicine
Virginia Tech
Blacksburg, VA 24061, USA

Brian R. MacPherson
Department of Anatomy
Faculty of Medicine, University of Alberta
509 Medical Science Building
Edmonton, Alberta, T6G 2H7, Canada

CHAPTER 8

The colon: Normal ultrastructure and pathological patterns

MARGARET B. LISTROM & CECILIA M. FENOGLIO-PREISER

1. Introduction

This chapter will focus on the large intestine, which begins at the ileocecal valve and includes the cecum, appendix, ascending, transverse, descending and sigmoid colon and rectum. The histologic and ultrastructural organization of the colon reflects its main function, the absorption of water and elimination of undigested material, a process facilitated by mucus secretion.

The purpose of this discussion is to bring together the published material on the ultrastructure of the normal and diseased colon. Transmission electron microscopy (TEM), scanning electron microscopy (SEM), immune electron microscopy (IEM), and autoradiography have made major contributions to our understanding of the normal and diseased human colon, but the observations remain widely scattered in the literature.

One of the main functions of the colon is to absorb salt by way of Na^+, K^+-dependent ATPase with water following passively. The colonic mucosa can also absorb proteins from the bowel lumen, a process that can be followed utilizing peroxidase labelled protein molecules. The labelled molecules are contained in surface vesicles within Golgi-associated bodies and subsequently found in the lamina propria (1). Potassium and bicarbonate are secreted from the crypts into the lumen (2). In addition, the secretion of mucus lubricates and protects the colon from the solid luminal contents.

2. Structure of the normal colon

The flat colonic mucosa is punctuated by regularly arranged crypts of Lieberkühn openings. The crypts extend from the lumen to the muscularis mucosae and are clustered into groups of five or six, separated by deep furrows (Figs. 8-1E, F).

The crypts are lined by simple columnar cells and goblet cells. Undifferentiated cells, immature absorptive cells, and endocrine cells are present in the basal portions of the crypts (3, 4).

2.1. *Epithelial cells*

SEM visualization of the colonic surface accentuates the fact that the grossly apparent flat mucosa is arranged in regular interconnected territories containing 10 to 100 parallel crypt mouths measuring 25–50 μm in diameter, separated by deep clefts (Figs. 8-1E, F) (3, 5). They are surrounded by elevations of luminal cells. Mucin secretions flow from the crypt openings and cover the surface (Figs. 8-1A, B), which is composed of absorptive cells and irregularly interspersed mature goblet cells recognizable by droplets of extruded mucus and irregular microvilli (3, 4). The crypts are lined by overlapping absorptive and goblet cells. Individual absorptive cells are polygonal in shape and covered by regular, evenly dispersed microvilli with knobby tips (3). Freeze-fracture replicas of the apical

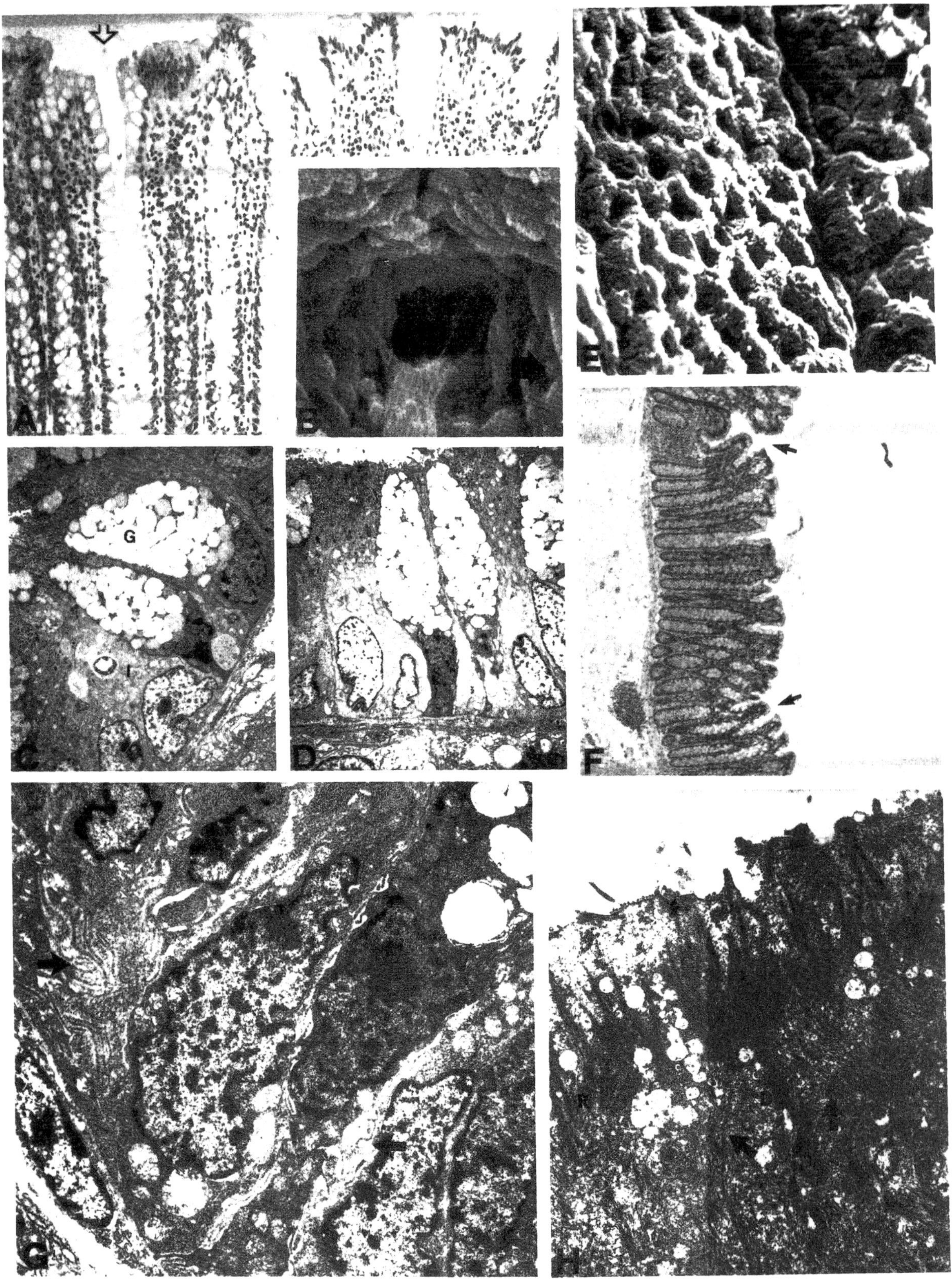

plasma membrane of absorptive cells confirm the regular arrangement of microvilli.

Normal colonic epithelium consists of: (a) undifferentiated cells, (b) mature goblet cells, (c) immature absorptive cells, (d) mature absorptive cells, and (e) enterochromaffin cells (4), with a regular pattern of cell differentiation proceeding from the base of the crypt to the luminal free surface (4).

The undifferentiated cells of the lower half of the crypts are similar to those in the small intestine. The crypt sides are composed of mucus-secreting cells, immature absorptive cells, and rare endocrine cells. The surface epithelium between the crypts consists of goblet cells and mature absorptive cells.

Basal immature cells are characterized by numerous ribosomes and few organelles; the nuclei are convoluted and contain a prominent nucleolus. As the cells differentiate, the cytoplasm becomes denser due to the development of rough endoplasmic reticulum (RER). Small apical secretion droplets, similar to those of mature goblet cells, appear (4).

Most of the cells lining the middle and upper crypt are immature absorptive cells with prominent (4) apical junctional complexes (Figs. 8-1C, D). The immature cells contain small vesicles with fibrillar glycoproteins that contribute to the glycocalyx. A supranuclear Golgi zone is present (4). These cells are in the second replicative zone, i.e., the zone in which partially differentiated cells can still undergo mitoses (4). In addition, they exhibit a complex network of 10 nm filaments encompassing the nucleus and forming extensive bundles in the terminal web (Fig. 8-1H). These are linked to intermediate filaments which are then anchored to lateral desmosomes (6).

The mature absorptive cell, which absorbs water and electrolytes, has numerous, short, regularly-spaced microvilli, and the cells are joined by junctional complexes and lateral desmosomes. The intercellular spaces are variably dilated depending on the physiologic activity of the colon (4). Interlocking cell processes cross the intercellular spaces and form labyrinths (Fig. 8-1G). The apical cytoplasm contains a terminal web composed of microfilaments and core rootlets. Mitochondria, lipid droplets, lysosomes, and apical vesicles are present (7). The round nucleus is basally located.

The well-developed microvilli are covered by a glycocalyx that invests the luminal surface of the epithelial cells. Hairlike, loosely packed projections extend perpendicularly from the plasma membrane into the lumen and radiate from the microvilli. This contrasts with the compact feltwork often seen in the small bowel. The filamentous, glycoprotein coat is constantly renewed and has a turnover time of 16–24 hours (8). Small, round 30–100 nm vesicles, called C-bodies, are embedded among the filaments (8, 10) (Fig. 8-2A). These extracellular bodies coat the microvilli and are surrounded by a unit membrane (9–11). They are interconnected to each other and the microvilli by the filaments of the glycocalyx (10). The number of C-bodies is inversely proportional to the number of vacuoles present in the cell (8). Similar bodies are seen in the apical cytoplasm of mature absorptive cells (12). These are called R-bodies and measure 106–2,500 nm in diameter (11). R-bodies have also been seen in the undifferentitiated cells. As the cells mature, there is a concomitant migration of the R-bodies to the apex of the cell. The fusion of the R-body membrane with the plasma membrane results in the formation and secretion of C-bodies (9).

The glycoprotein coat can be visualized ultrastructurally using periodic acid-silver-methenamine staining (13). The mucin droplets stain throughout the crypt, but as the cells mature,

←

Figure 8-1. (A). The crypts are parallel to each other and mucinous secretions flow from the crypt openings (arrow). (B). SEM of crypt mouth with extruded mucin (arrow = goblet cell). (C). Bottom of crypt. (D). Side of crypt. I, immature absorptive cell; G, goblet cell (see text for detail). (E). Normal colonic mucosa. The crypt openings are clustered into territories separated by a deep furrow running transversely across the right side of the picture (SEM). (Reprinted with permission from *Gastroenterology* 69, 1975). (F). The furrows are indicated by arrows (H & E X 2.5). (G). Extensive cytoplasmic interdigitations (arrows) of the basal portions of colonic epithelial cells. (H). Apical portions of two absorptive cells. An extensive network of intermediate filament (arrows) is seen associated with core rootlets (R) and desmosomes (D).

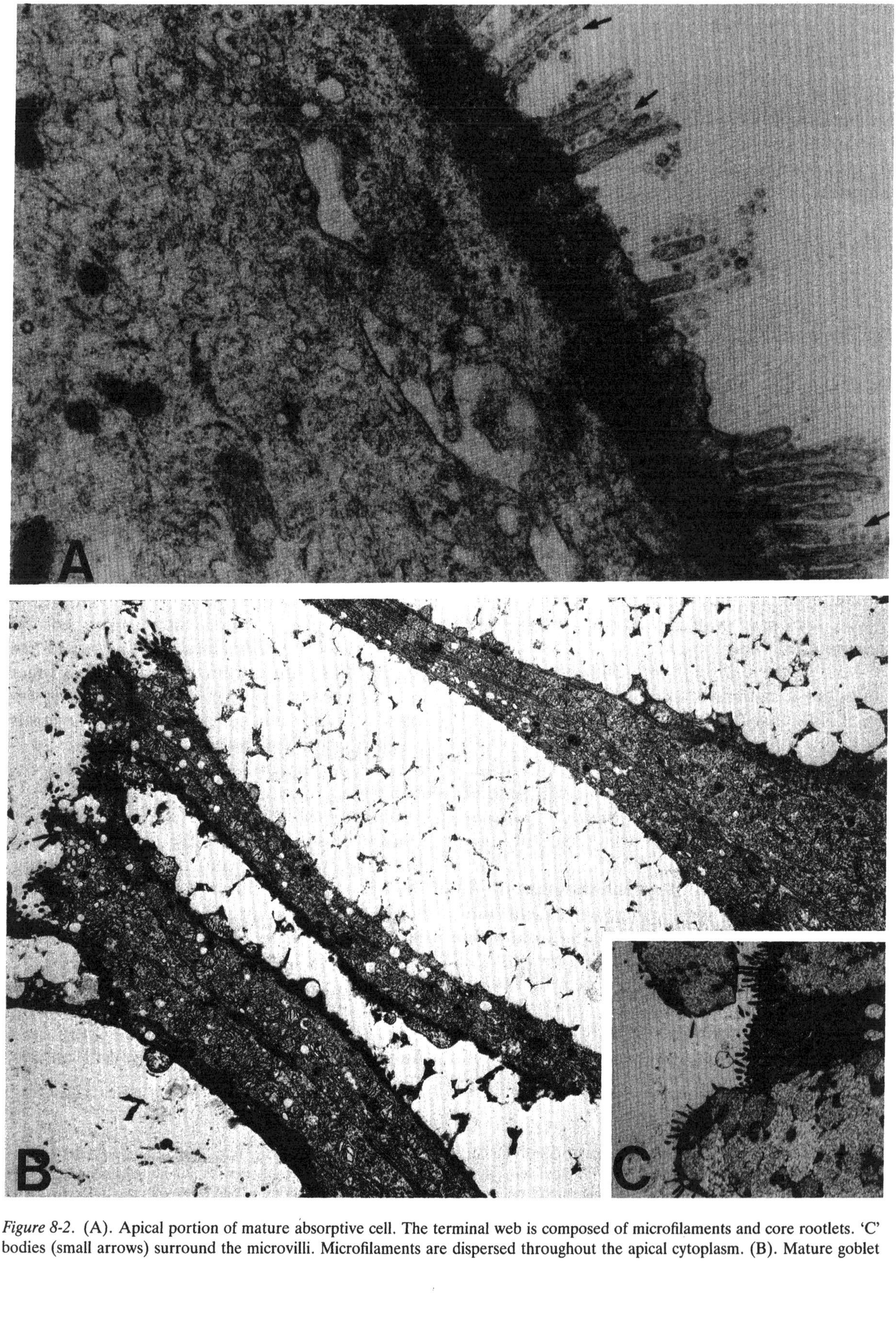

Figure 8-2. (A). Apical portion of mature absorptive cell. The terminal web is composed of microfilaments and core rootlets. 'C' bodies (small arrows) surround the microvilli. Microfilaments are dispersed throughout the apical cytoplasm. (B). Mature goblet

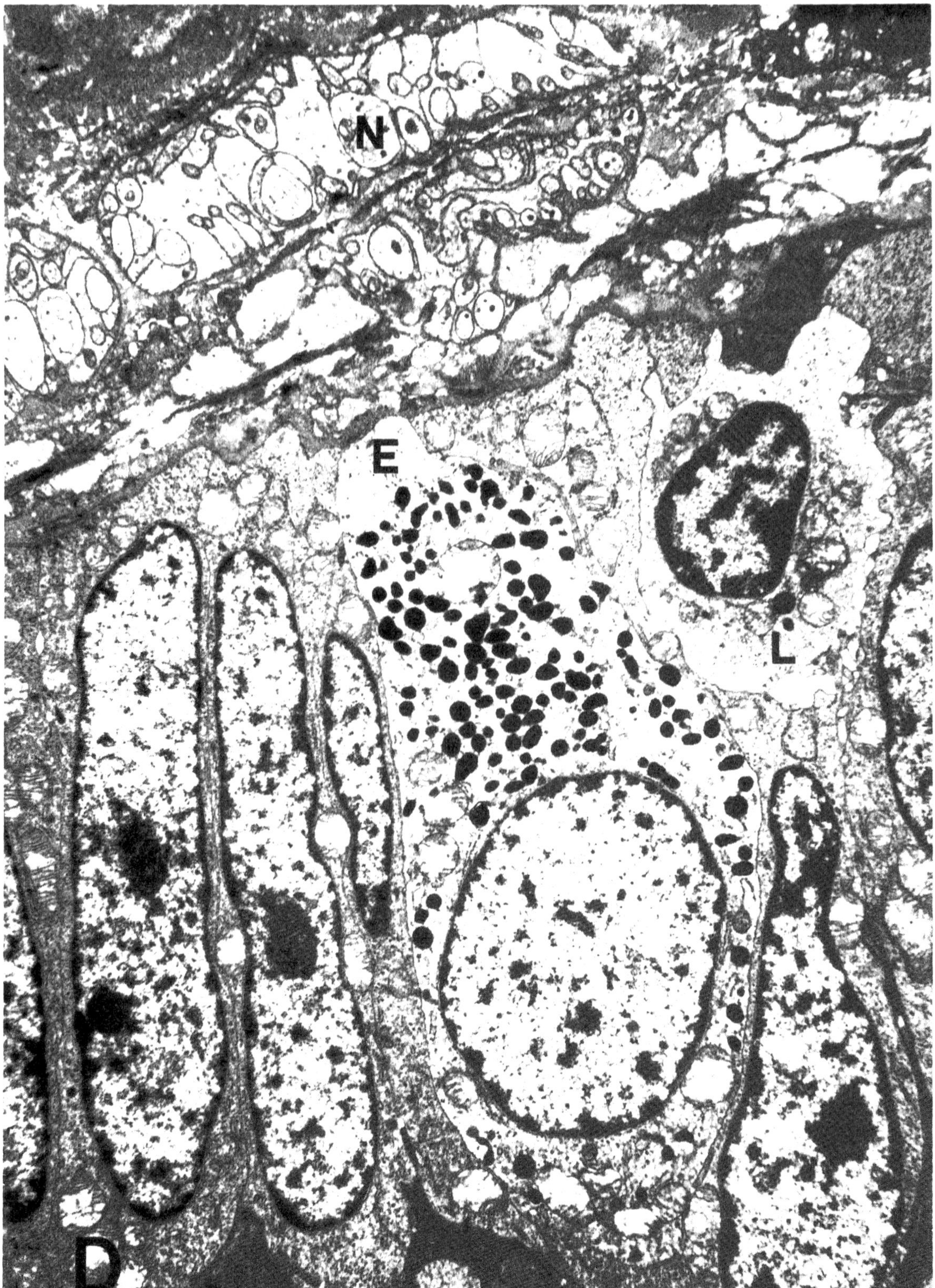

cells with distended theca and overhanging margins of absorptive cells. (C). Early phase of mucin secretion. Residual microvilli are seen on the surface of the goblet cell. (D). Endocrine cell (E) abutting the basement membrane of crypt. The cell is in close proximity to a nerve (N). An intraepithelial lymphocyte is also present (L).

silver staining increases. The apical vesicles in the absorptive cells also stain (13). Immunocytochemically some of these vesicles contain gamma-globulin (12). Immune labelling with ferritin-labelled antibody is also present along the luminal surface, suggesting that the glycocalyx coat is partly composed of secretory immunoglobulin (12).

Goblet cells are unevenly distributed and plentiful in the large intestine. They typically have a dense basal cytoplasm and a distended theca of the mature cell. The theca is a swollen supranuclear mucinous secretory mass compressing the nucleus. These cells are most abundant in the sigmoid (Figs. 8-2B, C). Evidence of early mucinous secretion can be seen at the base of the crypts in undifferentiated cells and immature goblet cells (4). In this location, the mucin granules vary considerably in size and shape and contain meshlike fibrils (12). The cytoplasm contains abundant ribosomes attached to the outer surface of the membrane of elongated cisternae of RER. These cisternae are parallel to the sides of the cell or to the contours of the nucleus. The Golgi apparatus is elaborate and located above the nucleus. The microvilli are irregular, shorter, and denser than those of the absorptive cell. As the cells become more distended with mucins, the microvilli become sparser. The lateral cell surfaces are interlocked by cytoplasmic processes. Junctional complexes are present at the cell apex, and desmosomes are located at various points along the lateral surface.

The goblet cell secretes the sulfated carbohydrate-protein complex synthesized in the RER, and glycosylated, sulfated, and packaged in the Golgi cisternae. The mucin granules migrate to the apical membrane where they are exocytosed. The synthesis and secretion of mucus occurs continuously throughout the life of the cell.

The cytoplasmic microfilaments of the epithelial cells form a loose, stratified network that retains the same basic organization as that found in the small intestine (14). Actin filaments (6–7 nm) extend laterally from the microvillous cores and interact with the tonofilaments of the desmosomal plaques. The apical terminal web of the absorptive cells stabilizes the epithelial surface.

Although the different cell types can be distinguished ultrastructurally, some authors feel that the columnar cells are goblet cells that have discharged their mucous granules, since both absorptive and goblet cells have the ultrastructural features of mucin secretion. After the mucus has been discharged, the mature goblet cell retains its electron density until it absorbs enough water to become electron-lucent. The entire epithelium lining the crypt rests on a continuous basal lamina (15).

2.2. *Endocrine cells*

Endocrine cells are small pyramidal cells with a broad basal cytoplasm containing dense cored granules. They are scattered among the nonendocrine epithelial cells. Ultrastructurally, there are at least 16 different types of granules recognizable in the gastrointestinal tract (16). These are described in detail in Chapter 13. Immunocytochemical methods are being used to correlate granular morphology with hormone production.

Ultrastructurally, some of the cells have a narrow apical pole with a tuft of microvilli extending into the glandular lumen. These are the "open" cells (10). Others are only in contact with the basal lamina and are "closed" (Fig. 8-2D) (5). The Golgi complex is supranuclear and the RER is sparse. The granules may be electron-dense or lucent (11).

Secretory products of these cells exert a local (paracrine) modulating effect on neighboring exocrine and endocrine cells. When these products cross the basal membrane, they enter the bloodstream reaching target organs where they exert endocrine effects. The neural effects of these cells are exerted as the hormones diffuse to synapses.

The colon has the least number of endocrine cells; they are primarily located in the proximal and distal colon (16). Substance P, 5-hydroxytryptamine (serotonin), somatostatin, glucagon, and glicentin as well as other hormones have been demonstrated in them.

2.3. *Other cells*

Paneth cells can be seen and are primarily located in the cecum and ascending colon (17). Their structure is similar to other zymogenic cells, with well-organized basal RER and large, electron-dense membrane-bound apical granules.

Intraepithelial lymphocytes may also be present either between the columnar epithelial cells or enfolded in the cytoplasm of an epithelial cell which provides a thin partition from luminal antigens (Fig. 8-2D). The surface epithelium overlying the lymphoid follicles contains microfolds, instead of microvilli, that may transport antigens across the cells' luminal surface to contact intraepithelial lymphocytes as well as those in the Peyer's patches (18). There, B and T cells and macrophages within the lymphoid follicles are stimulated by the antigens.

2.4. *Epithelial cell renewal*

The process of intestinal cell renewal involves proliferation, migration, differentiation, and sloughing or death of the epithelial cells. Dividing cells are located in the proliferative compartment of the crypts, which is larger than that seen in the small intestine. Autoradiographic techniques indicate that it involves the lowest two-thirds of the crypts (19). Cells migrate from the proliferative zone toward the gut lumen. This migration takes three to eight days (20). The normal colonic epithelium has a turnover time of approximately three to four days (9). All of the epithelial cells migrate together up the crypt wall onto the flat luminal surface, where they are eventually sloughed. Argentaffin cells probably do not divide or migrate. They probably arise from reserve cells.

Epithelial cells synthesize type IV collagen and laminin contributing to the formation of the underlying basal lamina. The fibroblasts within the lamina propria encircle the crypts and synthesize the underlying collagen table (4). Autoradiographic studies reveal that the flattened subepithelial fibroblast sheath proliferates near the base of the crypt and migrates upward together with the epithelial cells. These pericryptal fibroblasts are mitotically active at the base of the crypts and incorporate ^{3}H-thymidine. The polyploid pericryptal fibroblasts move upwards along the crypt and then laterally into the lamina propria.

The amount of collagen produced by fibroblasts is greater at the surface than at the base of the crypts. Under the surface, the fibroblasts are not mitotically active, confirming the fact that cell division and collagen synthesis do not take place together. The close relationship of the pericryptal fibroblast and the epithelium maintains the normal morphology and cellular dynamics of the colonic mucosa.

The fibroblasts have typical ultrastructural features with elongated nuclei and abundant RER, with many attenuated cell processes. Occasional myofibroblasts are seen; these cells have narrow bundles of microfilaments with peripherally located dense bodies. Many collagen fibrils are present in the subepithelial layer.

2.5. *Lamina propria*

The lamina propria separates the crypts of Lieberkühn. It consists of loose, reticular connective tissue containing fibroblasts, capillaries, and mononuclear cells, including macrophages, plasma cells, lymphocytes, and scattered mast cells. The area immediately under the epithelium primarily contains fibroblasts and macrophages. Macrophages are usually located in the upper 100 to 150 μm of the mucosa. These cells are identified by their abundant cytoplasmic vacuoles, lysosomes, and irregular nuclei. Mast cells are round with pleomorphic inclusions containing histamine. Eosinophils, when present, can be recognized by their crystalloid cytoplasmic granules (11). Plasma cells containing highly developed RER, and eccentrically located nuclei are also present. These inflammatory cells represent a barrier to entry by organisms and antigens to the interior of the body and are fewer in number in the intestines of germ-free animals (8).

Capillaries are present and their walls are usually fenestrated (10). Between the capillaries and the epithelium, there is an 8–12 μm reticular layer. The lymphatic plexus is associated with the muscularis mucosae, but no lymphatics exist at higher levels in the lamina propria (21). Lymphoid follicles characteristically split the muscularis mucosae and partially extend into the mucosa and submucosa.

2.6. *Submucosa*

The submucosa is subjacent to the muscularis mucosae and is composed of loose connective tissue, neurons, nerve fibers, blood vessels, lymph vessels, and scattered inflammatory cells. Ganglia of Meissner's plexus contain large cells with central nuclei and basophilic cytoplasm. These are further discussed in the section on the enteric nervous system.

2.7. Muscularis propria

The muscularis propria has an inner circular muscle layer and an outer continuous longitudinal layer gathered into bundles known as taeniae. Contraction of these bands results in outward bulges of colonic haustra. Auerbach's plexus is between the muscle coats. Individual muscle cells are electronically coupled through nexuses or gap junctions which are packed with hexagonal arrangements of 9 nm particles (22). The inner portion of the circular layer is made of a special muscle layer, three to eight cells thick (23). Abundant connective tissue is present inside this inner circular layer and between it and the outer layer. The specialized inner portion has wide areas of contact between cells (23). The cytoplasm of these cells contains abundant RER and a well-developed Golgi apparatus. Basal lamina is interposed between adjoining cells. At the border zone with the submucosa, there is a fibrous lamella rich in collagen and elastic fibers and lined by an incomplete layer of circularly oriented fibroblasts. Many nerve plexuses are present. It is thought that slow peristaltic contractions originate here. In fact, removal of the submucosa from the circular muscle layer abolishes all recorded slow waves (24).

Scanning electron microscopy of the outer longitudinal muscle layer reveals that the outer surface of cells is covered with collagen fibrils and a basal lamina rich in filamentous material (25). The outer surfaces of smooth muscle plasmalemma have many small vesicular invaginations (caveolae). Myofilaments form bundles running parallel to the cell axis. Some of the myofilaments leave the main bundles and coarse perpendicularly to the cell axis. Microtubules under the caveolae can be seen using the high resolution field emission scanning electron microscope (25). The arrangement of the collagen fibrils on the surface of the cells suggests that the basal lamina of the tissue plays a role in mechanical support against the stress of contracting and stretching (25).

2.8. Nervous system

The enteric nervous system is one of three divisions of the autonomic nervous system. It is composed to two major interconnected plexuses of ganglion cells and nerve fibers (22, 26). One of these is sympathetic and the other parasympathetic. These systems partially function independently of the central nervous system. The systems consist of ganglionic plexuses found in the submucosa (Meissner's) and between the two muscle layers (Auerbach's). Axons of parasympathetic, sympathetic, and sensory enteric plexuses contribute to the fiber network and form connections with the intrinsic neurons. Many types of nerves are present and are ultrastructurally differentiated by the types of vesicles seen in the nerve endings (Table 8-1; Fig. 8-3C) (27, 28). The types of nerves, their functions, and pathologic changes are discussed further in the chapter on the enteric nervous system.

3. Colonic polyps

A simple classification of epithelial colonic polyps includes hamartomatous, juvenile, hyperplastic, and neoplastic polyps (29). The polyps, present in patients with polyposis syndromes, resemble those found in patients without these syndromes.

Table 8-1.

Putative product	Morphology of axonal vesicles
Cholinergic	Small, agranular (50 nm)
Noradrenergic	Small, granular vesicles (50 nm)
Peptidergic "p-type" (nonadrenergic)	Large, opaque vesicles (90–150 nm), large granular vesicles
VIP	Large, granular vesicles (70–160 nm), pleomorphic granular vesicles, large opaque vesicles.
Dopamine?/ Catecholamines?	Small, flat, agranular vesicles (30–70 nm)

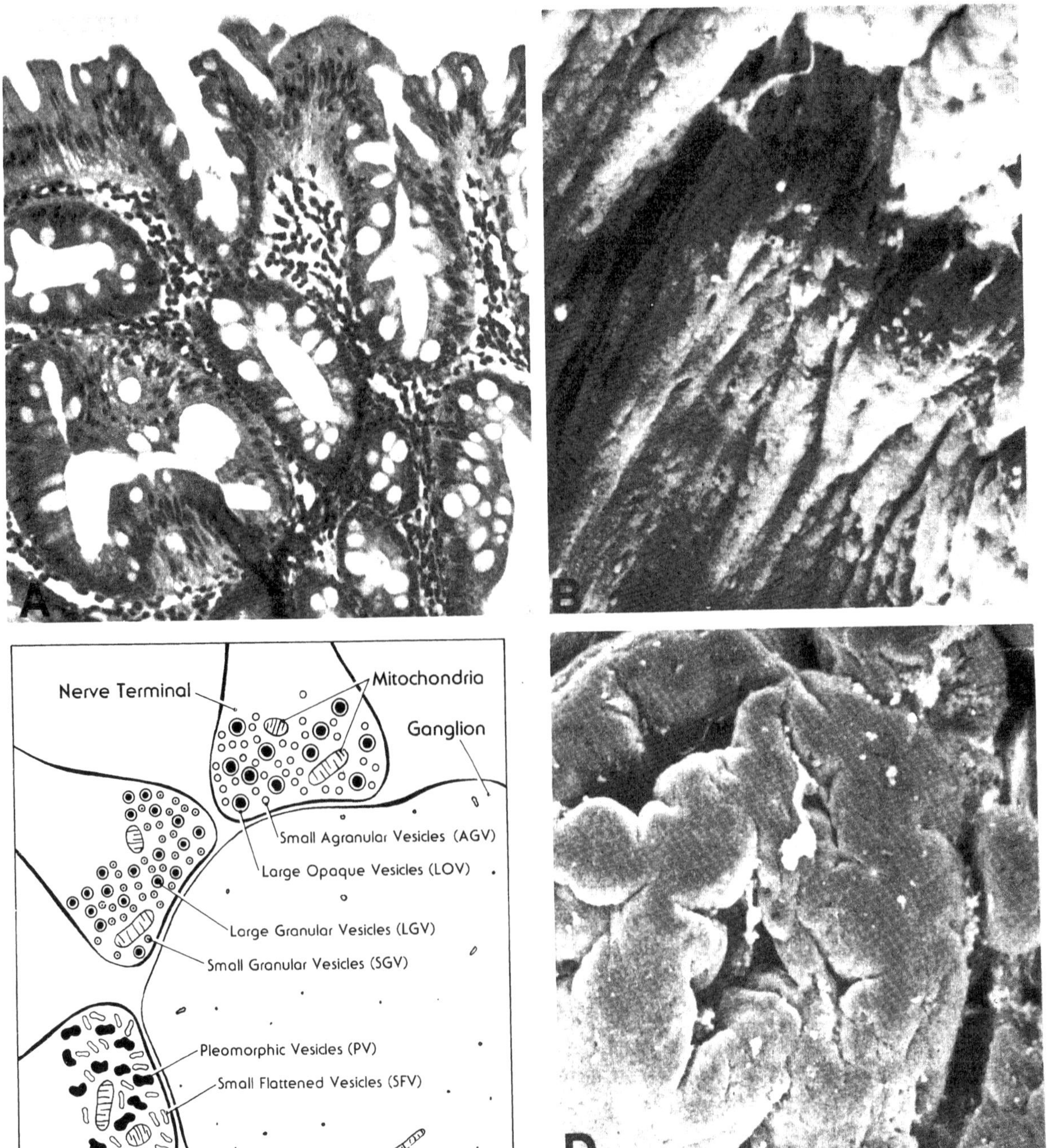

Figure 8-3. (A). Surface of hyperplastic polyp. The surface is serrated and star-shaped glands are lined by hypermature goblet cells. (B). SEM of the lateral border of hyperplastic polyp (see text for detail). (Reprinted with permission from *Gastroenterology* 69, 1975). (C). Schematic drawing of different types of vesicles found in nerve terminals within the enteric nervous system. (D). Surface of adenomatous polyp with cerebrifrom configuration and deep furrows. Normal territorial pattern is absent. (Reprinted with permission from *Gastroenterology* 69, 1975).

3.1. *Juvenile polyps*

Juvenile polyps are the most common type of hamartomatous polyp. They occur in children and are primarily located in the rectosigmoid. Histologically, these non-neoplastic polyps are composed of a stalk of normal colonic mucosa. The core of the polyp contains dilated glands, fibroblastic stroma, and many chronic inflammatory cells. The surface is usually ulcerated, and hemosiderin is deposited in the stroma. Ultrastructural examination has confirmed the histologic appearance and reaffirms the non-neoplastic nature of this polyp.

3.2. *Peutz-Jeghers polyps*

The Peutz-Jegher polyp is composed of mature epithelium arranged on an arborizing (20) proliferation of the muscularis mucosae. It is part of a familial Mendelian autosomal dominant syndrome that includes polyps in the stomach and small bowel, as well as mucocutaneous pigmentation and genital tumors. Ultrastructurally, the mucosa contains absorptive cells and mucus-secreting goblet cells (30). The absorptive cells have many mitochondria and a small number of lysosomes. The goblet cells are similar to those in the normal mucosa. Argentaffin cells are present in the basal portions of the polyp. The basal lamina is of normal thickness. Plasma cells, histiocytes, and lymphocytes comprise the inflammatory cell components. Normal smooth muscle cells are irregularly arranged in clusters of two or three cells.

3.3. *Hyperplastic polyps*

Hyperplastic polyps, the commonest type of colonic polyp (29), are characterized by focal enlargement of the crypts due to cellular crowding and excessive cytoplasm at the mouth of the crypts. There is marked infolding of the epithelium imparting a "sawtooth" appearance to the lumen (Fig. 8-3). They develop from a focal abnormality in cell replication resulting in an expansion of the proliferative zone (29). The surface of hyperplastic polyps is coated with mucinous secretions (3). The transition zone between the lateral surface of the polyp and the normal mucosa reveals an exaggerated territorial pattern of the crypts. The cell shapes are distorted due to stretching (Fig. 8-3B), and the microvilli are approximately two times their normal length. The tall microvilli of the intermediate cells at the base of the crypts have long rootlets. There are thick cordlike elevations of the surface of the polyp which taper into a flattened zone (Figs. 8-3A, B) (3). The number of surface goblet cells is decreased.

The crypt walls have over-distended goblet cells, which are about twice normal size. Occasionally, well-defined goblet cells are seen in the base (4). The intercellular space may also be over distended (4). In general, the hyperplastic epithelium has more mature cells in the upper part of the crypts. Abnormalities of the pericryptal sheath result in accelerated maturation of pericryptal fibroblasts (4). This is manifested by the thickened fibroblastic sheath.

3.4. *Adenomatous polyps*

These are benign neoplasms characterized by a glandular proliferation with closely packed tubules lined by crowded columnar cells containing large hyperchromatic nuclei. The growth pattern may be tubular, villous, or tubulovillous.

The surface of adenomatous polyps is flat with irregularly arranged, closely packed crypt openings that vary in size and shape (31). The surface is composed of irregularly anastomosing cerebriform cords of cells surrounded by deep furrows (3). This appearance corresponds to the raspberry-like configuration of the head of the polyp (Fig. 8-3D). Villous polyps are leaflike with deep furrows. The microvilli are short and blunted; mature goblet cells are decreased except in the hypersecretory type (3). Mild atypia is manifested by mild alterations in size and shape of cells; severe atypia is seen as extreme variations of cell size and shape.

Ultrastructurally, there is maturational arrest as the epithelial cells migrate toward the surface. The primary cell of adenomatous polyps is an intermediate, partially differentiated cell. The nucleus of this cell is oval and has evenly distributed chromatin (4, 32). The cytoplasm contains little RER and many free ribosomes; these are similar to cells seen in the lower part of the normal crypt. There are a few irregularly shaped microvilli. Goblet cells have variably dense mucin droplets and atypical dense bodies (32). The lateral cell borders are straight, the intercellular spaces are not distended, and junctional complexes are present (4). Malignant change is manifested by increased undifferentiated cells with enlarged nuclei containing prominent nucleoli. Several layers of these atypical cells may be seen (Figs. 8-4A–D). Cell processes pass through the basement membrane. The presence of undifferentiated and atypical cells distinguishes adenomatous polyps from hyperplastic polyps (Fig. 8-2). Focal collections of enterchromaffin cells can be seen at the surface of these polyps; in normal mucosa they are usually located at the base of the crypts (4).

In carcinoma, the majority of the glands contain high-grade dysplasia (33). The proliferative zone of the mucosa from patients with polyposis syndromes is displaced upwards, similar to that seen in colorectal carcinomas (20).

Villous adenomas associated with electrolyte loss have characteristic ultrastructural features. The cytoplasm of these cells have an extensive Golgi zone with occasional vacuoles that are either empty or contain granular material (32). The vacuoles are similar to the mucin droplets of goblet cells and are occasionally membrane bound. The cells resemble immature goblet cells.

In addition to the mucinous epithelial cell types, adenomas contain Paneth cells, squamous cells, neuroendocrine cells, and, rarely, melanocytes (34). Rather than representing a metaplastic process, these various cell types represent the pluripotentiality of endodermally derived cells.

4. Adenocarcinoma

Carcinomatous cells have a surface that is deranged and distorted (31), and the abnormalities seen in adenomatous polyps are exaggerated in carcinomas. The surface is usually pebbly due to the decrease in microvilli (35). Those microvilli that remain are club-shaped and of variable height (36). The mucosal territorial pattern is obliterated, and the absorptive cells do not exhibit their characteristic polygonal shape. Epithelial cells appear crowded and compressed and protrude from the surface in a disorderly fashion. The crypts are no longer parallel.

Malignant nuclei are large with prominent nucleoli (Figs. 8-4E–F). The cells have straight, lateral borders. Irregular microvilli with long rootlets are characteristic of colonic adenocarcinomas (35). Apical dense bodies (0.1–0.3 μm) are seen in most cells of colorectal tumors and have been previously described in the immature absorptive cells of transitional mucosa (36).

The cytoskeletal system is also significantly altered. In normal mucosa, the intermediate filaments form a fine network surrounding the nuclei and mucin droplets, and form extensive bundles in the terminal web and basal cytoplasm. In adenomas, tonofilament bundles cross the cytoplasm, and the cells have well-developed apical collections of intermediate filaments (37). In well-differentiated carcinomas, these apical intermediate filaments are associated with lateral desmosomes. In more poorly differentiated carcinomas, randomly placed surface and intracytoplasmic desmosomes are associated with the intermediate filaments. These desmosomes vary from immature ones to fully formed desmosomes. As the cell dedifferentiates the intermediate filament system becomes more disorganized and collapses around the nucleus (5). This leads to accumulations of disorganized intermediate filaments in the perinuclear area (38).

NM-dimethylforinamide (DMF) induces differentiation of human colon carcinoma cells in culture (38). This is manifested by an increase in the number of tonofilaments and increased desmosomes. There is a correlation between the number of these structures and the degree of differentiation (38).

Round basophilic inclusions, approximately 1 to 2 μm in diameter, originally described by Leuchtenberger, are present in both benign and malignant tumors of the colon (39, 40). These inclusions are also seen in inflammatory disorders such as ulcerative colitis, diverticulitis, and appendicitis (40). They are intracytoplasmic and extranuclear. Feulgen positivity indicates that they contain DNA (40). They probably do not represent degenerating cellular organelles, bacteria, or extraneous material (40).

The stromal response in invasive carcinomas is marked by activation of the stromal fibroblasts.

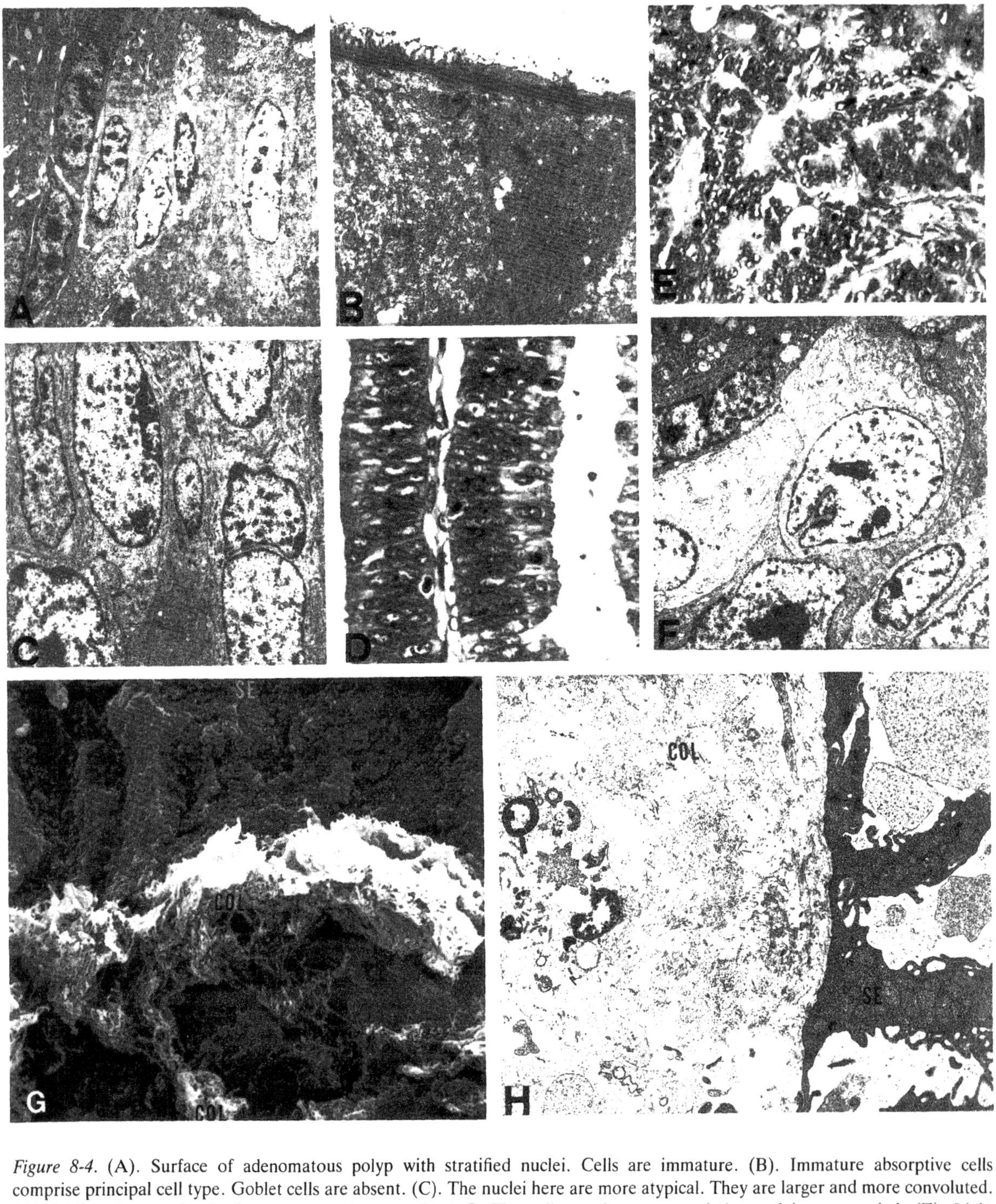

Figure 8-4. (A). Surface of adenomatous polyp with stratified nuclei. Cells are immature. (B). Immature absorptive cells comprise principal cell type. Goblet cells are absent. (C). The nuclei here are more atypical. They are larger and more convoluted. (D). Light micrograph of adenomatous polyp seen in A-C. The cells are immature and the nuclei are crowded. (E). Light micrograph of an invasive adenocarcinoma. (F). Corresponding electron micrograph. Notice loss of cell polarity and anaplastic nuclei. (G). Collagenous colitis. SEM of subepthelial collagen layer (COL). (H). Thickened collagen table (COL) underlies the epithelium. The surface epithelium (SE) is degenerating and desquamated. (Reprinted with permission from *Scandinavian Journal of Gastroenterology*. 1985; 20 (supplement 107).)

Many combine ultrastructural features of fibroblastic and smooth muscle cells, demonstrating their myofibroblastic nature (41). The stromal cells become activated when cancer cells are in the lamina propria, but true desmoplasia does not occur until submucosal invasion is present (41). In anaplastic carcinomas, almost all stromal cells are activated myofibroblasts with indented nuclei, dense bodies, and abundent RER. Elastic fibers can also be seen around invasive tumor cell clusters.

5. Epithelial changes adjacent to carcinoma (transitional mucosa)

Light microscopy usually shows an abrupt transition between benign and malignant epithelia. However, histochemical studies and ultrastructural studies reveal differences not appreciated histologically. This transitional mucosa is confined to an area within 2 cm of carcinomas (42). Mitotic indices are increased. This suggests an acceleration of cell renewal in the colonic mucosa even in the absence of dysplasia. Immature cells persist at higher levels in the crypts, and the differentiating zone of the crypt is extended (19,20).

SEM visualization of transitional mucosa reveals compressed crypts with clumps of mucus on the cell surfaces. The mucus is fragmented, leaving the underlying epithelium focally exposed to luminal contents. This change is also seen in polyps. Changes in the sialic acid concentration of the mucin probably decreases its layering abilities, accounting for this change.

Ultrastructural cytochemical analysis indicates that, although more mucin is produced, it is enzymatically and histochemically abnormal. There is a prevalence of sialomucins, whereas the mucin in normal colonic mucosa is predominantly sulfated mucin. Groups binding ruthenium red, a marker for sulfated mucopolysaccharides, are decreased. Staining with silver methenamine-PAS also reveals an increase in sialomucins accompanied by a decrease or absence of the sulphomucins. The midportion of the crypts show a heterogeneous pattern of silver-staining as compared to normal crypts. The upper crypts have larger mucus droplets and contain greater amounts of stainable glycoprotein. The luminal epithelium of normal mucosa contains mucin droplets surrounded by a dense rim of staining, whereas in the transitional mucosa, the mucin droplets stain diffusely.

Ultrastructurally, increased goblet cells and the appearance of vesiculated intermediate cells, which are not common in normal mucosa, are easily seen. The absorptive cells have many variably sized (0.15–0.3 μm) supranuclear, cytoplasmic vesicles budding into the luminal surface. These granules have a dense core and a clear halo and are continuous with the ER (33). They increase in number as the degree of dysplasia worsens (Table 8-2) (33). There appears to be a direct relationship between the number of vesicles and development of a fuzzy glycoprotein coat in transitional mucosa and early carcinogenesis.

6. Carcinoid tumors

Gastrointestinal carcinoids occur primarily in the appendix and distal ileum. The second most common site is the rectum. The behavior of these lesions is more aggressive when they occur in the colon (44). Their ultrastructural characteristics are described in the section on the appendix and small bowel.

6.1. Adenocarcinoid

This unusual tumor occurs primarily in the appendix but has also been reported in the colon

Table 8-2. Ultrastructural changes in ulcerative colitis and granulomatous colitis.

	Ulcerative Colitis	Granulomatous Colitis
Microvilli	↓ #, shorter	Normal
Glycocalyx extraneous coat	Thickness	Normal thickness
Swelling of ER	Present	Not seen
Vesicles	↑ # & size	Normal
Mitochondria	Swollen	Normal
Cell separation	Present	Present
Inflammatory cells	Intraepithelial	Subepithelial

(45). The tumors generally have the appearance of typical carcinoid tumors with an admixture of mucin-secreting cells and signet ring cells.

Ultrastructurally, epithelial nests containing mucin-producing cells are seen. Many of the mucin granules are confluent and a few are surrounded by limiting membranes (45). Other cells contain electron-dense pleomorphic granules ranging in size from 100 to 425 nm. The granules are membrane bound. The cytoplasm of these cells contains moderate amounts of RER, free ribosomes, and the Golgi apparatus is small.

Transitional cell forms (amphicrine cells) are seen with mucin secretion present in the neuroendocrine cells. This is evidence of a close relationship between intestinal neuroendocrine cells and mucus cells. Goblet cell carcinoids are probably derived form the endodermal crypt cell that is normally present in the gastrointestinal tract (46). In the colon these tumors are slightly more aggressive than appendiceal adenocarcinoids.

7. Ulcerative colitis

Grossly, ulcerative colitis is a diffuse disease with distal large intestinal predominance. Histopathologically, the crypts are destroyed and distorted by inflammatory and reparative processes. The activity of the disease is estimated by the amount of crypt infiltration by neutrophils. The severely inflamed lamina propria is composed of granulation tissue with increased numbers of IgG producing plasma cells.

By SEM the mucosa is edematous and covered by a cellular exudate. In mild cases the crypt orifices are still arranged at regular intervals. They are surrounded by an irregular cuff of heaped-up cells that vary in size, shape, and polarity. The altered polarity of the cells around the crypt mouths resembles that seen in tubular adenomas (47, 48). In severe lesions, there is irregular denudation of the mucosa leaving residual islands of surviving luminal cells. Disorganization of primary crypt architecture is characteristic of regenerated areas. In very severe cases, the crypt structure is replaced by a low villous appearance (49), with prominent cleft formation, crevasses, and cerebriform appearances (50).

Atrophic changes are characterized by decreased and irregularly-spaced crypts. Dysplasia is ultrastructurally manifest by marked variation in epithelial cell size and shape. Goblet cells are reduced in number as are the number of microvilli. The earliest ultrastructural change seen by TEM is degeneration of the reticular layer which underlies the surface columnar epithelium (6). Stromal fibroblasts produce abundant collagen. The degeneration gradually extends from small foci to involve more of the surface.

The glycocalyx coat is decreased and the C-bodies which normally accompany the microvilli are reduced in number. The cytoplasm of the epithelial cells become vacuolated and the number of mitochondria is reduced (6, 51). The subepithelial capillaries are decreased in number and lined by swollen endothelial cells with enlarged nuclei (6, 52). Many of these ultrastructural changes are not specific for ulcerative colitis and are seen in other acute and chronic inflammatory diseases (51).

Paneth cells are increased in number in ulcerative colitis. The cells are usually numerically unaltered in developmental disorders and increased in inflammatory and neoplastic diseases.

Ulcerative colitis has a high incidence of malignant change. Prolonged mucosal inflammation and repeated regenerations with expansion of the proliferative zone may lead to premalignant epithelial changes. The precancerous lesions include irregularity in cell size and shape with hyperchromatic nuclei and prominent nucleoli (47). Ultrastructural examination reveals a decrease in the number of goblet cells with degeneration of both the goblet and absorptive cells (47). Enlargement and irregularity of the nuclei as well as stratification are also present.

8. Crohn's disease

Although primarily a disease of the small intestine, the large bowel may be affected. The earliest recognizable histological lesions of Crohn's disease are "aphthoid" ulcers that represent small mucosal micro-ulcerations. These small ulcers are surrounded by normal appearing mucosa. There is a uniform increase in IgA, IgG, and IgM secreting plasma cells (53).

Ultrastructure of granulomatous colitis

(Crohn's disease) reveals no loss of microvilli. They are normal in size, quantity, and distribution. There is no inflammatory cell infiltrate in the intercellular spaces, although there is marked inflammation of the subepithelial layer (6). The fissures which are characteristic of Crohn's disease contain young collagen fibrils, erythrocytes, and inflammatory cells. The subepithelial basal lamina and the pericryptal fibroblast sheath are destroyed. Fibrosis of the lamina propria is present or absent depending on the duration of the disease. Most of the components of the lamina propria do not differ significantly from those in the normal colon. The submucosa is fibrotic and edematous and contains mast cells, neutrophils, and macrophages. Large macrophages, full of debris, can be seen in the lamina propria, especially in those cases without granulomas (53). Many of the lymphocytes have plasmacytoid features. The nuclei of epithelioid cells within the granulomas have a dispersed chromatin pattern and a prominent nucleolus. The cytoplasm contains variable amounts of RER, many mitochondria, and very little phagocytosed matter. Acid phosphatase has been localized to the clear vesicles also seen in the cytoplasm (53).

Mucus production is increased in the diseased colon; however, the cell coat is not increased in thickness (54). This may affect the barrier function of the intestinal epithelium by decreasing motility and, therefore, hindering the removal of foreign material (55). Detachment and loss of the cell coat is seen in some areas. The epithelial cytoplasm contains many dense vesicles suggesting increased losses and increased production of cell coat in diseased areas (55). The ultrastructural differences between ulcerative colitis and granulomatous colitis are summarized in Table 8-2.

9. Pseudomembraneous colitis

The administration of antibiotics, especially clindamycin, may be complicated by severe gastrointestinal symptoms. Pseudomembraneous colitis occurs in about 10% of patients receiving the drug. Histologically, the mucosa is congested, edematous, and necrotic. The acute inflammatory response results in the formation of pseudomembranes. The ultrastructural changes reflect the histological alterations (56). Mild cases show only focal loss of microvilli as well as edema. These changes are seen initially at the tops of the crypts. There is a normal pattern of crypt openings (56). By SEM the edema is manifested by flattening of the mucosa. The next phase is characterized clinically by diarrhea and morphologically by pronounced edema, blunted microvilli, and neutrophils. Severe colitis is associated with pseudomembrane formation. The surface is irregular and swollen folds of mucosa are separated by grooves. Cellular uniformity is lost (56). Goblet cells are decreased. Fibrin is deposited as a fine network which polymerizes into coarse beaded fibrils (56).

10. Collagenous colitis

Collagenous colitis, a recently described entity, is characterized by watery diarrhea (57). Histologically, there is a distinct 50 to 60 μm fibrous layer immediately beneath the surface epithelium (57). Normally this subepithelial layer averages 1.9 μm in thickness. Increased thickness of this layer is also seen in hyperplastic polyps, but averages 10 to 30 μm in thickness. The increase is secondary to an abnormality of pericryptal fibroblasts. There are also hyperplasia, degeneration, and desquamation of the luminal epithelial cells. A mild inflammatory response is present in the lamina propria (Figs. 8-4G, H).

Ultrastructurally, the epithelial cells contain large cytoplasmic vacuoles. The microvilli are decreased in number (57). In the subepithelial area, the basal lamina is of variable thickness. Underneath the basal lamina there is a regularly thickened collagen layer between 20 and 80 μm (57). The collagen fibrils are not oriented, and the fibrils maintain the usual periodicity of 64 nm. The subepithelial capillaries are surrounded by collagen fibrils and the endothelial cells are swollen. The basal lamina is reduplicated in areas.

11. Amyloidosis

This condition is characterized by an extracellular amorphous eosinophilic deposition of abnormal fibrillar protein between connective tissue fibers. When the GI tract is involved, symptoms include malabsorption, obstruction, ulceration, hemorrhage, and impaired motility (58). Histologically, one sees a diffuse thickening of basal lamina due to deposition of amyloid between endothelial cells and pericytes (59). The vessels usually affected are those located in the submucosa and muscularis mucosae (10, 59). The most commonly used staining technique to visualize amyloid is the dye Congo red which, under polarized light, reveals a green birefringence. This reaction is due to the cross-beta pleated configuration of amyloid fibrils.

Ultrastructurally, amyloid is composed of thin nonbranching fibrils which measure 10 nm in diameter (59). Collagen fibrils are distinguished from amyloid fibrils by their characteristic transverse bands which repeat every 64 to 70 nm.

12. Changes associated with laxative use

Prior to endoscopy and/or biopsy, many patients are subjected to purgation by various laxatives. The mucosa examined after mannitol induced diarrhea or after saline enema cannot be distinguished from normal. However, hypertonic phosphate enemas (Fleet's enema) or bisacodyl (Ducolax) invariably change the proctoscopic appearance and alter the histologic and ultrastructural features of the mucosa. The epithelium is pale and surface erosions are seen. The subepithelial area is edematous and the lamina propria is infiltrated by neutrophils. Inflammation produced by bisacodyl may resemble mild idiopathic ulcerative colitis.

Melanosis coli is a condition found in patients who utilize laxatives containing anthracene dyes. The macrophages in the lamina propria contain dense pigment granules composed of lipofuscin which is related to the dyes (10). Ultrastructurally, the granules are both a fine- and coarse-tangled network of membranes (60).

13. Bacterial infections

Various indigenous, nonpathogenic microorganisms are associated with the epithelial cells of the colon. One such organism is a spirochete, in the genus *Borrelia* (61), whose pathogenicity is obscure. The organism can be found in about 1% to 7% of colorectal biopsies in symptomatic patients. Flagellated organisms have also been seen (62). Both of these organisms attach themselves to the surface membrane of the absorptive cells. Usually the epithelial cells are normal even though the bacterial surface protein, which is similar to a lectin, may interact with a specific glycosylated component on the host cell (62). The spirochete may be eliminated by antibiotic treatment. There is one case report of intestinal spirochetosis which resulted in damaged epithelial cells and penetration of the spinochetes into the lamina propria (61).

Intestinal spirochetosis has been associated with carcinoma, Crohn's disease, ulcerative colitis, ischemic colitis, chronic diarrhea, diverticulitis, hemorrhoids, and adenomatous polyps. This association between spirochetosis and various gastrointestinal diseases probably reflects the fact that the biopsy specimens were taken for the underlying gastrointestinal disease. Ultrastructurally, numerous spiral microorganisms are present at the luminal surface of the epithelium. SEM of this entity has revealed a continuous layer of spirochetes masking the underlying structures and obstructing the gland openings.

Most cases of inflammatory infectious diarrhea are caused by bacterial enteric pathogens which tend to localize in the distal small intestine and proximal colon (63). The histologic patterns observed depend on the pathogenic mechanism, i.e., via toxin production or direct mucosal invasion.

Bacteria which produce diarrhea by enterotoxin production do not cause severe light micro-

Table 8-3. Colonic bacterial causes of diarrhea.

	Invasive	Enterotoxin production
Campylobacter jejuni		+
Yersinia enterocolitica	+	
E. coli	+	
Salmonella	+	
Shigella	+	
S. aureus enterocoliticus	+	
C. difficile		+

scopic or ultrastructural changes (64). These organisms are listed in Table 8-3. Neutrophil infiltration of the mucosa is the hallmark of acute colitis caused by invasive bacteria. TEM reveals organisms in the glycocalyx as well as in the surface cytoplasm, giving evidence for direct cellular invasion (64, 65). Biopsy specimens from patients with acute onset of diarrhea may suggest an infectious etiology, but the diagnosis must be confirmed by culture or serologic tests.

12.1. Whipple's disease and malacoplakia

Both these diseases are the result of bacterial infection (10). Whipple's disease is a systemic disorder primarily affecting the small bowel but also affecting the large intestine. The macrophages in the lamina propria contain bacilliform organisms seen by TEM within and between glycoprotein-laden PAS-positive macrophages. The bacilli are 2.50 by 0.15 μm in size. The PAS-positive granules are thought to be bacterial products. Malacoplakia is classically seen in the genitourinary system and only rarely in the colon (66). It is characterized by two types of cytoplasmic inclusion: PAS-positive granules, corresponding to phagolysosomes, and Michaelis-Guttman bodies, which develop later within lysosomes. Cells containing these structures are called Hansemann macrophages. *E. coli* or Klebsiella is phagocytosed by macrophages and is incorporated into phagolysosomes. Ultrastructural pathology reveals large granular, concentrically laminated, siderocalcific structures (Michaelis-Guttman bodies) (66), representing calcium encrusted phospholipid membranes. Defective phagocytic function of the macrophage is the currently favored histogenetic theory. A malfunction of the cytoskeleton of the macrophages is postulated, impairing the movement of lysosomes to phagocytic vacuoles and blocking lysosomal enzymatic degradation of the ingested bacteria.

13. Viral diseases

13.1. Viral gastroenteritis

Many of the viruses which cause viral gastroenteritis have been identified (from stool specimens) by immune electron microscopy, which has proven to be an invaluable tool to detect viruses which cannot be cultivated (66). Rotavirus, the Norwalk agent, and adenovirus have all been seen utilizing this technique (66–70). However, viruses must be intact particles to be detectable (Fig. 8-4D). Furthermore, 10^6 organisms per 1 ml of stool must be present (66). Finally, the technique is expensive and labor intensive. The number of specimens which can be processed in one day is fewer than with other techniques such as ELISA, RIA, or immunofluorescence (66). The viruses are detected by negative staining of a suspension of feces. The centrifuged pellets are stained with potassium phosphotungstate. Viruses are identified on the basis of their characteristic size and morphology (Table 8-4), but since many viruses are similar, immune electron microscopy can be used for specific identification. Immune electron microscopy is the negative staining of a reaction between the virus and a specific antibody (Fig. 8-4C). Optimal antibody-antigen concentration must be achieved for the aggregation to occur. If the aggregates are too large, they may cover the grid and result in a dark, uninterpretable image.

13.2. Cytomegalic virus (CMV)

CMV colitis is a severe and usually fatal complication found in immunocompromised patients. Although CMV has been recognized in ulcerative lesions of the colon for years, its etiologic relationship to the ulcer is controversial (71). CMV infection is prevalent in homosexual males. Most of the infections are subclinical; however, with the onset of acquired immune deficiency syndrome (AIDS), serious life-threatening infections can develop. In addition, renal transplant patients develop overwhelming CMV colitis.

Table 8-4. Viruses causing gastroenteritis/colitis.

Virus	Nucleic acid	Size
Rotavirus	ds RNA	65 nm
Enteric adenovirus	ss DNA	70 nm
Norwalk virus	ss RNA	27–32 nm
Herpes simplex virus	ds DNA	180–200 nm
Cytomegaloviruses	ds DNA	150–200 nm

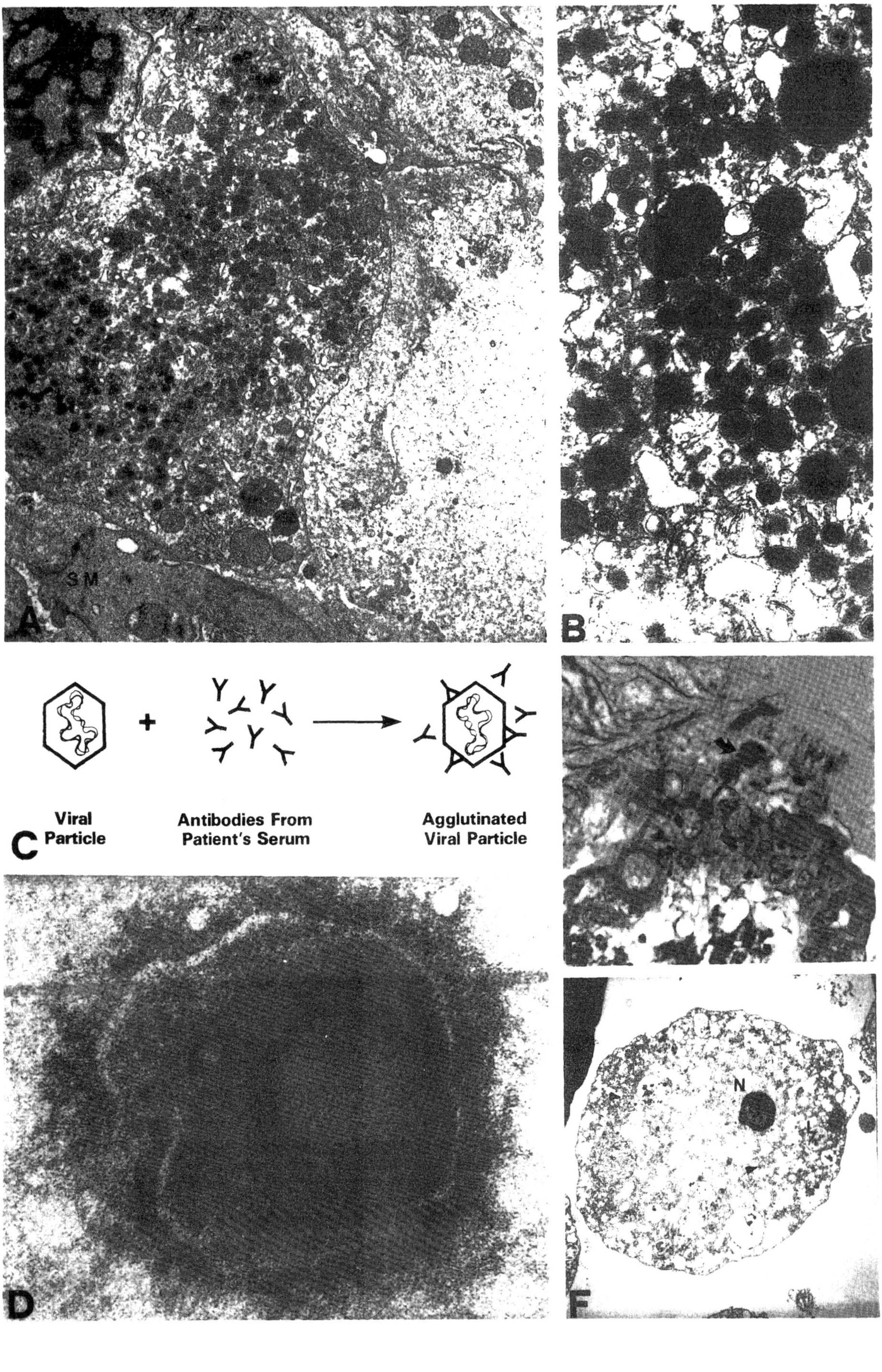
A
SM
B
C
Viral Particle
+
Antibodies From Patient's Serum
Agglutinated Viral Particle
D
F
N

Ultrastructurally intranuclear and intracytoplasmic viral particles are demonstratable in endothelial cells (71). Macrophages and pericytes also contain viral particles. These particles consist of irregular anastomosing strands of heterochromatin with small, round, enveloped viral particles approximately 100–140 nm in diameter. (Figs. 8-5A, B) (71). The cytoplasmic inclusions measure 17–340 nm. Endothelial and pericytic involvement is a feature of very severe infections.

13.3. Herpes simplex colitis

Herpes simplex causes intestinal ulceration in which the viral particles can be seen by electron microscopy (71). Ultrastructurally, the inclusions are present in the cytoplasm and nucleus of epithelial cells. The particles measure approximately 130 nm in diameter (72). Histologically, one sees multinucleate giant cells and intranuclear inclusions displacing the nucleus to the periphery or they are surrounded by a clear halo. Immunoperoxidase stains can be done to document the presence of herpes antigen.

14. Parasitic infection

14.1. Amebiasis

Few ultrastructural studies have been performed on amebic colitis since the diagnosis of amebiasis is made histologically. Histologically, changes are either diffuse or focal. Diffuse lesions resemble nonspecific inflammatory diseases, whereas focal lesions are ulcerated, flask-shaped, and shallow (73).

Ultrastructurally, the diffuse lesions are characterized by decreased numbers of microvilli and a sparse glycocalyx. The epithelial cells contain myelin figures, and the mitochondria and ER are swollen.

Ulceration is probably due to the release of lysosomal enzymes from the ameba (74). Plasma cells, polymorphonuclear leukocytes, erythrocytes, and eosinophils are present in the lamina propria (73).

The ameba has an outer fuzzy layer. It has large numbers of intracytoplasmic glycogen particles and numerous vacuoles. Red blood cells are occasionally seen in the cytoplasm. Clumps of chromatin bind to the inner nuclear membrane and there is a central karyosome (Figs. 8-5E, F).

14.2. Cryptosporidosis

Although the small bowel is the most prominent site of involvement, patients with colonic involvement have been reported. The clinical setting is usually that of a homosexual male with AIDS who presents with diarrhea and malabsorption. The infection is initiated by ingestion of oocysts. The organisms then undergo an asexual stage of development. The growing trophs are attached to the surface of the bowel epithelium and are surrounded by membranes. Whether the membranes originate from the host or parasite is controversial (75).

In the colon there may be minimal histologic changes, such as epithelial mucin depletion and focal infiltration by polymorphonuclear leukocytes. Oval basophilic, gram-negative organisms about 2 μm in diameter can be seen along the brush border. The organism is not highly toxic and does not destroy the host cell; however, there is damage to the epithelium with loss or elongation of the microvilli at the attachment site. The lack of microvilli probably impairs digestion and absorption, resulting in, or contributing to, diarrhea. The free parasites are crescent or ovoid in shape with multiple electron-dense cytoplasmic granules of various sizes. Most have fuzzy surfaces similar to epithelial cells. A single nucleus and nucleolus are present as well as two double-layered outer membranes (75).

←

Figure 8-5. (A). CMV colitis. Submucosal macrophage containing cytoplasmic and nuclear (arrow) viral particles. SM, smooth muscle cell. (B). Higher power of cytoplasmic enveloped viral particles. (Micrographs courtesy of Dr. Kathy Foucar, University of New Mexico). (C). Principle of immune electron microscopy. (D). Herpes virus, surrounded by lipid coat, visualized by negative staining (X 390,000). (E). Amebic colitis. *Entameba histolytica* (arrow) present in exudate covering surface epithelium. (F). Organisms with abundant cytoplasmic glycogen (arrow) and scattered phagolysosomes (L). Notice the characteristic nucleus (N).

15. Macrophage lesions

15.1. Colonic histiocytosis

Histiocytes are normally present in the colon. Although they contain PAS-positive material similar to that seen in Whipple's disease, ultrastructural studies have shown that there are no bacterial components in them (76). The osmiophilic inclusions are due to extrusion of goblet cell contents into the lamina propria which is then phagocytized by the macrophages. Macrophages within the lamina propria contain membrane-bound inclusions called phagosomes and contain material similar to goblet cell mucin.

15.2. Mercury poisoning

Characteristic granules are seen in the macrophages of the lamina propria. Amorphous electron-dense, granular material is randomly distributed in the cytoplasm. The granules are membrane bound by a single unit membrane. These deposits are mercuric sulfide crystals (77).

15.3. Storage diseases

It is not the purpose of this review to describe the details of lysosomal storage disease. Many excellent texts have been devoted to this subject (78). Since the sphingolipidoses are the most common lysosomal storage diseases, they have been studied extensively. The abnormally accumulated lipids are seen in ganglion cells as well as in histiocytes (see Table 8-5).

Ultrastructural pathology of the stored material in the cells is in two forms. The Gaucher or globoid-type inclusion is usually tubular (79) and is due to the accumulation of monohexosyl ceremide. The other type of inclusion is lamellar and

Table 8-5. Mesenchymal tumors of the colon.

Tumor	Incidence: clinical	Site of origin	Light microscopy	Ultrastructure
Smooth muscle tumors[1]	3% of total GI smooth tumors	Muscularis propria, blood vessels of bowel wall, or muscularis mucosae	Spindle cell neoplasms with palisading, cigar-shaped nuclei; perinuclear halo. Nuclear anaplasia and mitotic activity in malignant tumors	Fusiform dense bodies, microfilament bundles, pinocytic vesicles, absent cell junctions, absent basal lamina
Leiomyomas	0.1% of all colonic maligancies			
Leiomyoma peritonealis disseminata	Women of reproductive age or on oral contraceptives			
Schwann cell tumors[2] Neurofibromas Neurilemomas Malignant schwannomas	Occurs in 10% to 25% of patients with von Recklinghausen's disease. Occur in other patients much more rarely	Mucosa, submucosa, or muscularis propria	Spindle cell neoplasm with palisading, comma-shaped nuclei; no perinuclear halo	No fusiform dense bodies, no cell junctions; basal lamina, prominent RER
Ganglioneuromas	Associated with multiple polyposis syndromes or multiple endocrine neoplasia type 2B	Myenteric plexus	Schwann and interspersed ganglion cells, proliferating nerves	Proliferating nerve endings have large numbers of neurosecretory endocrine granules
Vascular Tumors[3]				
Hemangiomas	Rare, associated with visceral hemangiomas	Submucosal vascular plexus	Small or large blood filled vascular channels with variable amounts of connective stroma	Endothelial cells, by basal lamina, cytoplasm containing rod-shaped Weibel-Palade bodies, thick and thin filaments
Angiomatosis	Manifestation of widespread vascular disorders.			

Table 8-5. (Continued).

Tumor	Incidence: clinical	Site of origin	Light microscopy	Ultrastructure
Hemangiopericytoma	Rare		Branching blood vessels lined by endothelial cells. Pericytes surround the vessels separated from endothelial cells by reticulin layer. Pericyte oval cells with hyperchromatic nuclei with nucleoli	Pericytes and transitional cell between pericytes and mesenchymal cells
Angiodysplasia (82)	Primarily cecum and ascending colon	Submucosa	Large, blood-filled spaces, arteriovenous malformation	Not reported
Kaposi's sarcoma (82)	Associated with AIDS	Submucosa	Proliferating capillaries, spindle cell stroma	Three cell types: endothelial, perithelial, fibroblasts
Lymphangiomas (82, 88)	Rare: may be associated with protein losing enteropathy	Submucosa	Cystically dilated lymph vessels	Endothelial lined vessels without continuous basal lamina; cytoplasmic edges overlap; pinocytotic vessels; anchoring filaments
Adipose tissue tumors (82)				
Lipomas	Common	Submucosa	Benign adipocytes	Not reported
Liposarcoma	Very Rare		Lipoblasts characteristic cells	Primitive mesenchymal cells and lipoblasts with non-membrane bound lipid droplets. No desmosomes
Fibrous tumors fibrosarcoma (82)	Rare		Spindle cells with tapered nuclei, no palisading	No fusiform dense bodies, no filament bundles, no cell junctions, rare basal lamina, prominent RER
Malignant fibrous histiocytoma (82)	Rare		Spindle cells with storiform pattern, histiocytes, multinucleated giant cells	Fibroblasts, histiocytes primitive mesenchymal cells
Inflammatory pseudotumor (82, 89, 90)	Rare	Submucosa into muscularis propria	Myxoid stroma with reactive blood vessels inflammatory cells	New blood vessels lined by prominent endothelial cells collagenous stroma, inflammatory cells

[1] Data from references 82, 83, and 87.
[2] Data from references 82, 84, 85, and 87.
[3] Data from references 82, 85, and 87.

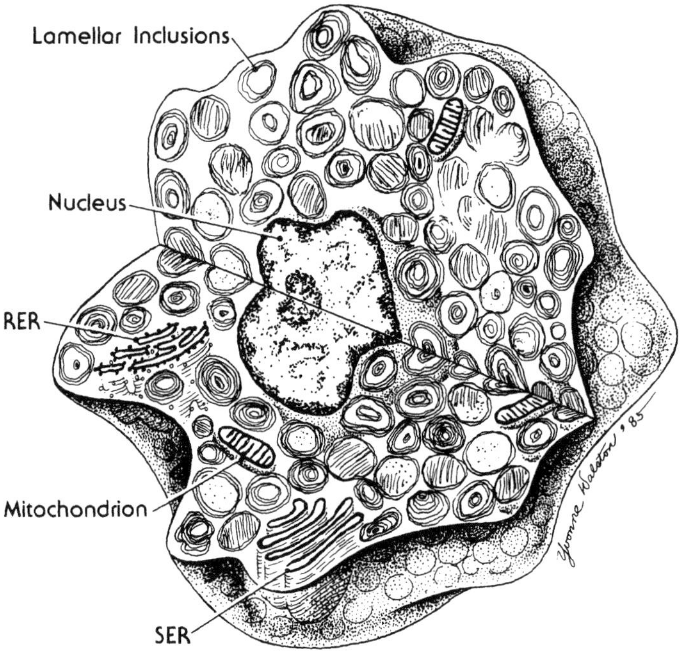

A Macrophage (X 5000)

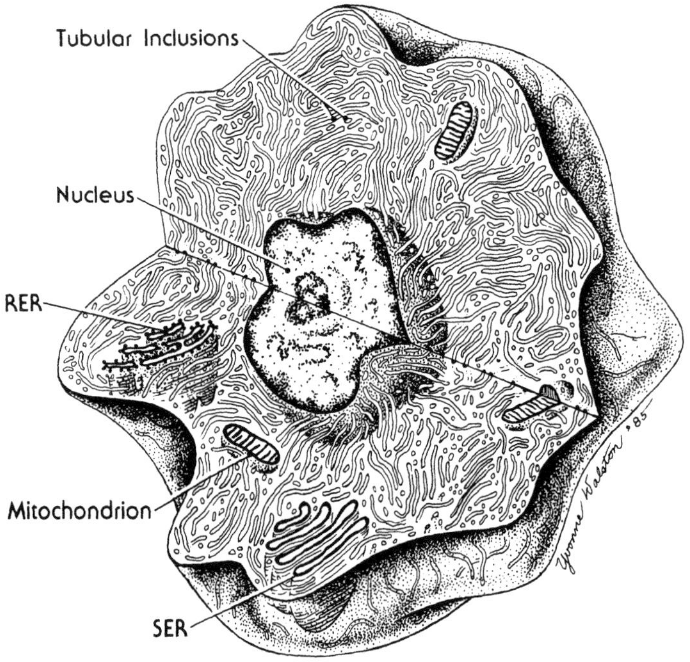

B Macrophage (X 5000)

Figure 8-6. (A). Drawing of typical lamellar-type inclusions seen in lysosomal storage disease. (B). Tubular type inclusion.

is a result of the accumulation of larger sphingolipid molecules (79). The periodicity of the lamellar inclusions varies depending on the specific disease (Figs. 8-6A, B).

16. Mucosal pseudolipomatosis of the colon (lipomatosis)

This uncommon entity has been thought to be due to fatty infiltration of the submucosa and mucosa. Recently an ultrastructural study has shed light on this phenomenon (80). The spaces apparently are unlined and usually empty except for some proteinaceous debris. The cytoplasmic projections adjacent to the spaces appear to belong to fibroblasts or lymphoid cells. There is no evidence that the "fat cells" are adipocytes or lymphatics. Inflammatory reactions around the spaces are not present (80). The authors have suggested that the lesion represents an artifact of endoscopy (80). Dilatation is probably secondary to increased intraluminal pressure (81).

17. Mesenchymal tumors of the colon

Tumors that arise from smooth muscle, fat, fibrous tissue, vessels, and nerves are uncommon to rare in the colon (82–90). The histologic and ultrastructural characteristics of these tumors are similar to those occurring in the soft tissues. Table 8-5 lists and briefly characterizes the mesenchymal neoplasms that occur in the colon.

Acknowledgment

Many thanks to Jeannine Dodge and Gwenn Cooka who painstakenly prepared this manuscript.

References

1. Esterhuizer AC, Lever JD, Palmer S: An electron microscopic, histochemical study in protein uptake and distribution in the colonic mucosa of the neonatal rat. *J Anat* 131: 705–715, 1983.
2. Welsh MJ, Smith PL, Fromm M, Frizzell RA: Crypts are the site of intestinal fluid and electrolyte secretion. *Science* 218: 1219–1221, 1982.
3. Fenoglio CM, Richart RM, Kaye GI: Comparative ultrastructural features of normal, hyperplastic, and adenomatous human colonic epithelium. *Gastroenterology* 69: 100–109, 1975.
4. Kaye GI, Fenoglio CM, Pascal RR, Lane N: Comparative electron microscopic features of normal, hyperplastic, and adenomatous human colonic epithelium: Variations in cellular structure relative to the process of epithelial differentiation. *Gastroenterology* 64: 926–945, 1973.
5. Takeuchi A, Zeller JA: Scanning electron microscopic observations on the surface of the normal and spirochete-infected colonic mucosa of the rhesus monkey. *J Ultrastruct Res* 40: 318–324, 1972.
6. Brown DT, Anderson BH, Wylie CC: Alterations in the organization of cytokeratin filaments in normal and malignant human colonic epithelial cells during mitoses. *Cell Tissue Res* 233: 619–628, 1983.
7. O'Connor JJ: An electron microscopic study of inflammatory colonic disease. *Dis Colon Rectum* 15: 265–277, 1972.
8. Neutra MR, Grand RJ, Trier JS: Glycoprotein synthesis, transport, and secretion by epithelial cells of human rectal mucosa. *Lab Invest* 36: 535–546, 1977.
9. Stone J, Mukherjee M, Hecker R: C bodies and R bodies in the epithelial cells of normal and diseased human rectum. *Arch Path Lab Med* 101: 437–441, 1977.
10. Dobbins WO III: Diagnostic pathology of the intestine and colon. In: Diagnostic Electron Microscopy. BF Trump, RT Jones (eds), John Wiley & Sons, p 253–337, 1978.
11. Toner PG, Carr KE, Wyburn GM: The Digestive System. An Ultrastructural Atlas and Review. New York: Appleton-Century-Crofts, p 107–204, 1971.
12. Schofield GC, Atkins AM: Secretory immunoglobulin in colon epithelial cells of the large intestine. *J Anat* 107: 49–504, 1970.
13. Dawson PA, Filipe MI: An ultrastructural application of silver methanamine to the study of mucin changes in the colonic mucosa adjacent to and remote from carcinoma. *Histochem J* 8: 143–158, 1976.
14. Hull BE, Staehelin LA: The terminal web. A reevaluation of its structure and function. *J Cell Biol* 81: 67–81, 1979.
15. Low FN, McClugage: Microdissection by ultrasonication: Scanning EM of the epithelial basal lamina of the alimentary canal of the rat. *Am J Anat* 169: 137–147, 1984.
16. Sjolund K, Sanden G, Hakanson R. Sundler F: Endocrine cells in human intestine: An immunocytochemical study. *Gastroenterology* 85: 1120–1130, 1983.
17. Symonds DA: Paneth cell metaplasia in diseases of the colon and rectum. *Arch Path* 97: 343–347, 1982.
18. Neutra MK, Padykula HA: The gastrointestinal tract. In: Modern Concepts of Gastrointestinal Histology. L Weiss (ed), Amsterdam: Elsevier Science Publishing Co, Inc, p 693–695, 1983.
19. Maskens AP, Deschner EF: Tritiated thymidine incorpo-

ration into epithelial cells of normal-appearing colorectal mucosa of cancer patients. *J Natl Cancer Inst* 58: 1221–1224, 1977.
20. Eastwood GL: Gastrointestinal epithelial renewal. *Gastroenterology* 72: 962–975, 1977.
21. Fenoglio CM, Kaye GI, Lane N: Distribution of human colonic lymphatics in normal, hyperplastic, and adenomatous tissue. *Gastroenterology* 64: 51–65, 1973.
22. Gabella G: Innervation of the gastrointestinal tract. *Int Rev Cytol* 59: 129–193, 1979.
23. Pellegrini MSG, Cortesini C: Ultrastructural peculiarities of the inner portion of the circular layer of colon. *Acta Anat* 120: 185–189, 1984.
24. Durdle NG, Kingma YJ, Bowes KL, Chambers MM: Origin of slow waves in the canine colon. *Gastroenterology* 84: 375–382, 1983.
25. Swada H: Scanning EM of guinea pig taenia coli. *Biomed Res* 2: 153–158, 1981.
26. Gershon MD, Ade SM: The nervous system of the gut. *Gastroenterology* 80: 1571–1594, 1981.
27. Burnstock G: Ultrastructural identification of neurotransmitters. *Scand J Gastroenterology* 70: 1–9, 1981.
28. Baumgarten HG, Holstein AG, Owman CH: Auerbach's plexus of mammals and man: Electron microscopic identification of 3 different types of neural processes in myenteric ganglia of the large intestine from rhesus monkeys, guinea pigs, and man. *Z Zellforsch* 106: 376–397, 1970.
29. Fenoglio-Preiser CM, Hutter RVP: Colorectal polyps: Pathologic diagnosis and clinical significance. *CA-A Cancer Journal for Clinicians* 35: 322–359, 1985.
30. Estrada R, Spjut HJ: Hamartomatous polyps in Peutz-Jeghers syndrome. A light, histochemical, and electron microscopic study. *Am J Surg Pathol* 7: 747–754, 1983.
31. Traynor OJ, Costa NC, Blumgart LH, Wood CB: A SEM study of ultrastructural changes in the colonic mucosa of patients with large bowel tumors. *Br J Surg* 68: 701–704, 1981.
32. Balazs M: Electron microscopy of polyps of the colon. *Exp Path* 19: 112–121, 1981.
33. Mughal S, Filipe MI: Ultrastructure of the normal mucosa-adenoma cancer sequence in the development of familial polyposis coli. *J Natl Cancer Inst* 60: 753–756, 1978.
34. Bansal M, Fenoglio CM, Robboy SJ, King DW: Are metaplasias in colorectal adenomas truly metaplasias? *Am J Pathol* 115: 253–265, 1984.
35. Cooke T, Kirkham N, Stainthorp DH, Inman C, Goeting N, Taylor I: Detection of early neoplastic changes in experimentally induced colorectal cancer, using SEM and cell kinetic studies. *Gut* 25: 748–755, 1984.
36. Imai H, Stein AA: Ultrastructure of adenocarinoma of the colon. *Gastroenterology* 44: 410–417, 1983.
37. von Bassewitz DB, Roessner A, Grundmann E: Intermediate-sized filaments in cells of normal human colon mucosa, adenomas, and carcinomas. *Path Res Pract* 175: 238–255, 1982.
38. Christensen TG, Burke B, Dexter DL, Zamcheck N: Ultrastructural evidence of Dimenthylforinamide-induced differentiation of cultured human colon carcinoma cells: Increased expression of desmosomes. *Cancer* 56: 1559–1565, 1985.
39. Balazs M: Electron microscopy of polyps of the colon: II. Inclusion bodies of Leuchtenberger. *Exp Path* 19: 164–171, 1981.
40. Fisher ER, Sharkey DA: The ultrastructure of colonic polyps and cancer with special inference to the epithelial inclusions of Leuchtenberger. *Cancer* 24: 160–169, 1962.
41. Ohtani H, Sasano N: Stromal changes in human colorectal adenomas and carcinomas. Ultrastructural study of fibroblasts, myofibroblasts, smooth muscle cells. *Virch Arch (Path A Anat)* 401: 209–222, 1983.
42. Riddell RH, Levin B: Ultrastructure of the transitional mucosa adjacent to large bowel carcinoma: *Cancer* 40: 2509–2522, 1977.
43. Dawson PA, Filipe MI, Bussey HJ: Ultrastructural features of the colonic epithelium in familal polyposis coli. *Histopathology* 1: 105–113, 1977.
44. Sanders RJ: Carcinoids of the Gastrointestinal Tract. Springfield, Il: CC Thomas, Publisher, p 1–110, 1973.
45. Cooper PH, Warkel RL: Ultrastructure of goblet cell type of adenocarcinoid of the appendix. *Cancer* 42: 2695, 1978.
46. Isaacson P: Crypt cell carcinoma of the appendix (so-called adenocarcinoid tumor). *Am J Surg Pathol* 5: 213–224, 1981.
47. Jenkinson JA, Dawson IMP: The value of electron microscopic studies in diagnosing malignant change in ulcerative colitis. *Gut* 12: 110–118, 1971.
48. Siew S: The application of scanning electron microscpy in the clinical investigation of the human colon. *Scanning EM* 4: 1911–1929, 1983.
49. Riddell RH, Eisentat L, Golomb H, Brynes R, Levin B: A low power SEM study of normal and colitic human large bowel. In: 33rd Ann Proc Electron Microsc Soc Amer. GW Bailey (ed), Baton Rouge, LA: Claitor's Publishing Div p 118–119, 1975.
50. Shields HM, Bates ML, Goldman H, Zuckerman GR, Mills BA, Best CJ, Bair FA, Goran DA, Deschryver-Kecskemiti K: Scanning electron microscopic appearance of chronic colitis with and without dysplasia. *Gastroenterology* 89: 62–72, 1985.
51. Gonzales-Licea A, Yardley JH: Nature of the tissue reaction in ulcerative colitis, light and EM findings. *Gastroenterology* 51: 825–840, 1966.
52. Dobbins WO III: Colonic epithelial cells and polymorphonuclear leukocytes in ulcerative colitis. *Dig Dis* 20: 236–252, 1975.
53. Otto HF, Gebbeis JO: Electron microscopic, ultracytochemical, and immunohistological observations in Crohn's disease of the ileum and colon. *Virchows Arch* 391: 189–208, 1981.
54. Rickert RR, Carter HW: The gross, light microscopic, and scanning electron microscopic appearance of the early lesions of Crohn's disease. *Scan Electron Microsc* 2: 179–186, 1977.
55. Dvorak AM, Dickerson GR: Crohn's Disease: Transmission electron microscopic studies; I: Barrier function. Possible changes related to alterations of cell coat,

mucous coat, epithelial cells and Paneth cells. *Hum Pathol* 2: 561–571, 1980.
56. Siew S, Tedesco FJ: Scanning EM of human colonic biopsies as an aid in the diagnosis of clindamycin-associated colitis. *Scan Electron Microsc* 2: 187–194, 1977.
57. Fausa O, Foerster A, Hovig T: Collagenous colitis. A clinical, histological, and ultrastructural study. *Scand J Gastroenterol* 20: 8–23, 1985.
58. Vernon SE: Amyloid colitis. *Dis Colon Rectum* 25: 728–739, 1982.
59. Shousha S, Lowdell CP, Bull TB, Parkins RA: Secondary amyloidosis of the gastrointestinal tract, an electron microscopic study. *Hum Pathol* 16: 596–601, 1985.
60. Schrodt GR: Melanosis coli: A study with the electron microscope. *Dis Colon Rectum* 64: 277–283, 1983.
61. Antonakopoulos G, Newman J, Wilkinson M: Intestinal spirochaetosis: An electron microscopic study of an unusual case. *Histopathology* 6: 477–488, 1982.
62. Neutra MR: Prokaryotic-enkaryotic cell junctions. Attachment of spirochetes and flagellated bacteria to primate large intestinal cells. *J Ultrastruct Res* 70: 186–203, 1980.
63. Scully RE, Mark EJ, McNeely BV: Case records of the Massachusetts General Hospital; Case 39-1985. *N Engl J Med* 313: 805–815, 1985.
64. Sack RB: Acute diarrheal diseases in humans caused by bacteria. In: Virus Infections of the Gastrointestinal Tract. DAJ Tyrrell, AZ Kapikain (eds), Marcel Dekker, Inc, p 239–270, 1979.
65. Van Spreeuwel JP, Durrsma GC, Meijer CJLM, Bax R, Rosekrans PCM, Lindeman J: Campylobacter colitis: Histological, immunohistochemical and ultrastructural findings. *Gut* 26: 945–951, 1985.
66. Kewon KJ, Harell GS, Lee AS, Growley LG: Malakoplakia, an electron microscopic study: Demonstration of bacilliform organisms in malacoplakia macrophages. *Gastroenterology* 66: 28–45, 1984.
67. Kjeldsberg E: 40–43 nm virus-like particles associated with gastroenteritis. *Ultrastruct Pathol* 1: 457–459, 1980.
68. Davies HA: Electron microscopy and immune electron microscopy for detection of gastroenteritis viruses. In: Virus Infections of the Gastrointestinal Tract. DAJ Tyrrell, AZ Kapikain (eds), Marcel Dekker, Inc, p 37–45, 1979.
69. Grohmann GS, Greenberg HB, Welch BM, Murphy AM: Oyster-associated gastroenteritis in Australia. The detection of Norwalk virus and its antibody by immune electron microscopy and radioimmunoassay. *J Med Virol* 6: 11–19, 1980.
70. Yolken RH: Rotaviruses and other viruses causing gastroenteritis. In: *Texbook of Human Virology*. RB Belshe (ed) Littleton, Mass: PSG Publishing Company, p 795–809, 1984.
71. Foucar E, Makai K, foucar K, Sutherland DER, Van Buren CT: Colonic ulceration in lethal cytomegalovirus infection. *Am J Clin Pathol* 76: 788–801, 1981.
72. Boulton AJM, Slata DN, Hancock BW: Herpes virus colitis: A new cause of diarrhea in a patient with Hodgkins disease. *Gut* 23: 247–249, 1982.
73. Pittman FE, El-Hashini WK, Pittman JC: Light and electron microscopic observation of colonic mucosa and exudate in acute amebic colitis. Studies of human amebiasis. *Gastroenterology* 65: 588–603, 1973.
74. DeGirolani PC, Dunn JC, Federman M: Infections caused by intestinal protozoa. *Pathol Annu* 20: 463–506, 1985.
75. Pohlenz J, Benrick WJ, Moon HW, Cheville NF: Bovine cryptosporidosis: A transmission and scanning electron microscopic study of some stages in the life cycle and of the host parasite relationship. *Vet Pathol* 15: 417–427, 1978.
76. Lou TY, Tepliz C, Thaya WR: Ultrastructural morphogenesis of colonic PAS-positive macrophages. *Hum Pathol* 2: 421–439, 1971.
77. Wands JR, Weiss SW, Yardley JH, Maddrey WC: Mercury poisoning — laxative. *Am J Med* 57: 92–101, 1974.
78. Stanbury JB, Wyngaarden JB: Disorders characterized by evidence of abnormal lipid metabolism. In: The Metabolic Basis of Inherited Diseases, 4th ed. JB Stanbury, JB Wyngaarden, DS Fredrickson (eds), New York; McGraw-Hill Book Co, p 544–841, 1978.
79. Takahashi K, Naito M: Lipid storage diseases. II. Ultrastructural pathology of lipid storage cells in splinogolidoses. *Acta Pathol Jpn* 35: 385–408, 1985.
80. Snover DC, Sandstad J, Hutton S: Mucosal pseudolipomatosis of the colon. *Am J Clin Pathol* 84: 575–580, 1985.
81. Haboubi NY, Horn RP, Hasleton PS, Ali HH, Anfield C, Hobbiss J, Schofield PF: Pneumatosis coli: A case report with ultrastructural study. *Histopathology* 8: 145–155, 1984.
82. Fenoglio-Preiser CM, Pascal RR, Perzin KH: Tumors of the Large and Small Intestine, *AFIP Fascicle* 2nd series (in press).
83. Weiss RA, MacKay B: Malignant smooth muscle tumors of the gastrointestinal tract. An ultrastructural study of 20 cases. *Ultrastruct Pathol* 2: 231–240, 1981.
84. Schaldenbrand JD, Appelman HD: Solitary solid stromal gastrointestinal tumors in von Recklinghausen's disease with minimal smooth muscle differentiation. *Human Pathol* 15: 229–232, 1984.
85. Gould VE, Kraft JR: Case 14. *Ultrastruct Pathol* 5: 359–568, 1983.
86. Waldo ED, Vuletin JC, Kaye GI: The ultrastructure of vascular tumors. Additional observations and a review of the literature. *Path Annu* 12: 279–308, 1977.
87. Enzinger FM, Weiss SW: *Soft Tissue Tumors*, St. Louis: CV Mosby Co, p 442, 1983.
88. Leak LW, Burke JF: Fine structure of lymphatic capillaries and adjoining connective tissue area. *Am J Anat* 118: 785–881, 1966.
89. Navas-Palacios JJ, Colina-Ruizdelgado F, Sanchez-Larrea MD, Cortes-Cansino J: Inflammatory fibroid polyp of the gastrointestinal tract. An immunohistochemical and electron microscopic study. *Cancer* 51: 1682–1690, 1983.
90. Williams RM: An ultrastructural study of a jejunal inflammatory fibroid polyp. *Histopathology* 5: 193–203, 1981.

Authors' addresses:
Margaret B. Listrom
Assistant Professor of Pathology
University of New Mexico School of Medicine
Albuquerque, New Mexico 87131
and
Attending Pathologist
Veterans Administration Medical Center
Albuquerque, New Mexico 87131

Cecilia M. Fenoglio-Preiser
Professor of Pathology
University of New Mexico School of Medicine
Albuquerque, New Mexico 87131
and
Chief, Laboratory Service
Veterans Administration Medical Center
Albuquerque, New Mexico 87108

CHAPTER 9

Ultrastructure of the rectum with special emphasis on tumors

LOKENDRA N. CHOWDHURY & HYMIE KAVIN

1. Introduction

There is considerable overlap between the normal ultrastructural morphology of the rectum and other parts of the large intestine. Ultrastructure of the colon is described in detail in Chapter 8. The present discussion will focus on some specific aspects of rectal ultrastructure. Emphasis will be placed on those aspects of ultrastructural morphology bearing a close relationship to tumorigenesis. Since epithelial neoplastic diseases of the rectum are the most common tumors of the gastrointestinal tract, the present discussion will concentrate on the mucosa.

The mucous membrane of the rectum differs in some respects from that of the mucosa of the large intestine. The crypts are shallower in the rectum and have more goblet cells. The stem cells which are found at the base of the crypts, unlike those of the ascending colon, often become vacuolated by mucus relatively early on their way toward the crypt mouth. However, before reaching the surface the vacuoles may be lost and the cells become columnar cells with microvilli. Occasional cells on the surface of the mucosa contain a few small vacuoles. The crypts of Lieberkühn disappear at the anorectal junction where circumanal glands are located. The latter glands are lined by stratified columnar epithelium and have a branched tubular configuration.

2. Transmission electron microscopic (TEM) appearance of normal rectal mucosa

Compared with the colon, the crypt-lining cells of the rectum have more mucous vacuoles and the surface epithelium shows increased numbers of mucous cells. Rectal mucosa also contains an increased number of enteroendocrine cells at the base of the crypts.

2.1. *Surface epithelium*

The surface epithelium is lined by both columnar and mucous cells. The columnar epithelial cells which lie on the basal lamina are elongated with basally placed nuclei. The cytoplasm shows numerous mitochondria, Golgi bodies, RER, and ribosomes. The apical part of the cytoplasm contains numerous 200–300 nm vesicles, microfilaments, a well-developed terminal web, and core rootlets (Fig. 9-1). These cells are capable of mucus production (1). Membrane bound bodies (R bodies) surrounded by a limiting membrane are also seen in these cells and probably represent degenerated mitochondria. The cells represent senescent mucous cells (2). The dark cells represent the immediate stage following complete discharge of mucus (Fig. 9-1).

2.2. *Mucous cells*

Mucous cells are most numerous in the rectal mucosa and comprise almost the entire crypt epithelium. These cells show RER, Golgi complexes, and mucous vacuoles (Fig. 9-2). Histochemically the mucous cells in the rectum contain predominantly sulfomucin which stains blue with the alcian-blue PAS method (2, 3). As the cell becomes distended with mucus, individual mucous vacuoles lose their membranes and form larger vacuoles which are eventually discharged into the lumen. Most mucous cells also show dense

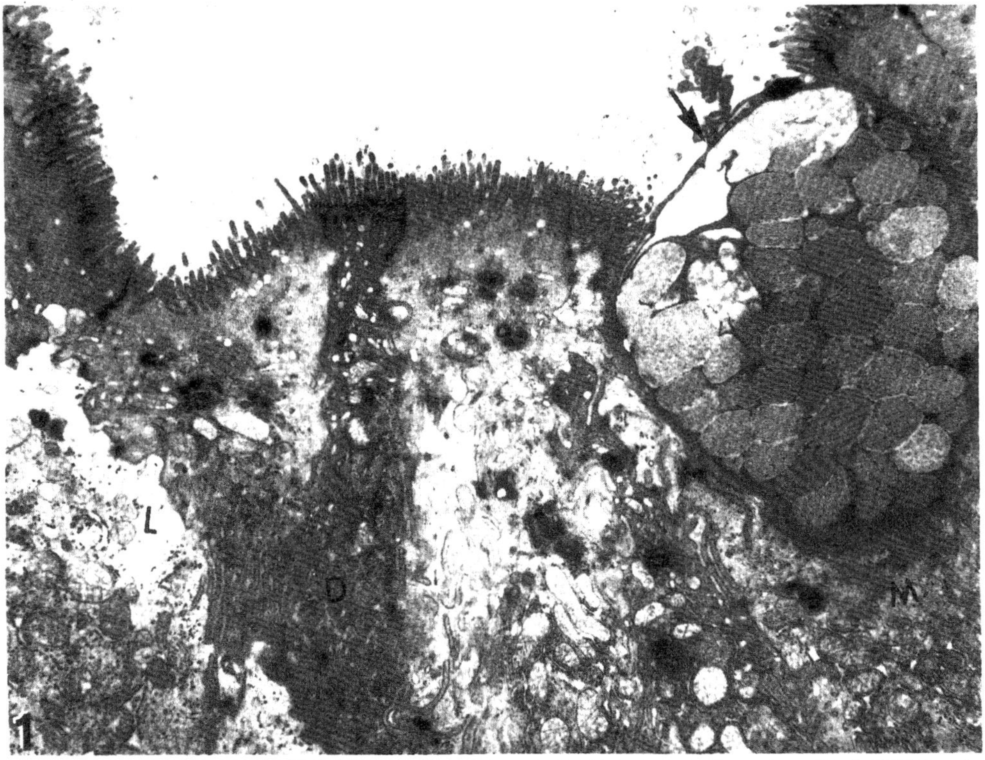
1
L
D
M

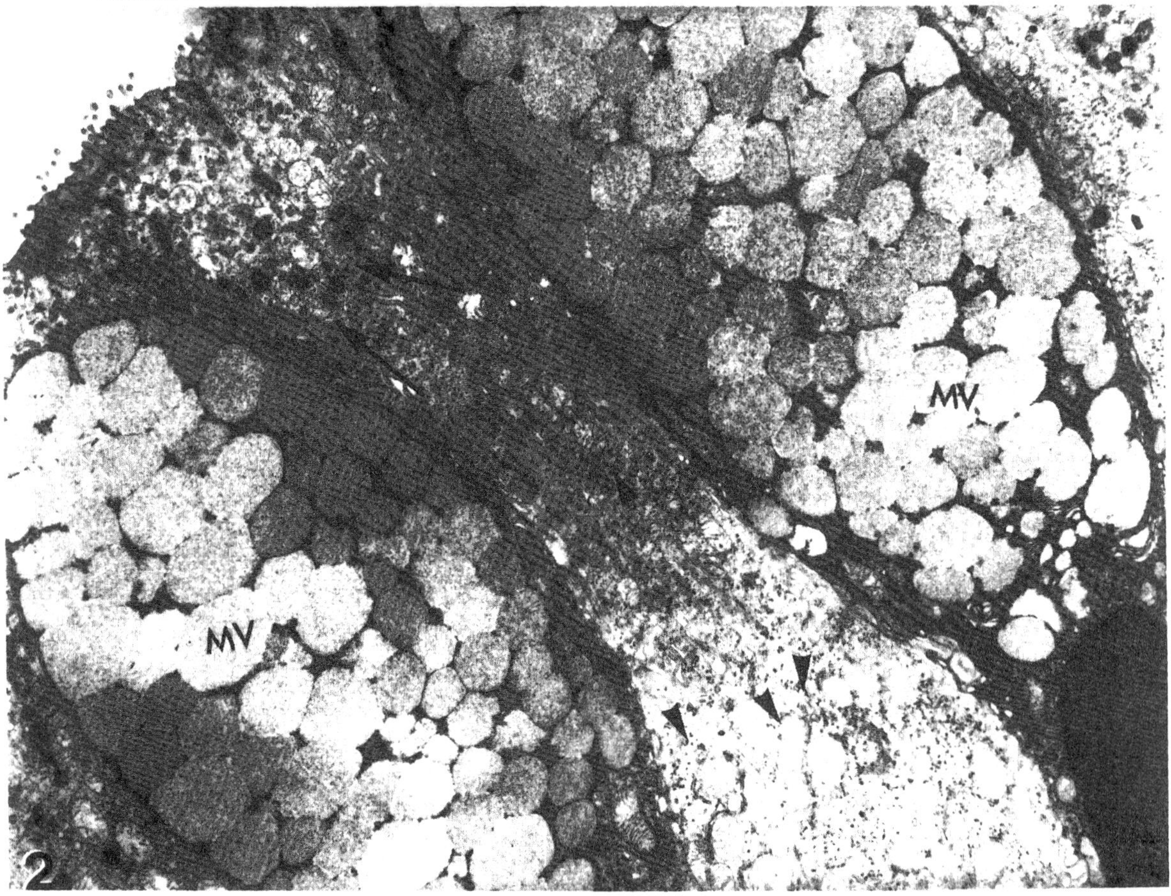
2
MV
MV

vesicles located at the apical regions. These vesicles are relatively less electron dense in the rectum compared to those in the ascending and transverse colon. They are surrounded by a unit membrane and vary in size from 300–500 nm.

The lateral plasma membranes show some interdigitation. The apical plasma membranes have microvilli which may be flattened when the cells are hyperdistended with mucus.

2.3. Enteroendocrine cells

Enteroendocrine cells are most numerous in the rectum and may be seen in clusters deep in the mucosa lying on the basement membrane. These cells may be argentaffin positive and have shown immunoreactivity against glucagon, glicentin, and bovine and human pancreatic polypeptide antisera (4). The cells characteristically show electron-dense granules which vary in size from 200–600 nm. Although they may vary in shape, they are generally round. The granules have a central electron-dense core surrounded by a limiting membrane with a clear halo between the dense core and the membrane. The cytoplasm of the endocrine cell contains sparse mitochondria, microfilaments, and a few segments of RER and ribosomes (Figs. 9-3, 9-4).

2.4. Undifferentiated cells

Undifferentiated cells are present at the basal aspect of the crypts. The cytoplasm of these cells has a paucity of organelles with a few ribosomes and mitochondria. The nuclear membrane shows prominent indentations and nucleoli are also seen (Figs. 9-3, 9-4). These cells differentiate into mucous cells which, after discharging their contents, become the columnar cells on the surface. Although the surface columnar cells have been frequently referred to as absorptive cells, their ultrastructural morphology and histochemical and autoradiographic studies indicate that they are derived from mucous cells (1–3).

There is controversy regarding the genesis of the enteroendocrine cells. Pearse (5) postulated that these cells are derived from the neural crest. Later studies, however, favored an endodermal origin from the undifferentiated crypt epithelial cells (6, 7).

It would appear, therefore, that the undifferentiated cells at the base of the crypts are the progenitors for different types of cells in the rectal mucosa. In addition, differentiation of the various cell types occurs at different levels of the crypt. Enteroendocrine cells differentiate at the base of the crypts and the mucous cells at higher levels (8).

2.5. Plasma membrane

The lateral plasma membrane shows extensive interdigitations. The apical plasma membrane shows microvilli surrounded by the glycocalyx. Glycocalyceal bodies (30–100 nm) are closely associated with the microvilli. Apical vesicles surrounded by a unit membrane contain glycocalyx-type material indicating that they are formed through pinocytosis. Leuchtenberger bodies represent lymphocytes migrating in between the epithelial cells (9).

3. Scanning electron microscopic (SEM) appearance of the surface of rectal mucosa

The SEM appearance of normal rectal mucosa has been previously described (10–17). The utility of SEM in the study of the rectal mucosa was first described by Kavin et al. (16). The surface morphology shows a regular geometric pattern separated by parallel rows of furrows and ridges. The surface may appear to be divided into multiple polygonal units each comprising a central hole approximately 18 μ in diameter which is the mouth of a crypt. The polygonal units are

←

Figure 9-1. TEM surface epithelium rectal mucosa. Arrow points to discharge of mucus from the surface. L, light cells; D, dark cells; M, mucous cells. X 2,775.

Figure 9-2. TEM mucous cells in the mid-crypt of normal rectal mucosa. Cells are distended by large mucous vesicles. Cell in between shows abundant mitochondria (arrowheads), RER, ribosomes, and few dense vesicles (arrows). MV, mucous vesicles. X 8,208.

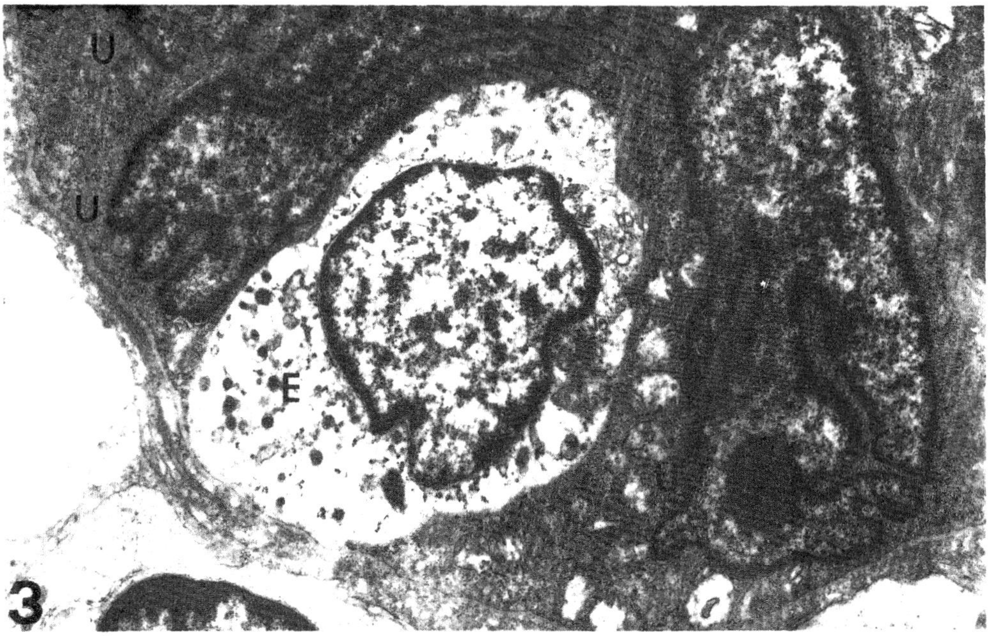
U
U
E
3

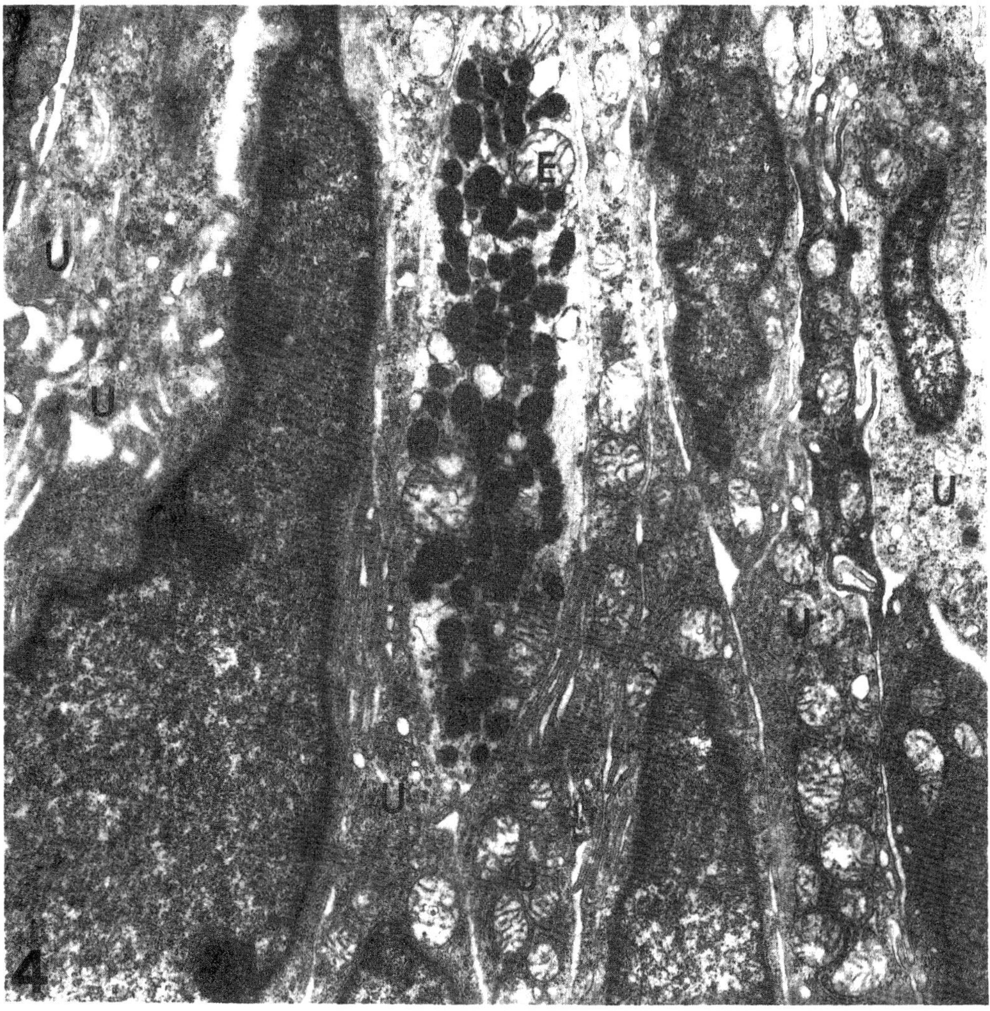
E
U
U
U
U
U
4

delineated by sharp but shallow depressions or furrows. Two types of cells can be readily identified. A larger number of cells showing hemispherical projections on the surface represent the columnar epithelial cells; these cells at higher magnifications show microvilli on the surface (Figs. 9-5, 9-6). The mucus secretory or goblet cells are irregularly scattered in between the surface columnar cells and show variable surface features. While some have smooth surfaces, others contain a few microvilli. The general surface appearances of rectal mucosa may also show considerable variation (14). At one extreme, the crypt openings may project above the adjacent intercrypt area; often the crypt openings may be difficult to see, and in some the crypt pattern may not be demonstrable. The cellular morphology, however, in all these instances remains very similar.

4. Lamina propria

Ultrastructural examination of the lamina propria of the rectum may provide additional information not readily seen by light microscopic evaluation. Particularly important in this regard is the fine structural morphology of the macrophage. Malacoplakia of the colorectal areas (18) can be confirmed by the ultrastructural demonstration of the characteristic bacillary bodies. Electron microscopic examination of the lamina propria macrophages may show the characteristic pigment-laden macrophages of neomycin toxicity (19). Various lipid storage disorders and some of the neuronal storage diseases may be readily identified by ultrastructural examination of the lamina propria macrophages. Moreover, in these disorders a rectal biopsy would be the least invasive and most easily obtainable material for EM evaluation. Likewise, examination of the rectal mucosa may be most helpful in the diagnosis of amyloidosis by the demonstration of characteristic amyloid fibrils (20, 21). Among the infectious diseases, rectal biopsy examination may be useful for identification of cryptosporidiosis.

5. Submucosa

Ultrastructural examination of the submucosa and other layers of the rectum provides important information in relationship to neoplastic processes. Neoplastic lesions of mesenchymal origin may show differentiation into various cell lines including smooth muscle cells, fibroblasts, and myofibroblasts. Tumors of vascular or vasoformative tissues, neural tissues lymphoid structures, mesothelial elements also occur, and ultrastructural evaluation of these lesions can be of considerable help in delineating the histogenesis and, perhaps, the clinical behavior of these lesions.

6. Cell kinetics and morphogenesis of colorectal cancers

Experimental carcinogenesis studies using cholanthracenes, aromatic amines, hydrazines, nitrosamines, and other miscellaneous agents have shown that colorectal cancer may be produced in the following ways:

1. A chemical agent may cause specific "mutation." The involved cells will not express this mutation.
2. Expression of malignancy will be caused by a second assault.

This two step model of carcinogenesis illustrates the significance of cell proliferation. Mathematical consequences of the proliferative phenomena (22) and their significance can be readily appreciated in experimental colorectal cancer models where these lesions occur as adenomas, dysplasias, etc. (23–25).

Crypt cell kinetic studies and SEM observations in experimental carcinogenesis using the dimethylhydrazine model in rats (10) have shown that the most prominent feature of the precancerous colon is the development of microadenomas which increase in size and number during carcinogenesis. Microadenomas and early carcinomas occur most frequently in the distal colon. In man, using in vitro thymidine labelling, an increase in the crypt proliferative zone has been

←

Figure 9-3. TEM undifferentiated cells (U) and an enteroendrocrine cell (E) of normal crypt base X 5,265.
Figure 9-4. Enteroendocrine cell (E) among undifferentiated cells at the base of a normal crypt. X 16,758.

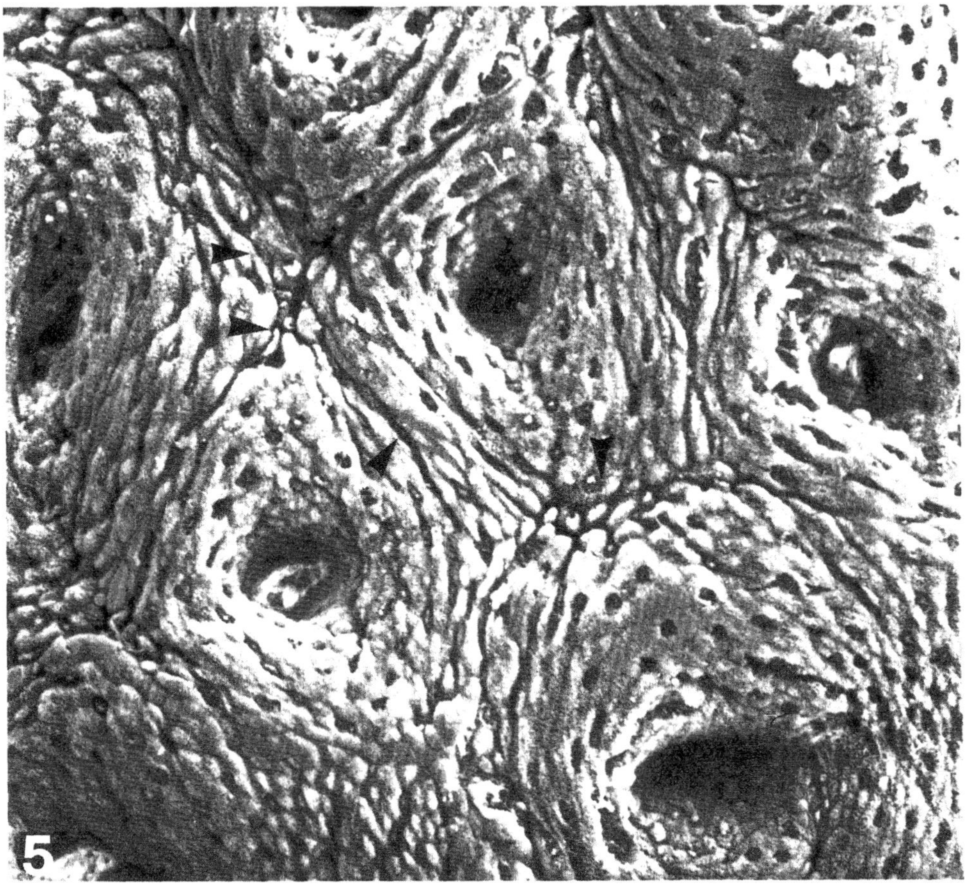
5

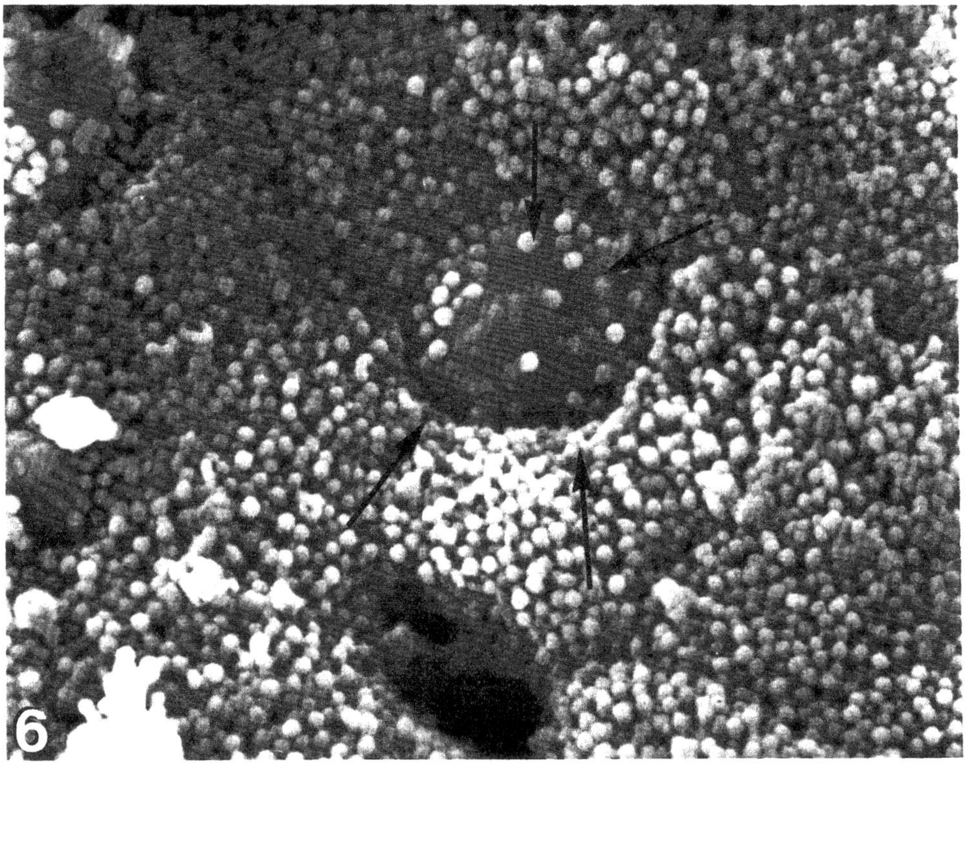
6

observed in the transitional mucosa close to the colorectal cancers. This proliferative pattern of response is also seen in patients with polyposis coli and chronic ulcerative colitis (24–26).

Using cell kinetic studies it was possible to determine that during the progressive stages of abnormal development, epithelial cells acquire an increased ability to proliferate and accumulate in the mucosa (26–29). Failure of the cells to repress DNA synthesis during migration to the surface of the mucosa (30, 31) has been shown to occur in polyposis syndromes, in cancer-prone individuals without polyposis, as well as in other conditions associated with a high incidence of cancer such as in ulcerative colitis. In experimental animals carcinogen-induced proliferative lesions have been demonstrated increasingly in the distal colonic segments and in the rectum. DNA synthesis occurring in cells that normally would be terminal or end cells is seen in adenomas, atypias, dysplasias, and malignancies in the colorectum. Computer applications to demonstrate this pattern of failure of repression of DNA synthesis could help segregate high risk groups from low risk population groups (29).

Cytoproliferative abnormalities precede the actual development of neoplasia and are considered as markers of increased risk for colorectal cancer. The upward extension of the proliferating cells can follow two different phases: a phase I abnormality is indicated by scattered proliferating cells in the upper third of the crypts but persistence of these cells in the lower third of the crypts. In phase II abnormality there is a shift of the area of maximal proliferation to the middle and/or upper portions of the crypts. The development of phase II abnormality is characterized morphologically by tubular and villous adenomas and carcinomas. Kanemitsu et al. (31) showed that cell proliferation kinetics in ulcerative colitis, polyps, and cancer are markedly abnormal. In ulcerative colitis the proliferation zones extend to the surface of the crypts. In addition, the labelling index of ^{3}H-thymidine was double that of normal mucosa. The turnover time of the epithelial cells in this disease is significantly shortened. The high incidence of colorectal cancer in ulcerative colitis (32, 33) is possibly due to the fact that the active DNA-synthesizing cells in the upper crypt are acted on more easily by fecal carcinogens (malignant transformation may occur in tissues with a higher proliferation rate). Polyps and cancer showed ^{3}H-labelled cells restricted to the upper one-third of the crypt or along the whole length of the crypt.

7. Ultrastructure of adenomatous and malignant epithelium

7.1. *Transitional mucosa*

Ultrastructural and histochemical studies of the colorectal mucosa adjacent to a colorectal carcinoma (1, 3, 34) have demonstrated significant biochemical, histochemical, and structural differences from corresponding normal mucosa. The transitional mucosa adjacent to the tumors shows a relative increase of sialomucins as compared to sulfomucins which form the predominant component of the mucinous secretion in the normal rectal mucosa. Structurally there is a relative increase of mucous cells in the transitional mucosa with persistence of immature and basaloid types of cells at the middle and upper portions of the crypt. In addition, the cells show an increased number of mucin vesicles and electron-dense bodies with prominent elements of endoplasmic reticulum and of Golgi systems. Both elements of smooth and rough endoplasmic reticulum may be increased as well (Figs. 9-7, 9-8). The predominant cell type, rarely seen in normal mucosa, is characterized as an "intermediate cell," which possibly is a type of stem cell capable of differentiating into other cell types. The presence of a complex Golgi zone structure is indicative of excess glycoprotein synthesis by these cells. The

←

Figure 9-5. SEM normal surface epithelium of rectum. Geometric pattern formed by shallow furrows (arrowheads). Wide crypt mouths are clearly shown. X 585.

Figure 9-6. SEM normal surface epithelium of rectum. Columnar epithelial cells show prominent microvilli; mucous cells have fewer microvilli (arrows). X 12,000.

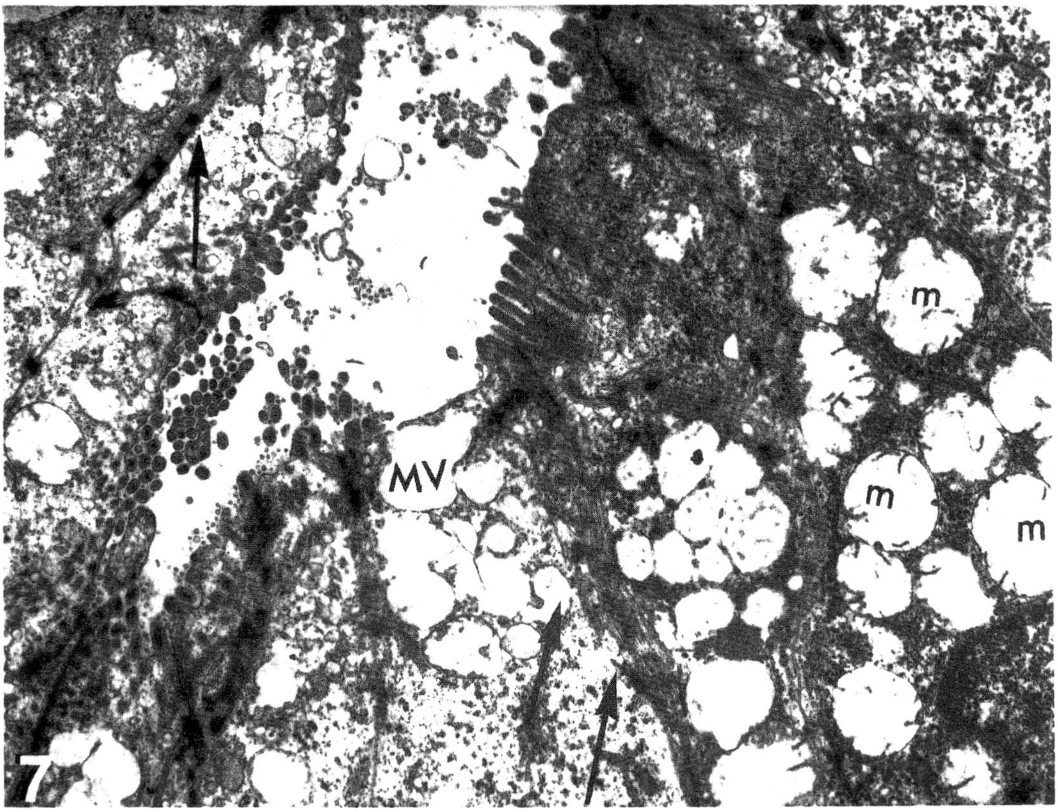
MV
m
m
m
7

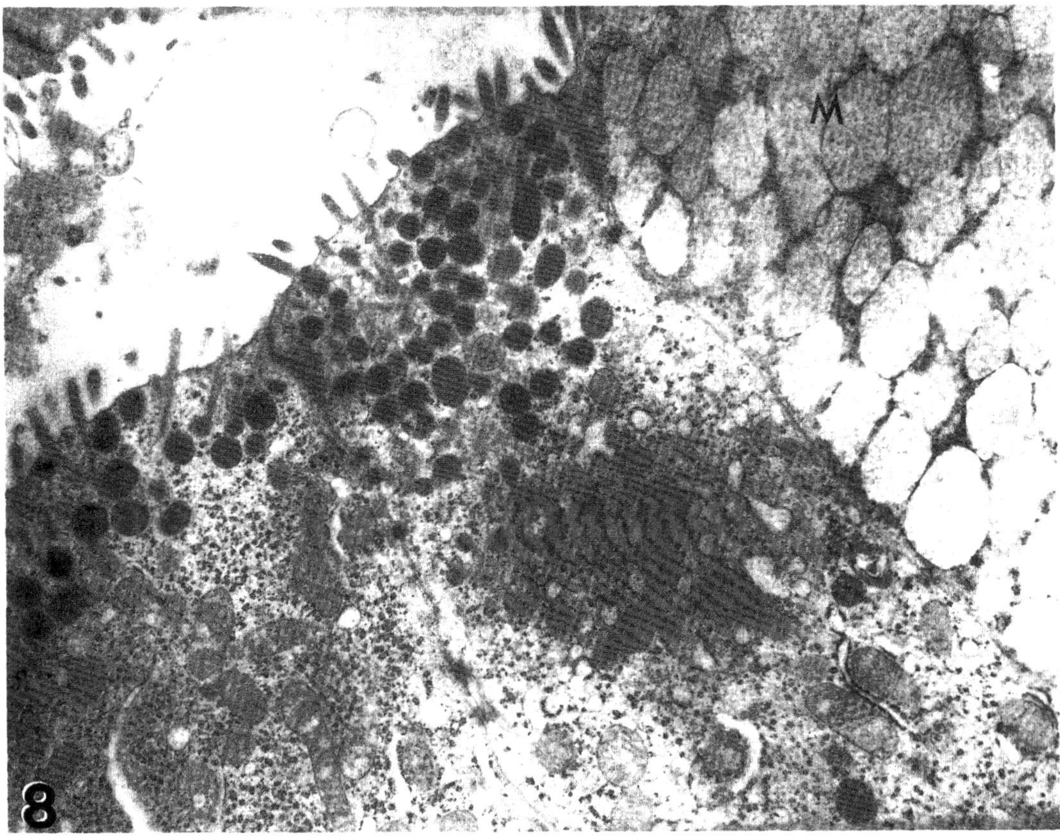
M
8

variable electron density of the mucin vesicles and the histochemical changes in the mucin are due to changes in glycoprotein synthesis. Experimental studies using dimethylhydrazine-induced cancer in rats show predominance of sialomucins in the dysplastic segments of the mucosa (35). These changes are also seen in ulcerative colitis with carcinoma but not in ulcerative colitis alone.

By scanning electron microscopy there may be architectural distortion of the transitional mucosa with loss of primary cryptal arrangement and replacement by irregular nodulelike formations; however the crypt orifices can still be identified. The surface epithelium is irregular in shape. Microvilli are usually less prominent. The cryptal arrangement and surface morphology vary considerably from patient to patient. Patterns with total architectural distortion of crypts with incoordinated epithelial cell proliferation may also be seen.

7.2. Hyperplastic and adenomatous epithelium

The ultrastructural morphology of hyperplastic and neoplastic colorectal mucosa has been amply discussed (12, 14, 15, 36, 37). Hyperplastic epithelium shows a normal pattern of maturation of cells along the crypts (Fig. 9-9). The morphologic features of the progressively maturing cells may be seen even at lower levels of the crypts. At the base of the hyperplastic cryptal epithelium, cells of intermediate differentiation may predominate. Adenomatous epithelium, on the other hand, rarely shows maturation of cells beyond the intermediate cells at this level. The predominant cell type in hyperplastic epithelium is the mucous cell. Most hyperplastic polyps will show cells with elongated microvilli in the midportion of the crypts, unlike the adenomatous polyps, where the cells at this site may show only few microvilli. In hyperplastic polyps, the lateral margins of the epithelial cells in the midcrypt may show convolutions. By contrast, in adenomatous polyps the cells at this level appear essentially similar to those at the base.

Adenomas may show significant variation in the degree of cellular differentiation at the surface. The predominant cell at the surface is still the intermediate cell with irregular short microvilli and a small number of apical vesicles. Microvilli may be absent in other cells. The microvillous border has increased numbers of small round vesicles (20–30 nm). The convolutions at the lateral margins and the desmosomal structures also vary with the degree of cellular differentiation, and the desmosomes appear to become more complex. Enteroendocrine cells may be found at the apex of adenomatous crypts (Figs. 9-10, 9-11).

By SEM the hyperplastic lesions show deep mucosal ridges which give rise to an irregular lobulation. The basic cryptal architecture is maintained and the crypt openings are regularly spaced. However, the openings are more widely separated (Fig. 9-12). The cells appear uniform in size and shape. SEM of adenoma shows marked distortion of the cryptal structure. Crypts are irregularly dispersed. The crypt openings vary considerably in size and shape.

7.3. Dysplasia in colorectal epithelium

Most carcinomas of the rectum arise in adenomas. The cell kinetics and morphogenesis of the proliferative lesions leading to adenomas and dysplasia have already been discussed in Section 6. Both genetic and environmental factors play significant roles in the development of these lesions. Grading of adenomas on the basis of epithelial atypias (dysplasias) has shown that their malignant potential increases with increasing degrees of dysplasia. Small adenomas under 1 cm in diameter usually show only mild atypia and have a low risk of malignancy. With severe atypia the incidence of malignancy rises to 27%. All carcinomas arising from colorectal epithelium presumably pass through various stages of epithelial atypia (38, 39). A small proportion of large bowel and rectal carcinomas arise as a complication of long-standing inflammatory diseases, especially in

←

Figure 9-7. TEM transitional mucosa. Complex junctions (arrows). m, mitochondria; MV, mucous vesicles. X 15,960.
Figure 9-8. TEM transitional mucosa. Cells show numerous mitochondria, RER, and dense bodies. Mucous cell (M) distended with mucus. Mucin granules are present in the apical cytoplasm of other cells. X 7,448.

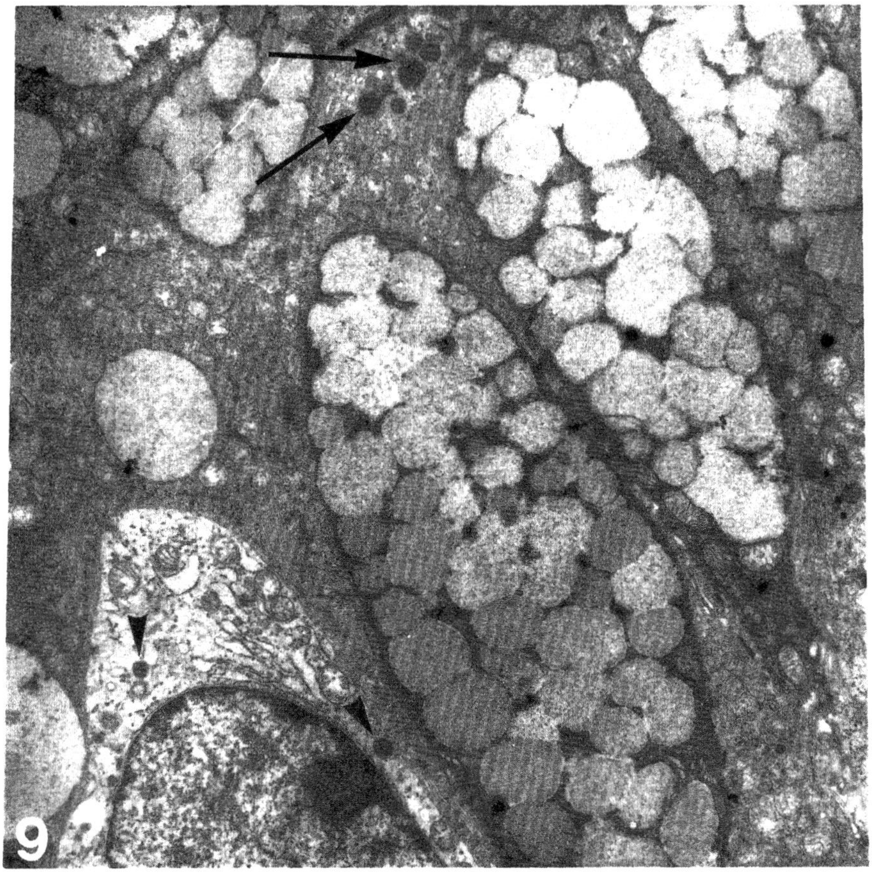
9

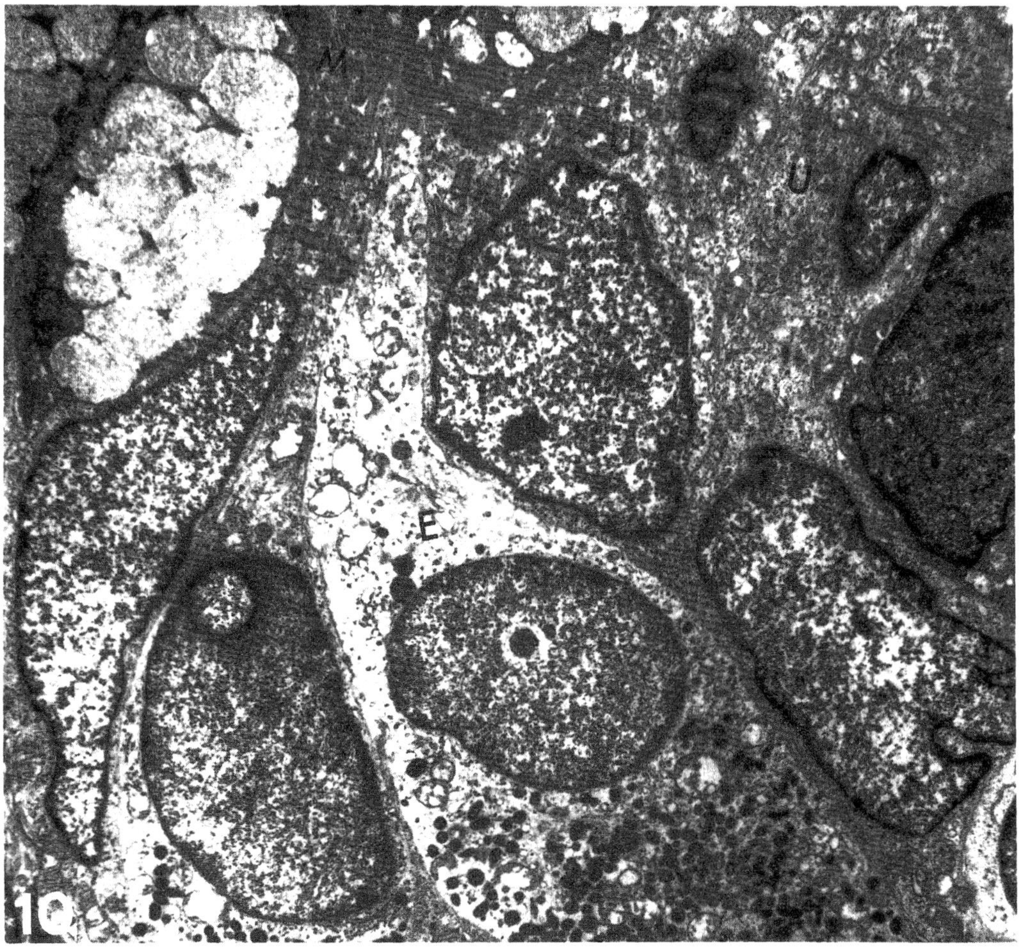
M
E
10

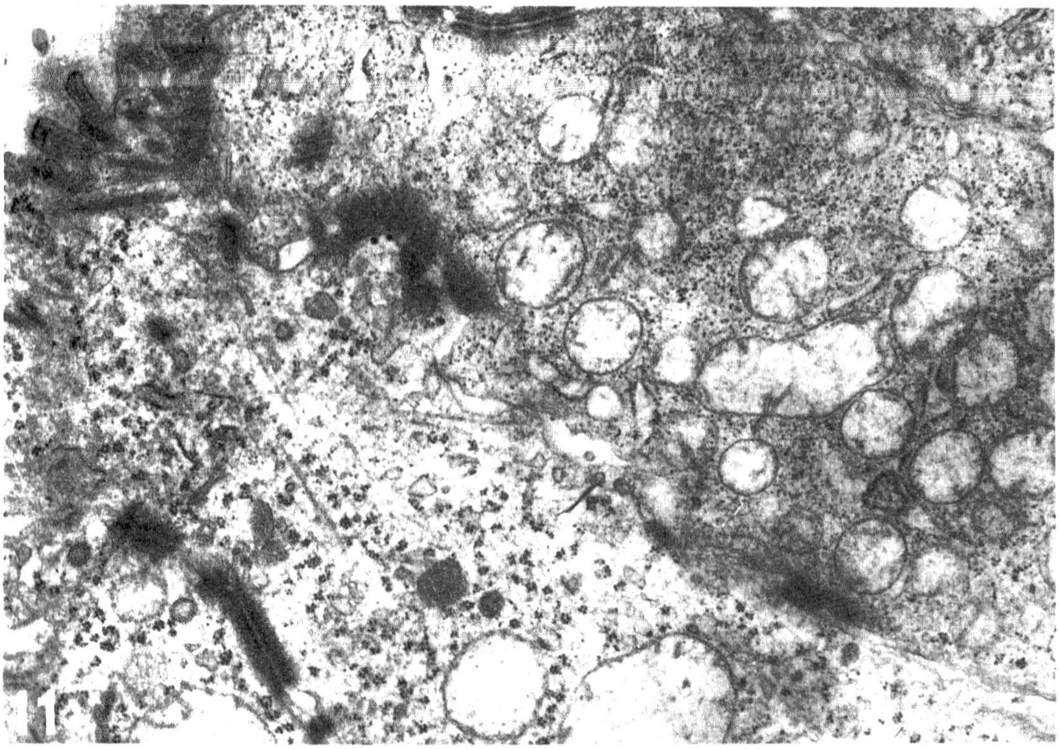

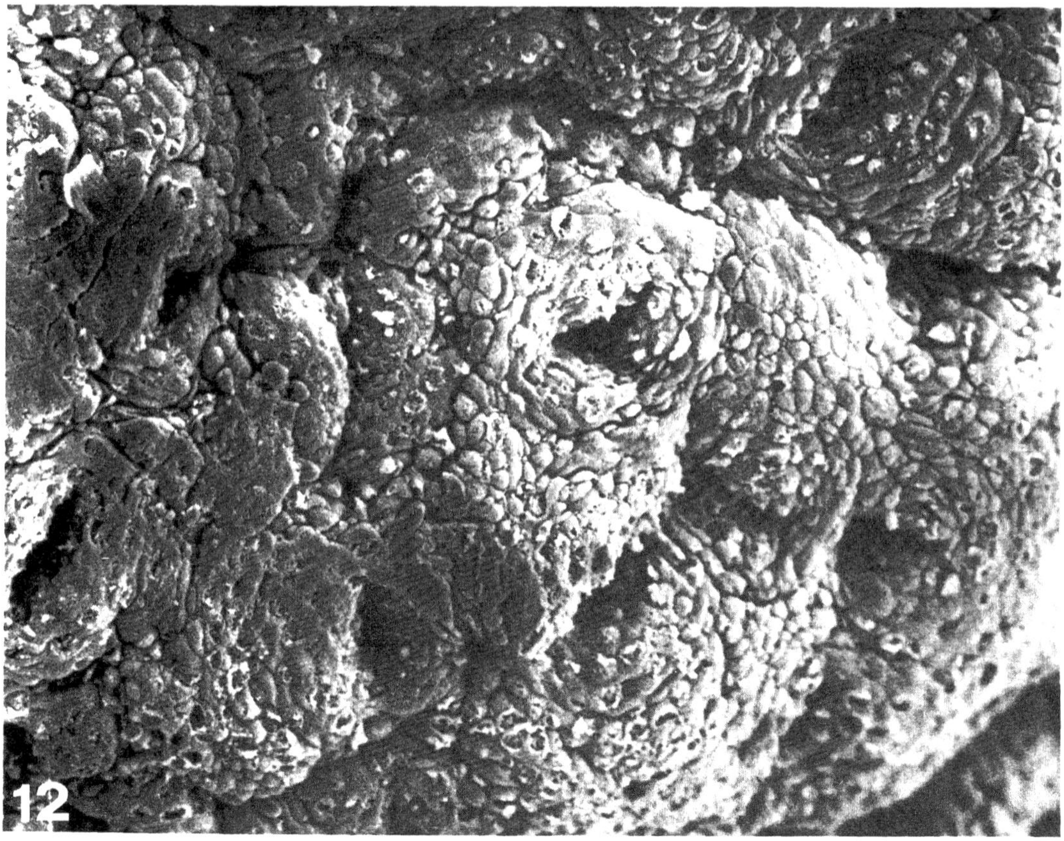

Figure 9-9. TEM hyperplastic polyp epithelium midcrypt. Proliferation of mucous cells seen; most cells show distension by mucous vesicles; a few cells show dense mucin granules (arrows). Arrowhead points to a cell showing endocrine differentiation. X 7,980.
Figure 9-10. TEM adenoma at mid-crypt. Undifferentiated cells (U), mucous cells (M), and enteroendocrine cells (E). X 6,384.
Figure 9-11. TEM adenoma at crypt surface. Prominent junctions and cells with 'intermediate' differentiation are seen. X 10,520.
Figure 9-12. SEM hyperplastic epithelium. Deep mucosal ridges are present. Crypt openings are relatively regular. X 400.

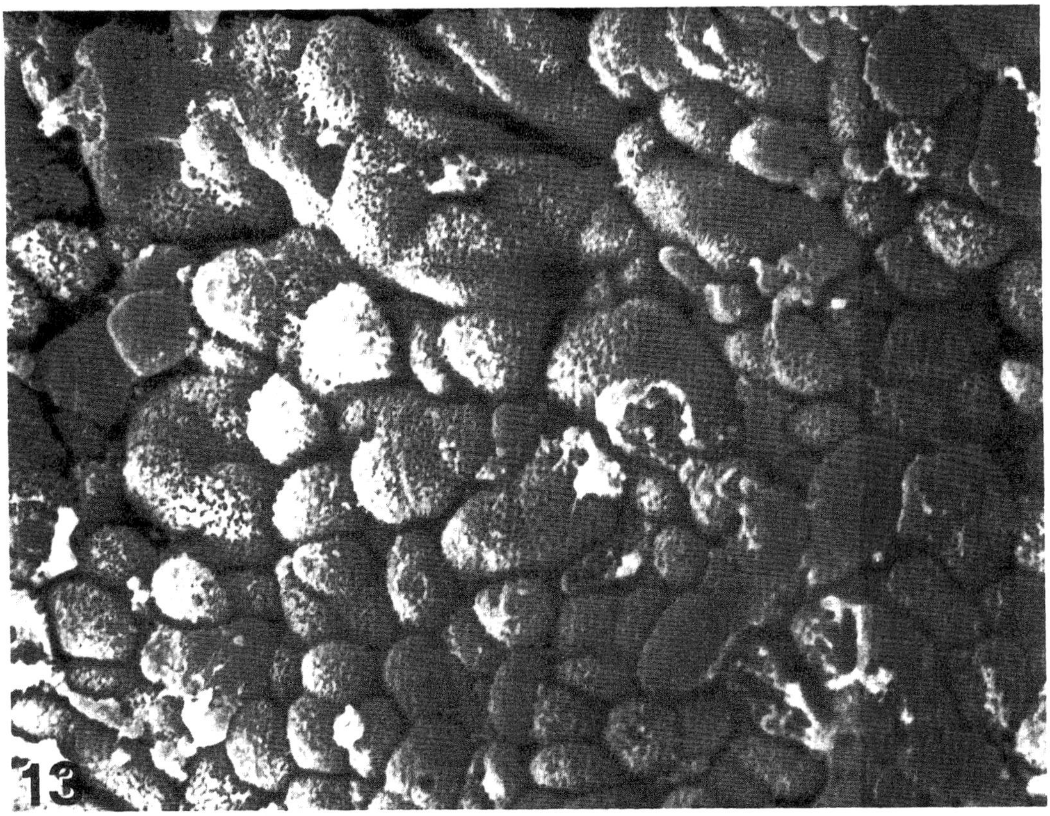

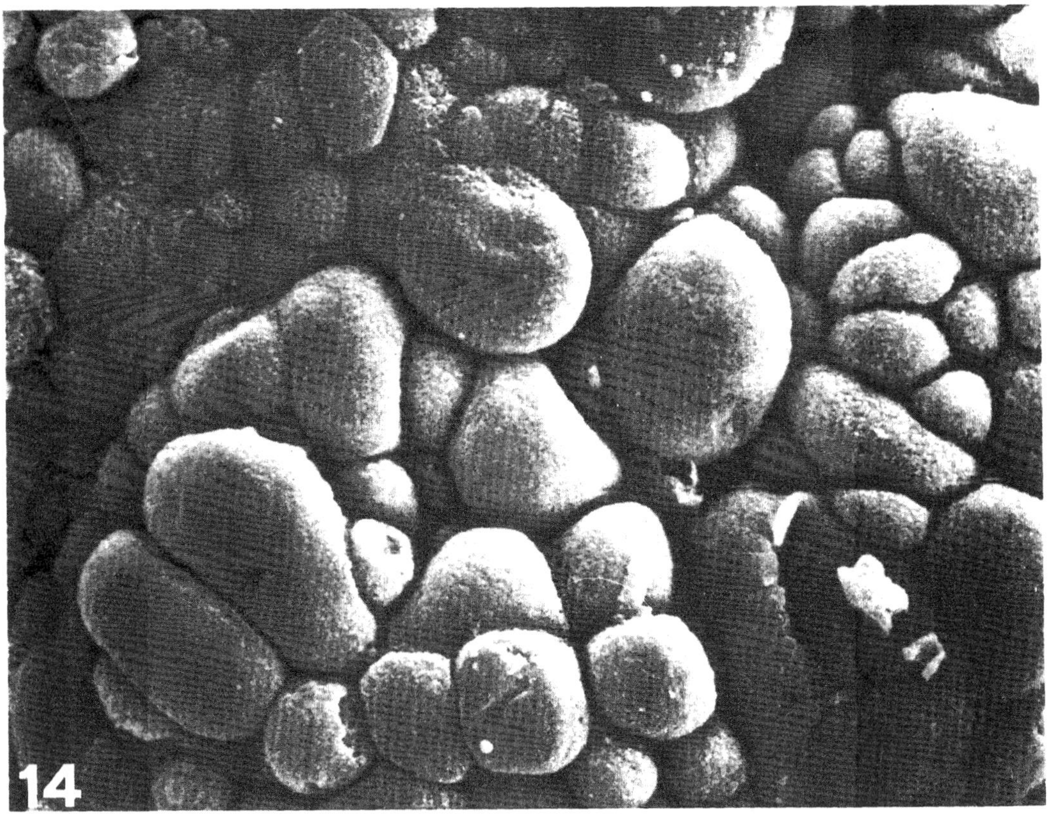

Figure 9-13. SEM moderate dysplasia in ulcerative colitis. Cells show marked variation in size and irregular microvillous formation. X 1,880.

Figure 9-14. SEM severe dysplasia in ulcerative colitis. Marked variation in size of cells. Microvilli are sparse. X 3,000.

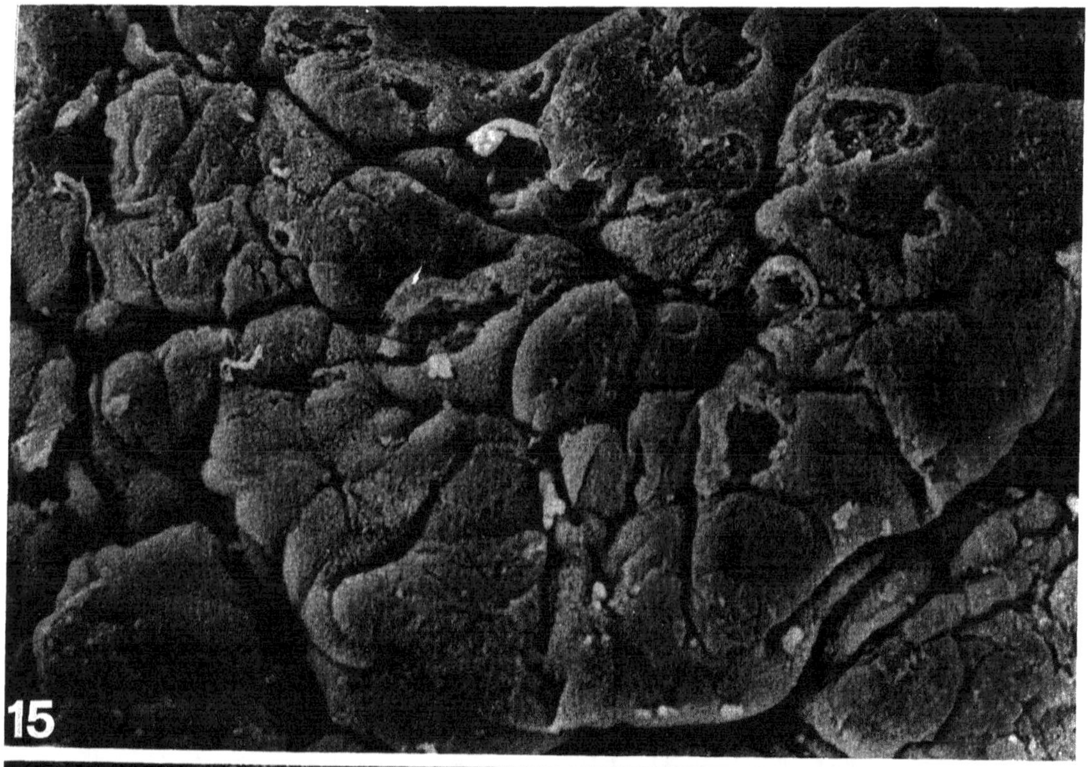

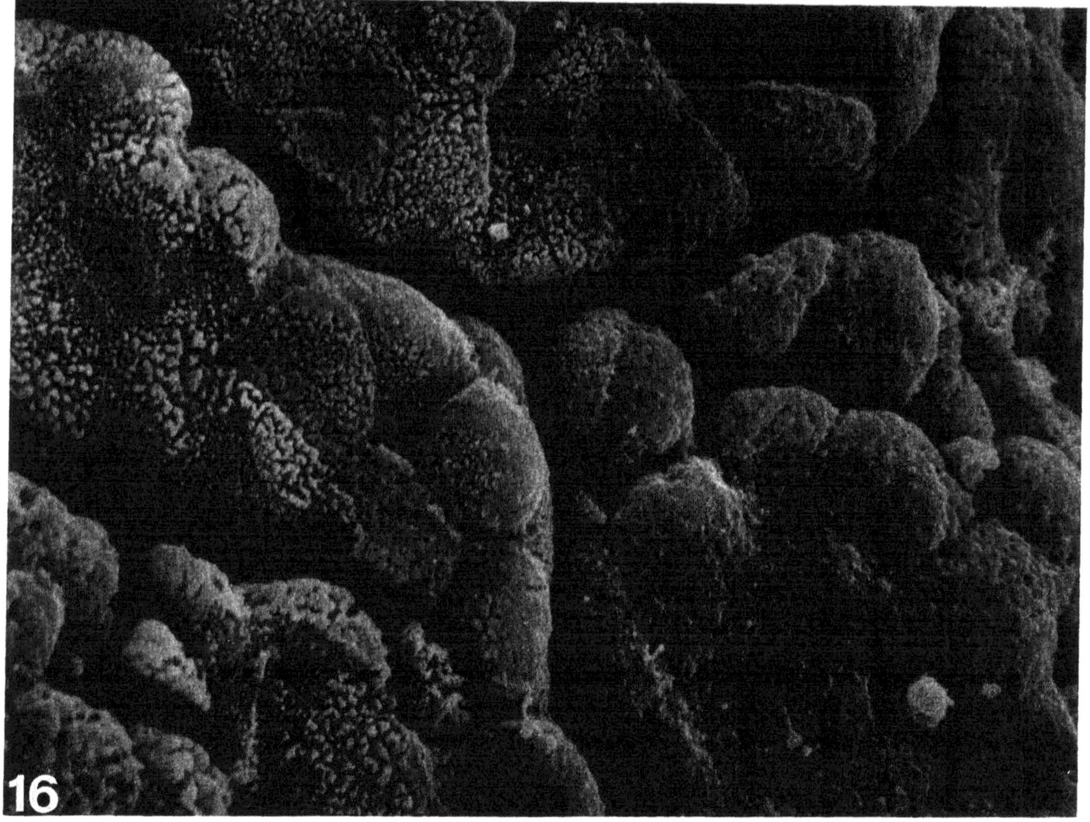

Figure 9-15. SEM carcinoma of rectum. Complete loss of surface architecture and irregular large lobular formation are seen. Occasional crypt formation noted. X 1,198.

Figure 9-16. SEM carcinoma of rectum. Complete loss of surface architecture and irregular large lobular formations are noted. Microvilli are seen in some of the cells (arrows). X 6,600.

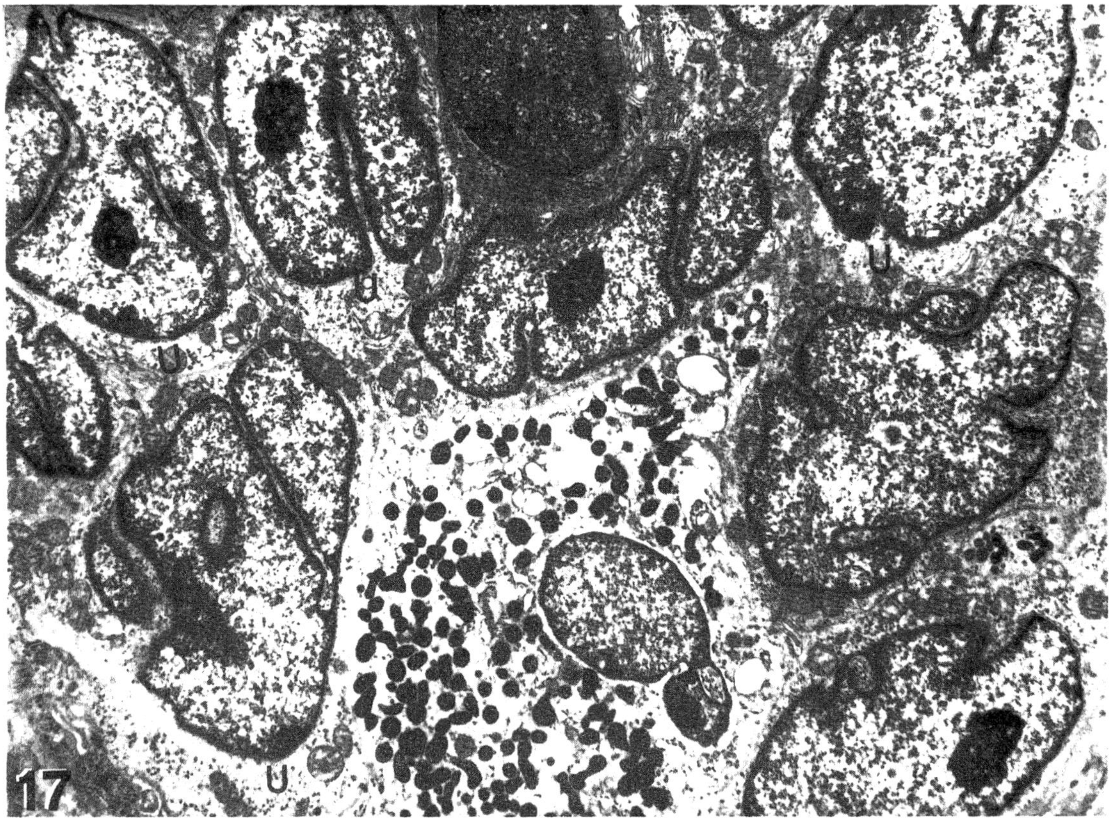

17

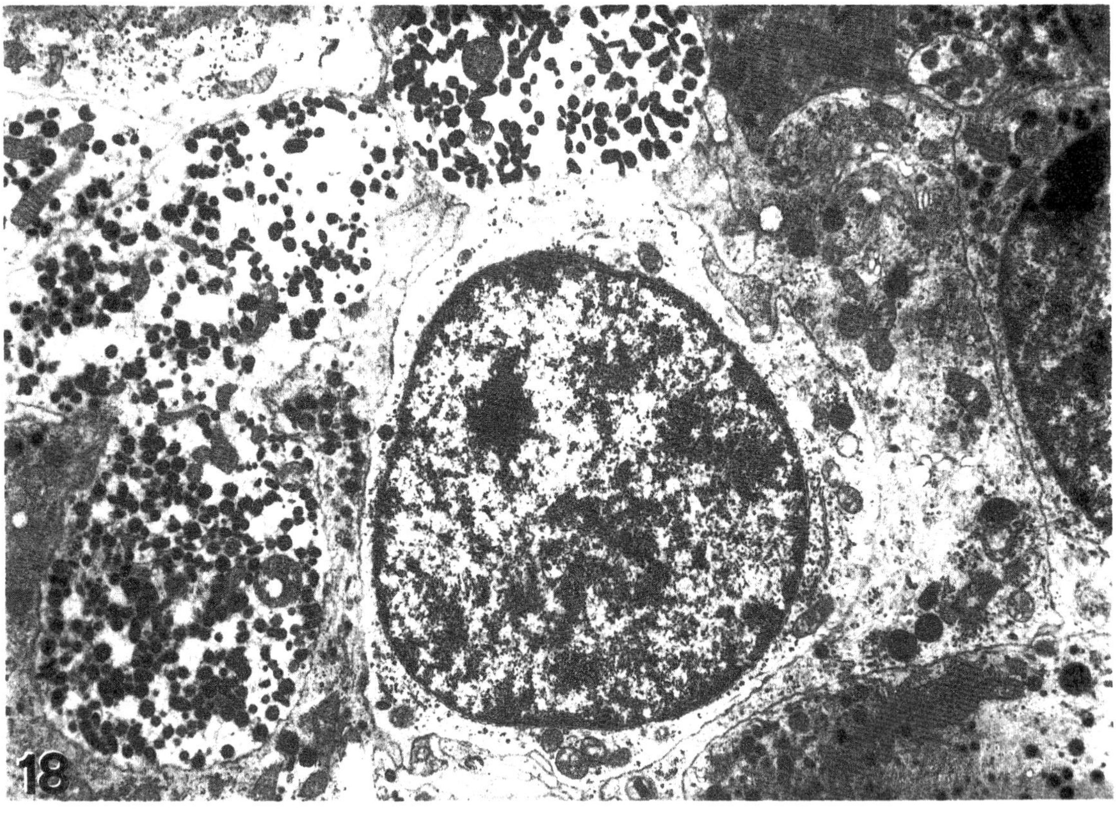

18

patients with extensive ulcerative colitis of more than 10 years' duration (40). Rectal dysplasia has been well documented histologically (41). The finding of high grade (severe) dysplasia in a rectal biopsy from a patient with long-standing ulcerative colitis indicates a high probability of a synchronous carcinoma in the colon.

Under transmission electron microscopy, dysplasia is indicated by a predominance of undifferentiated and intermediate cell types with varying degrees of nuclear atypia. In severely dysplastic mucosa, the cells are almost entirely made up of undifferentiated cells with nuclei showing marked indentations and dense peripheral heterochromatin, and nucleoli may also be seen. Under scanning electron microscopy, the epithelial cells show marked variation in size and shape, and the microvilli are sparse or may be absent. The number of mucous cells is greatly reduced. Architectural abnormalities with formation of irregular convolutions are frequent (Figs. 9-13, 9-14). Studies using scanning electron microscopy with morphometric analysis appear to contribute to the grading of dysplasia in chronic colitis (17).

7.4. Malignant epithelium

There is complete distortion of cryptal architecture. Irregular lobulation may be present. The epithelial cells form dense collections which obliterate the crypt openings. The cells vary markedly in size and shape, and the cell surface is also very irregular (Figs. 9-15, 9-16).

By transmission electron microscopy, rectal adenocarcinomas show epithelial cells which vary significantly in the degree of differentiation and mucin production (Fig. 9-17). The nuclei are large with marked nuclear indentations and dense intranuclear chromatin aggregates. The heterochromatin may be prominent in some cells while in others prominent nucleoli are observed. Dense mucin vesicles may be seen in some of the cells. Most tumors show complex desmosomal attachments except in poorly differentiated carcinomas where desmosomes are sparse. The Golgi system is usually prominent. The cells show groups of microvilli. However, the microvillous arrangement may be very irregular with microvilli projecting out in different directions. Occasional intracytoplasmic microvillous formations may also be seen. These possibly represent neoformation of the villous structures. Most, however, are due to intracytoplasmic invaginations of the surface microvilli.

Adenocarcinomas may show varying proportions of enteroendocrine cells. Composite tumors with different proportions of epithelial and enteroendocrine cells (Fig. 9-17) have also been well documented (42–44). Malignant epithelial tumors of enteroendocrine cell origin are not uncommon in the rectosigmoid. Ultrastructurally the tumor cells (Fig. 9-18) are characterized by dense core secretory granules which vary in size and shape (200–600 nm). These tumors belong to the generic group of APUDomas. While tumors arising in the proximal gastrointestinal tract may produce a variety of polypeptide substances identifiable by immunohistochemical methods, the great majority of rectal tumors are not associated with endocrine secretory products (45). However, a few tumors showing monoamine neurotransmitter-like substances and immunoreactivity to polypeptide hormones such as glucagon, glicentin, and pancreatic polypeptide have been reported (4, 46).

Acknowledgments

The authors express sincere thanks to Donna Ellzey for typing the manuscript and to Jean Makowsky and Martha Newbill for photographic assistance.

←

Figure 9-17. TEM carcinoma of rectum. Proliferation of undifferentiated cells with large indented nuclei (U). Two enteroendocine cells with granules also shown. X 7,980.

Figure 9-18. TEM tumor of rectal enteroendocrine cells (carcinoid). Numerous dense core (neurosecretory) granules are prominent. X 12,160.

References

1. Dawson PA and Filipe MI: An ultrastructural application of silver methenamine to the study of mucin changes in the colonic mucosa adjacent to and remote from carcinoma. *Histochem J* 8: 143–158, 1976.
2. Shamsuddin AM, Phelps PC, Trump BF: Human large intestinal epithelium: Light microscopy, histochemistry and ultrastructure. *Human Pathol* 13: 790–803, 1982.
3. Dawson PA, Filipe MI: An ultrastructural and histochemical study of the mucous membrane adjacent to and remote from carcinoma of colon. *Cancer* 37: 2388–2398, 1976.
4. Fiocca R, Capella C, Buffa R, Fontana R, Solcia E, Hage E, Chance RE, Moody AJ: Glucagon-, glicentin-, and pancreatic polypeptide-like immunoreactivities in rectal carcinoids and related colorectal cells. *Am J Pathol* 100: 81–92, 1980.
5. Pearse AGE: Neural crest origin of the endocrine polypeptide (APUD) cells of the gastrointestinal tract and pancreas. *Gut* 12: 783–788, 1971.
6. Cheng H, Leblond CP: Origin, differentiation and renewal of the four main epithelial cell types of the mouse small intestine. *Am J Anat* 141: 537–562, 1974.
7. Chang WWL, Leblond CP: Renewal of the epithelium in the descending colon of mouse. I. Presence of three cell populations, vacuolated columnar, mucous and argentaffin. *Am J Anat* 131: 73–99, 1971.
8. Lorenzsonn V, Trier JJ: The fine structure of human rectal mucosa: The epithelial lining of the base of the crypt. *Gastroenterology* 55: 88–101, 1968.
9. Fisher ER, Sharkey DA: The ultrastructure of colonic polyps and cancer with special reference to the epithelial inclusion bodies of Leuchtenberger. *Cancer* 15: 160–170, 1962.
10. Cooke T, Kirkham N, Stainthorp DH, Inman C, Goeting N, Taylor I: Detection of early neoplastic changes in experimentally induced colorectal cancer using scanning electron microscopy and cell kinetic studies. *Gut* 25: 748–755, 1984.
11. Traynor OJ, Costa NL, Blumgart LH, Wood CB: A scanning electron microscopy study of ultrastructural changes in the colonic mucosa of patients with large bowel tumors. *Br J Surg* 68: 701–704, 1981.
12. Siew S: Scanning electron microscopy of neoplastic lesions of the human colon. In: SEM/1976/V Proceedings of the Workshop on Biomedical Applications of the SEM. Chicago: IIT Research Institute p 11–17, 1976.
13. Barkla DH, Tutton PJM: Surface changes in the descending colon of rats treated with dimethylhydrazine. *Cancer Res* 37: 262–271, 1971.
14. Riddell, RH, Eisenstat L, Levin B, Golomb H: Surface ultrastructure of human adenomatous and hyperplastic polyps. In: SEM/1976/V Proceedings of the Workshop on Biomedical Applications of the SEM. Chicago: IIT Research Institute, p 19–24, 1976.
15. Phelps PC, Toker C, Trump BF: Surface ultrastructure of normal, adenomatous and malignant epithelium from human colon. In: SEM/1979 Chicago: IIT Research Institute, p 169–175, 1979.
16. Kavin H, Hamilton DG, Greasley RE, Eckert JD, Zuidema G: Scanning electron microscopy: A new method in the study of rectal mucosa. *Gastroenterology* 59: 426–432, 1970.
17. Shields HM, Bates ML, Goldman H, Zuckerman GR: Scanning electron microscopic appearance of chronic ulcerative colitis with and without dysplasia. *Gastroenterology* 89: 62–72, 1985.
18. Di Silvio T, Bartlett EF: Malacoplakia of the colon. *Arch Pathol* 92: 167–171, 1971.
19. Dobbins WO III, Herrero BA, Mansbach CM: Morphologic alterations associated with neomycin induced malabsorption. *Am J Med Sc* 255: 63–77, 1968.
20. Gear EV Jr, Dobbins WO III: Rectal biopsy: A review of its diagnostic usefulness. *Gastroenterology* 50: 195–210, 1966.
21. Glenner GG, Ein D, Eaves ED: Creation of amyloid fibrils from Bence Jones proteins in vitro. *Science* 174: 712–714, 1971.
22. Moolgavkar SH, Knudson AG: Mutation and cancer: A model for human carcinogenesis. *J Natl Cancer Inst* 66: 1037–1052, 1981.
23. Bleiberg H, Mainguet P, Cheretien J and Dupont-Mairesse N: Cell renewal in the human rectum. 'In vitro' autoradiographic study in active ulcerative colitis. *Gastroenterology* 58: 851–855, 1970.
24. Deschner EE, Lewis CM, Lipkin M: In vitro study of human rectal epithelial cells. I. Atypical zone of H3-thymidine incorporation in mucosa of multiple polyposis. *J Clin Invest* 42: 1922–1928, 1963.
25. Deschner EE, Lipkin M, Solomon C: Study of human rectal epithelial cells in vitro. II. H3-thymidine incorporation into polyps and adjacent mucosa. *J Natl Cancer Inst* 36: 849–857, 1966.
26. Bleiberg H, Mainguet P, Galand P: Cell renewal in familial polyposis: Comparison between polyps and healthy mucosa. *Gastroenterology* 63: 240–245, 1972.
27. Cole JW, McKalen A: Studies on the morphogenesis on adenomatous polyps in the human colon. *Cancer* 16: 998–1002, 1963.
28. Lipkin M: Phase I and Phase II proliferative lesions of colonic epithelial cells in diseases leading to colon cancer. *Cancer* 34: 878–888, 1974.
29. Lipkin M, Blattner W, Fraumeni JF Jr, Lynch H, Deschner EE: Tritiated thymidine (phi P, phi h) labeling distribution as a marker for hereditary predisposition to colon cancer. *Cancer Res* 43: 1899–1904, 1983.
30. Lipkin M: Growth kinetics of normal and premalignant gastrointestinal epithelium. In: Growth Kinetics and Biochemical Regulation of Normal and Malignant Cells. B Drewinko, RM Humphreys, (eds), Baltimore: Williams and Wilkins Co p 562–589, 1977.
31. Kanemitsu T, Koike A, Yamamoto S: Study of the cell proliferation kinetics in ulcerative colitis, adenomatous polyps and cancer. *Cancer* 56: 1094–1098, 1985.
32. Goldgaber MG, Kirsner JB: Carcinoma of the colon in ulcerative colitis. *Cancer* 17: 657–665, 1964.
33. Morgan CN: Malignancy in inflammatory diseases of the large intestine. *Cancer* 28: 41–44, 1971.

34. Riddell RH: Ultrastructure of the "transitional" muscosa adjacent to large bowel carcinoma. *Cancer* 40: 2509–2522, 1977.
35. Filipe MI: Mucus secretion in rat colonic mucosa during carcinogenesis induced by dimethylhydrazine. A morphological and histochemical study. *Br J Cancer* 32: 60–77, 1975.
36. Kaye GI, Fenoglio CM, Pascal RR, Lane N: Comparative electromicroscopic features of normal, hyperplastic and adenomatous human colonic epithelium. *Gastroenterology* 64: 926–945, 1973.
37. Hayashi T, Yatani R, Apostol J, Stemmermann GN: Pathogenesis of hyperplastic polyps of the colon: A hypothesis based on ultrastructure and in vitro cell kinetics. *Gastroenterology* 66: 347–356, 1974.
38. Fenoglio CM, Lane N: The anatomical precursor of colorectal carcinoma. *Cancer* 34: 819–834, 1974.
39. Muto T, Bussey HJR, Morson BC: The evolution of the cancer of the colon and rectum. *Cancer* 36: 2251–2270, 1975.
40. Butt JH, Morson BC: Dysplasia and cancer in inflammatory bowel disease. *Gastroenterology* 80: 865–868, 1981.
41. Riddell RH, Goldman H, Ranshoff DF, Appleman HD: Dysplasia in inflammatory bowel disease. Standardized classification with provisional clinical applications. *Hum Pathol* 14: 931–969, 1983.
42. Ulich TR, Cheng L, Glover H, Yang K, Lewin KJ: A colonic adenocarcinoma with argentaffin cells. *Cancer* 51: 1483–1489, 1983.
43. Hernandez F, Reid J: Mixed carcinoid and mucus secreting intestinal tumors. *Arch Pathol* 88: 489–496, 1969.
44. Bates HR, Belten LF: Composite carcinoid tumor (argentaffinoma-adenocarcinoma) of the colon: Report of two cases. *Dis Colon Rectum* 10: 467–470, 1967.
45. Orloff MJ: Carcinoid tumors of the rectum. *Cancer* 28: 175–180, 1971.
46. Alumets J, Falmer S, Grimelius L: Immunocytochemical demonstration of eukephalin and B-endorphin in endocrine tumors of the rectum. *Acta Pathol Microbiol Scand* (A)88: 103–109, 1980.

Author's address:
Hymie Kavin, M.D.
Division of Gastroenterology
Michael Reese Hospital and Medical Center
Lake Shore Drive at 31st Street
Chicago, Illinois 60616, USA

CHAPTER 10

The musculature and innervation of the alimentary canal

FRANCESCO AMENTA & CARLO CAVALLOTTI

1. Introduction

The alimentary canal is traditionally divided into two basic parts, the esophagus and the gastrointestinal tract. As is well known, both portions are composed of four concentric layers: the *mucosa*, lined by epithelial cells, contains a supporting layer of connective tissue, the lamina propria, and the muscularis mucosae, which is present in most but not all of the alimentary canal; the *submucosa*, a second layer of loose connective tissue; the *tunica muscularis*, which is smooth muscle throughout the digestive tube, except for the superior one-third of the esophagus, which contains striated muscle; and the *serosa* which is the most external of the four layers. The tunica muscularis, in turn, is divided into an inner band of circular muscle and an outer band of longitudinal muscle. Thus, muscle is actually located in two areas of the canal wall, the muscularis mucosae and the tunica muscularis (1,2).

An *intramural nerve plexus* begins in the wall of the esophagus and extends throughout the alimentary canal to the anus. This plexus consists mainly of two layers of neurons and associated fibers. The inner layer is called *Meissner's plexus*; it is located in the submucosa. Between the circular and the longitudinal muscle layers lies *Auerbach's plexus*. Meissner's plexus, which is less extensive than Auerbach's plexus, is mainly sensory in nature; it receives impulses from the surface epithelium and from stretch receptors buried within the gut wall. Auerbach's plexus, on the other hand, is mainly motor in function. Stimulation of this plexus increases gut activity, causing the following principal effects: (a) increased tone of the gut wall through tonic contractions; (b) increased intensity of rhythmic contractions; (c) increased rate of rhythmic contractions; and (d) increased velocity of conduction of excitatory waves along the gut wall. The alimentary canal is richly innervated by sympathetic and parasympathetic nerves which can modify the overall activity of the entire tube or just one portion of it (3,4).

2. Musculature

2.1. Cellular arrangement in smooth muscle

Smooth muscle cells are roughly parallel to each other and they are densly packed together. They form sheetlike layers which, in the case of the circular and longitudinal muscle layers, are perpendicular to each other. The stomach, which is the only organ in the entire alimentary canal to have three layers or coats of muscle cells, has each adjacent layer at right angles to its neighbor. Within each layer, the muscle cells are grouped together by septa. The septa are parallel to the muscle cell and they are only a few hundred microns long. They are then replaced by other septa which are located in different positions.

There are abundant collagen fibrils present within gut smooth muscle. It is of interest that smooth muscle contains three to four times as much collagen as a comparable amount of striated

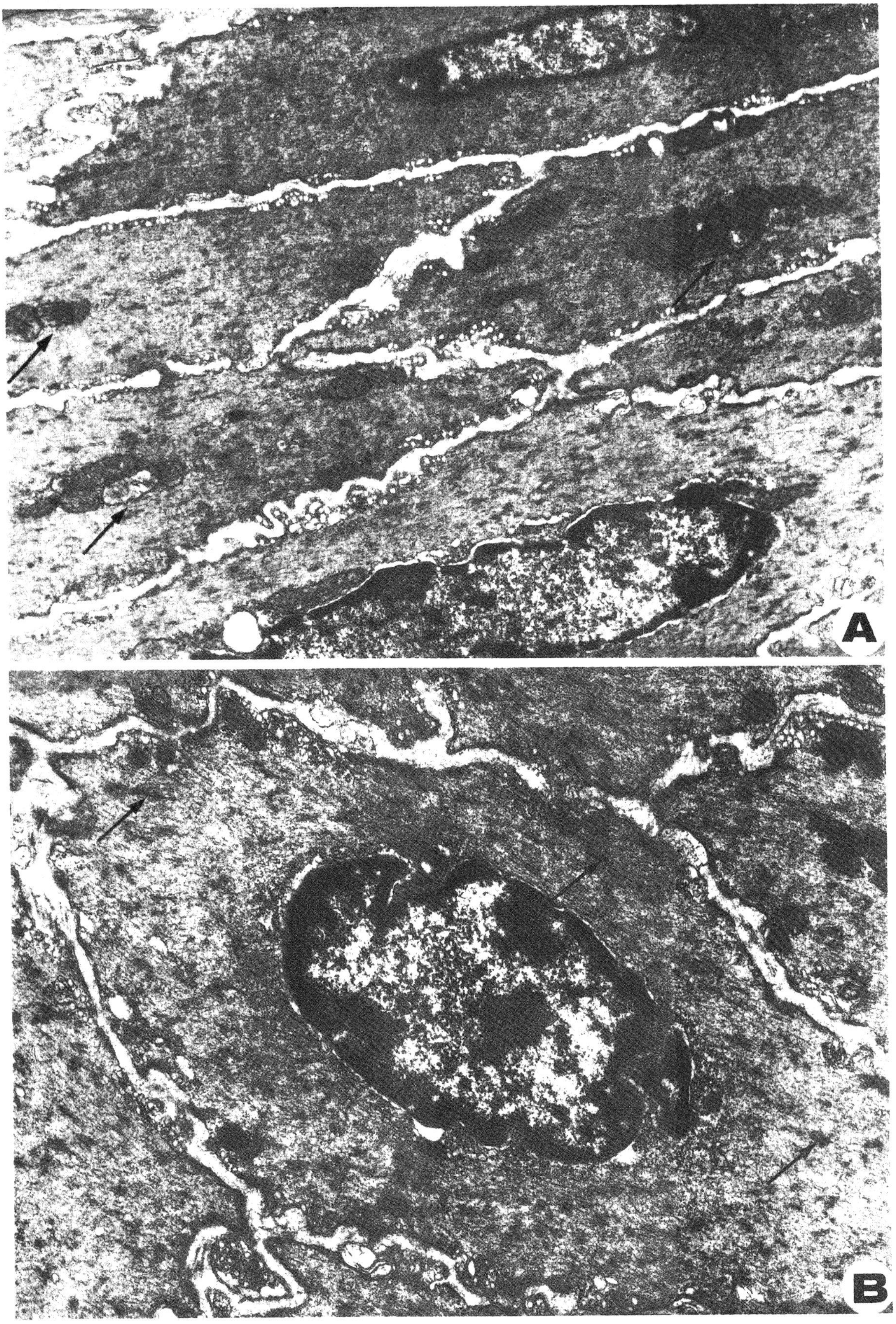
A
B

muscle; it has therefore been suggested that this increase in collagen serves as an intramuscular tendon (5) because smooth muscles lack true tendons.

Considerable interest has been aroused by the interstitial cells found in the intestinal musculature. They were originally considered to be modified sympathetic neurons which relayed motor impulses to the smooth muscle cells. This original hypothesis seemed to be substantiated by the observation that these cells readily stained with silver salts or with methylene blue (6). Electron microscopy, however, demonstrated their fibroblastic nature (6,7). While several investigators have proposed various hypotheses concerning the function of these cells (possible undifferentiated or immature smooth muscle cells, possible electrical coupling, etc.), most authors feel their function is not yet known (8,9).

2.2. *Fine structure*

Smooth muscle cells are spindle-shaped cells with a single nucleus. At rest they are 500 to 700 μm long, but in isotonic contraction they contract to 25% of their resting length. Of interest are the flasklike invaginations called *caveolae* which stud the surface of visceral muscle cells. While caveolae may occur singly, they usually are arranged in rows which are parallel to the long axis of the muscle fiber. Each individual caveola is perpendicular to the surface of the cell. Since these invaginations greatly increase the amount of cell membrane (by about 50% to 70%) (10), there has been much interest as to their possible function; there is general agreement that caveolae are smooth muscle organelles; but, unfortunately, their precise role is still unknown (11–13).

Where the cell surface is free of caveolae, so-called *dense bands* of electron-dense material adhere to the inner layer of the cell membrane (Fig. 10-1). Connected to these dense bands are filaments of different diameters (14). It thus appears that the dense bands are points to anchorage for the internal filaments which are located in visceral muscle. Since a dense band of one cell may be aligned with a dense band in a neighboring cell, these two structures probably form an intermediate junction, something which will be considered in more detail in the next section. The prominent Z lines of striated muscle are lacking in smooth muscle. Histochemical and immunochemical investigations have detected alpha-actin which is a component of Z lines. This has led to speculation that the smooth muscle dense bands are the equivalent of the striated muscle Z lines.

Electron microscopy has shown the presence of numerous microtubules, as well as centrioles, Golgi apparatus, lysosomes, mitochondria, and sarcoplasmic reticulum. Comparison of developing smooth muscle cells to mature smooth muscle cells shows a predominance of sarcoplasmic reticulum, which is replaced by smooth reticulum cisternae as the cell ages. It is also interesting to note that there is a relatively greater proportion of extracellular calcium which enters the smooth muscle during contraction, compared to striated muscle. This does not mean that calcium is not stored within the smooth muscle cell, because it is now known that calcium is stored in the sacroplasmic reticulum (15–16).

2.3. *Contractile filaments*

Unfortunately, the ultrastructure of visceral muscle is less well understood than that of skeletal muscles. Actin, a myofilament which is approximately 7 nm in diameter, and myosin, a thicker myofilament of about 15 nm in width, can be extracted from smooth muscles. Estimates of the ratio of actin to myosin filaments in smooth muscle are approximately 12:1. A third type of filament, about 10 nm in diameter, is also present.

The actin filaments are rather easily preserved during fixation. They are usually arranged in reg-

←

Figure 10-1. Mouse duodenum. Electron micrographs of the internal (circular) muscle layer. (A). Micrograph showing the arrangement of smooth muscle cells. Muscle cells are separated by extracellular spaces containing glycoproteins and thin bundles of collagen. The arrows indicate groups of mitochondria frequently oriented parallel to the length of smooth muscle cell. X 6,500. (B) Transverse section of smooth muscle. The arrows indicate bands of electron-dense material that very likely represent points of anchorage for contractile filaments. X 11,000. (Courtesy of Profs. C. Fruschelli and R. Gerli, Siena.)

ular fascicules of unknown length. Actin filaments penetrate into the amorphous material of the dense bands associated with the cell membrane and the dense bodies (7, 14).

The thicker myosin filaments are much more difficult to preserve so, consequently, their presence in electron micrographs varies with the fixative used. They have tapered ends and an irregular cross section (19). Due to technical problems, the cross bridges which are quite apparent in voluntary muscle are much more difficult to see, but they probably have a helical arrangement along the filament (20). A bare area in myosin filaments (29, 30) suggests that the molecular arrangement is different from that found in voluntary muscle myosin filaments (22).

The composition of actins isolated from several types of smooth and striated muscles is similar; we cannot say the same about myosin. Antibodies formed against smooth muscle myosin are non-reactive with skeletal muscle (23); there are chemical differences as well when smooth muscle myosin is compared to striated muscle myosin and when myosin filaments from different species are compared (24).

The third type of filament now seen in smooth muscle is present in all smooth muscle but in concentrations which vary considerably even within the same population of cells. These filaments are quite numerous following injury and in developmental situations such as tissue cultures (25). These filaments are formed by four filamentous subunits. These intermediate filaments probably form a type of intracellular skeleton, similar to the cytoskeleton that intermediate filaments form in obliquely striated muscles of invertebrates (26).

2.4. *Smooth muscle cell junction*

Smooth muscle cells have cell-to-cell junctions due to modifications of the cell membrane (5) which contain numerous dense bands that receive myofilament attachments. The dense bands from one cell occasionally match with a similar area in a neighboring smooth muscle cell to form a symmetrical junction called an *intermediate junction*. The intermediate junctions form a direct mechanical linkage between two or more smooth muscle cells. Since myofilaments are connected to each dense band, it appears that the intermediate junction is a method of transferring forces from a bundle of myofilaments in one cell to another bundle of myofilaments in another cell (27).

A second type of junction found in many smooth muscles is the *gap junction* or *nexus*. This type of junction is oval- or disk-shaped and lies in close approximation to an identical area in a neighboring cell. The space between two cells in the gap junction area is about 2 nm and is filled with regularly arranged subunits. The gap junction allows the passage of certain ions and small molecules, resulting in an electric and/or metabolic communication between cells (28). Experimentally, this coupling can be interrupted by increasing the intracellular concentration of calcium or by lowering the pH.

It is interesting to note that certain smooth muscles lack gap junctions. For example guinea pig ileum and duodenum are richly supplied with gap junctions, while these structures are virtually absent in the rectum. The lack of gap junctions in certain muscles implies that these muscles exchange ions and molecules by other means.

3. Innervation

3.1. *Extrinsic and intrinsic innervation of the alimentary tract*

The intramural nerve plexuses found in the gut control the motility of the alimentary canal by their motor and sensory function. The gut has nerve fibers of extrinsic origin as well which reach the alimentary tract by means of the vagus nerve, the mesenteric nerves, and the pelvic nerves.

Most of the studies on the vagus nerve have been done on the cat. It has been seen that the majority of the cervical and abdominal vagal nerve fibers (75% to 80% in the neck, 90% in the abdomen) originate from the vagal ganglia and are thus postganglionic (29). These postganglionic parasympathetic fibers are, for the most part, unmyelinated fibers. It is of interest to note that small animals, such as the cat, rabbit, and rat, have only 1% of their abdominal vagal fibers myelinated, while larger animals, such as sheep and humans, have a much larger percentage of their abdominal vagal nerve fibers ensheathed.

The vagus nerve may also contain an adrenergic component which originates from the superior cervical ganglion and, possibly, from the stellate ganglion (30).

The mesenteric nerves originate from the celiac ganglia, the superior mesenteric ganglia, and the inferior mesenteric ganglia. These nerves, which are postganglionic, sympathetic, adrenergic fibers, are closely associated with the large arteries which supply the stomach and intestine. Studies have shown that each artery is paired with from three to six small nerves which, however, lie at some distance from the arterial adventitia. The adventitia itself is richly supplied with adrenergic fibers which provide motor control of the vascular musculature; these fibers constitute the so-called perivascular nerve fibers (3) (Fig. 10-4).

Fluorescence microscopic studies of the cells comprising the celiac ganglia, the superior mesenteric ganglia, and the inferior mesenteric ganglia show their adrenergic nature. In fact, the mesenteric nerves are the source of the adrenergic supply to the gut.

The final source of extrinsic innervation of the alimentary canal is the pelvic plexus. This plexus is located on the lateral sides of the rectum and gives rise to nerve fibers which supply both the distal portion of the alimentary canal and the urogenital organs. Since there are cholinergic, parasympathetic fibers originating from the sacral portion of the spinal cord and adrenergic, sympathetic fibers which originate from the lumbar portion of the spinal cord, it is difficult, if not impossible, to separate the sympathetic and parasympathetic pathways. It is probable that some functional interaction may occur between the two.

3.2. Intramural plexuses and intramural ganglia (Fig. 10-2)

The two intramural plexuses, Meissner's plexus (or submucosal) and Auerbach's plexus (or myenteric), begin in the esophagus and extend the length of the alimentary canal without interruption. Small nerves emerge from each of the plexuses to communicate with either the layers of the wall or with the other plexus. These communications form a meshlike configuration which, interestingly, characterizes each segment of the alimentary tract and, to some extent, different animal species. The significance of these unchanging patterns is not known.

There is, however, a great variation in the number of neurons in each ganglion in Auerbach's plexus. The number of neurons per unit area is roughly related to the thickness of the muscle in that area; for example, there is a greater density of neurons beneath the taeniae of the cecum in comparison to the rest of the cecum. With an increase in body size there is a much larger increase in the number of neurons per unit area (7). The rodent small intestine contains approximately 1.85×10^6 neurons in the myenteric plexus, while the feline small intestine contains about 6×10^6 for both the myenteric and the submucosal plexuses together (31).

Neuronal density in the submucosal plexus varies greatly from species to species. In the guinea pig and the rat, the submucosal neurons are about half as numerous as in Auerbach's plexus; in the cat, however, they are two to three times more numerous (32).

Intramural ganglia contain two types of cells, the glia and neurons. The ganglia of the gut resemble the nervous tissue of the CNS very much; that is, there are scant elements of connective tissue present within them. In fact, neurons, glial cells, and their processes are closely packed together within each ganglion. Studies show that there is an apparent space of only 20 nm between the membranes of neighboring structures. Apparently, the glia fill in any excessive space within the ganglion. Each neuron of an intramural gut ganglion is surrounded by a basal lamina as well as by glial cells. The effect is then an effective anatomical separation of one neuron from another. It is not surprising that relaxation and contraction of the smooth muscle cells of the gut change considerably the shape of the intramurales ganglia. This mechanical activity likewise alters the shape of the individual neuronal and glial cells.

It is interesting that large areas of the surface of the perikaryon of Auerbach's plexus are in direct contact with the basal lamina, thereby exposing it to the extracellular space and to connective tissue.

External to the ganglia and wrapping connec-

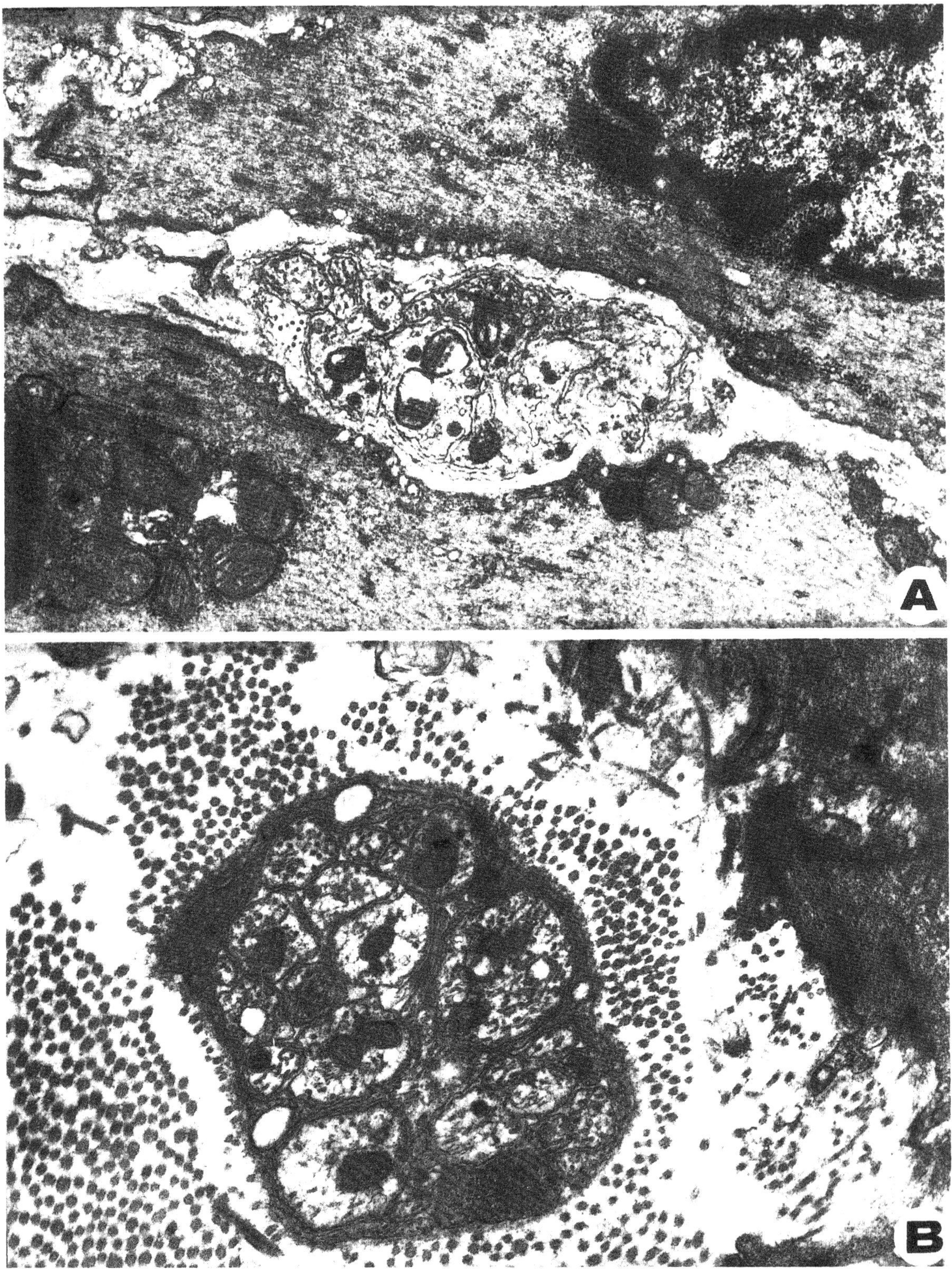

Figure 10-2. Mouse small intestine: Auerbach's plexus. (A). Electron micrograph showing an autonomic nerve ending placed between two smooth muscle cells. Granular and agranular vesicles can be seen. X 14,000. (B). Electron micrograph showing nerve fibers with several kinds of vesicles. X 30,000. (Courtesy of Profs. C. Fruschelli and R. Gerli, Siena.)

tive tissue are interstitial cells. These cells, which are flattened and have long laminar processes, cover partially or entirely the intramural ganglia. Where only a partial covering occurs — and this is the usual case — large areas of the ganglion directly face the visceral muscle cells (33).

3.3. Neuronal processes, nerve endings, and synapses

As might be expected, there is an interface between nerve cells and smooth muscle cells. Nerve endings within the alimentary canal are not, as is true elsewhere, anatomical endings of a fiber but rather varicosities along their length. The varicosities, which may number as many as 26,000 per nerve ending, allow an initimate association between the nervous system and the smooth muscle. The study of varicosities of adrenergic fibers using fluorescence microscopy is easy, and investigations involving the iris and rat hind limb show that varicosities of an adrenergic neuron contain about 100 times more norepinephrine than those in the corresponding cell body. Recent studies have indicated that the nonadrenergic axons of the gastrointestinal tract are also varicose (34).

Varicosities may contain mitochondria and microtubules. However, the more common components are synaptic vesicles, which are the basis for electron microscope morphological classification (Fig. 10-2). Since this work is still in an early phase and it is limited to a few portions of the gastrointestinal tract of just a few species, many investigators question its significance.

If we examine Auerbach's plexus in humans, guinea pigs, and rhesus monkies, we can see three types of synaptic vesicles (35). The first type, likely cholinergic, has numerous agranular vesicles, 35 to 60 nm in diameter, as well as a few larger granular vesicles. Some of these varicosities (Fig. 10-2) synapse on ganglionic neurons. The second type of synaptic vesicle is consituted of intensely osmiophilic granules mixed together with an equal number of agranular vesicles and a few large granular vesicles. Unlike the cholinergic vesicles, these adrenergic endings never form synapses. Their adrenergic identity is based upon the fact that they degenerate after 6-hydroxydopamine treatment, but their granule density increases after 5-hydroxydopamine treatment. The third and most common type of ending, once again, contains a mixture of agranular and granular vesicles. These types of endings are called *p-type varicosities*, where *p* stands for *polypeptide* (6).

Since there has been some controversy concerning the value of the above morphological classification, a newer and more elaborate taxonomy involving the guinea pig has been proposed. In this system, eight or more types of endings have been identified in Auerbach's plexus (36). It is likely that each type of ending is a different type of neuron and the different vesicles occurring within each ending represent different neurotransmitters. This system, too, has its critics; obviously, more work needs to be done in this area.

An interesting study by Campbell and Gibbins (37) has been carried out on p-type nerve endings of the gut. They have divided these endings into two separate groups, the small p-type endings, which have small clear vesicles and larger granular vesicles, and the large p-type endings, which have large granular vesicles. The small p-type vesicles have been identified as noncholinergic and nonadrenergic inhibitory (purinergic and/or peptidergic) endings, while the transmitter contained in the large p-type endings remains unknown.

In 1976 Cook and Burnstock (36) classified myenteric plexus neurons into nine different taxa based on their size, location, relation to satellite cells, and distribution of organelles; but there was no attempt to relate cell body ultrastructure to possible neural transmitters. In 1982, Burnstock (38) proposed grouping nerve profiles found in the gastrointestinal tract into eight classes based upon the characteristic structure of the predominant vesicle type found within that particular profile. These groups are as follows (see Fig. 10-3):

3.3.1. Small agranular vesicles (AGV) (Fig. 10-3A)

These profiles contain mainly small agranular vesicles which are 45 to 60 nm in diameter. There are, however, a few large granular vesicles present as well. This is the most frequent nerve

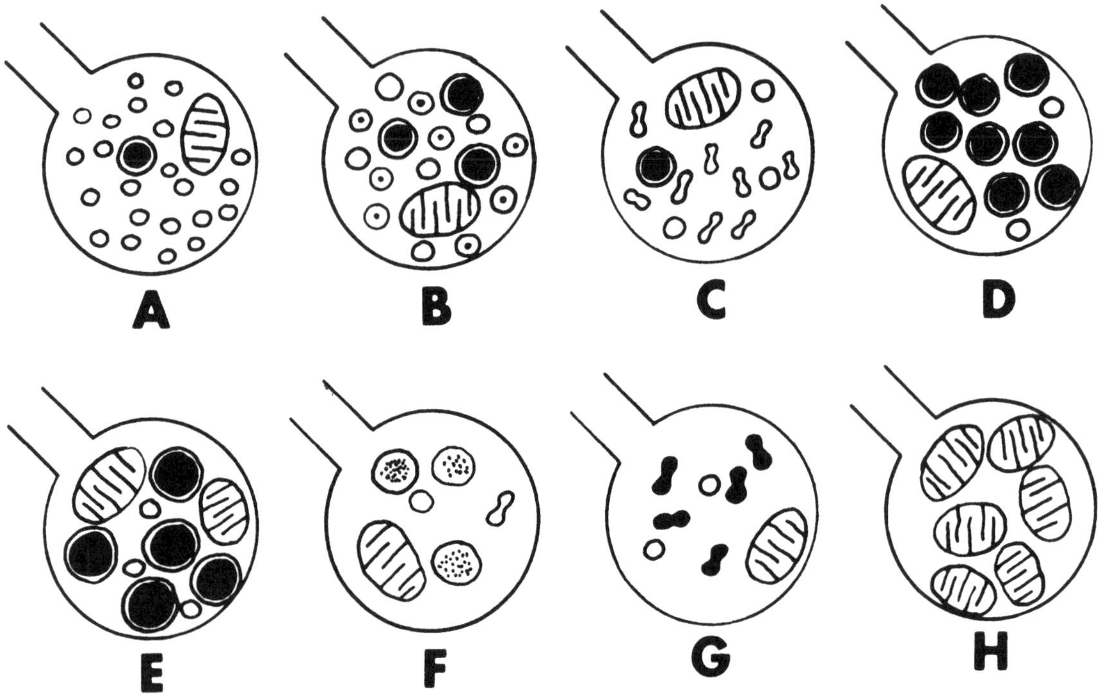
A
B
C
D
E
F
G
H

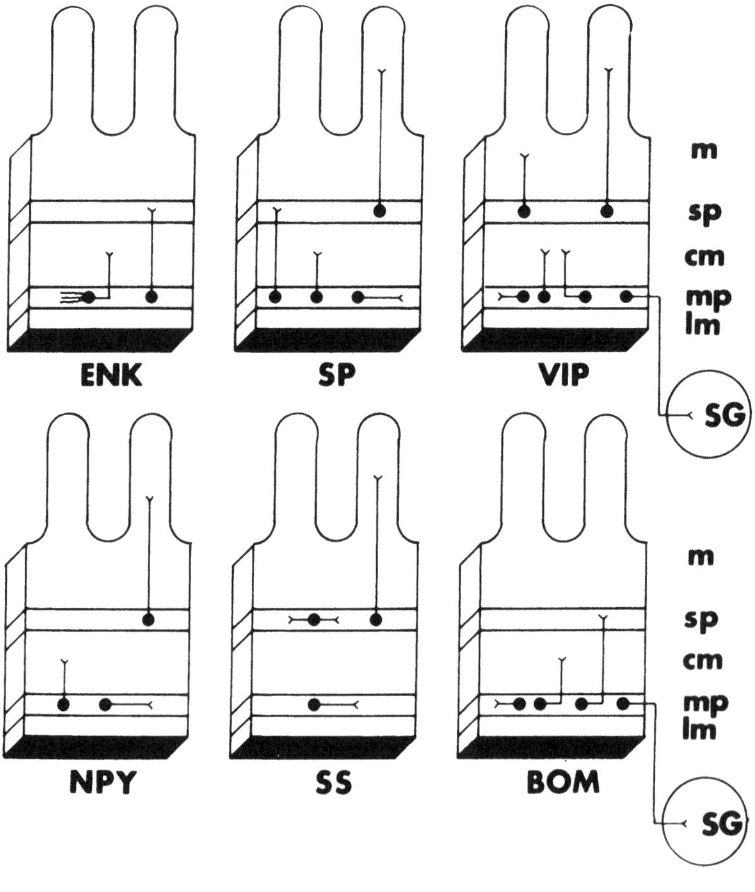
m
sp
cm
mp
lm
ENK
SP
VIP
SG
NPY
SS
BOM

profile, forming over 50% of the total number of nerves present.

3.3.2. Small granular vesicles (SGV) (Fig. 10-3B)

These profiles contain mainly small granular vesicles which are 45 to 70 nm in diameter. There are also some large granular vesicles, a few AGV and SFV.

3.3.3. Small flattened vesicles (SFV) (Fig. 10-3C)

The key morphologic components in this class of nerve profiles are flattened vesicles which measure 30 to 40 nm by 60 to 80 nm. There are also AGV and large granular vesicles present.

3.3.4. Large granular vesicles (LGV) (Fig. 10-3D)

The prominent features of this nerve profile are the large granular vesicles which measure 80 to 120 nm in diameter. These vesicles have an electron-dense core which is surrounded by a vesicle membrane, giving a halo effect. There are AGV as well.

3.3.5. Large opaque vesicles (LOV) (Fig. 10-3E)

While LOV profiles superficially resemble LGV profiles, they can readily be identified because the core is larger, 80 to 180 nm, and less electron-dense and the halo is less distinct. Some AGV are present as well.

3.3.6. Heterogeneous granular vesicles (HGV) (Fig. 10-3F)

These nerve profiles are classified by granular vesicles 70 to 140 nm in diameter having cores which are unevenly granulated. There is a halo, but it is difficult to see at times. Occasionally the vesicle membranes are incomplete. There are AGV and SFV present as well.

3.3.7. Pleomorphic vesicles (Fig. 10-3G)

This group contains granules with various shapes and sizes. The core is electron-dense, finely granular, and surrounded by a narrow halo. There are a few AGV sometimes present. This type of nerve profile is located in only certain regions of the gut.

It is important to note that each of the eight nerve profiles described above always contains mitochondria.

3.3.8. Small mitochondria (SM) (Fig. 10-3H)

This type of nerve profile is relatively rare in the alimentary canal. It consists of tightly packed clusters of small, round mitochondria. This type of profile may be difficult to distinguish from AGV.

3.4. Enteric neurotransmitters

Early histopharmacological studies proposed a comfortable, but now considered to be simplistic, understanding of gastrointestinal tract innervation. It was thought that cholinergic excitatory nerves located within the wall of the gut were under direct control of preganglionic cholinergic vagal and sacral parasympathetic fibers. This system was antagonized by the postsympathetic adrenergic inhibitory fibers. Falck's method of fluorescence histochemistry devised in 1962 allowed direct visualization of adrenergic varicosities. When this method was applied to the gut wall, investigators noticed that the terminal varicosities of the majority of adrenergic fibers were found in Auerbach's and Meissner's plexuses and

←

Figure 10-3. Top. Schematic representation of nerve profiles in Auerbach's plexus according to the classification of Burnstock (38). A, small agranular vesicles (AGV); B, small granular vesicles (SGV); C, small flattened vesicles (SFV); D, large granular vesicles (LGV); E, large opaque vesicles (LOV); F, heterogeneous granular vesicles (HGV); G, Pleomorphic vesicles; H, small mitochondria (SM).
Bottom. Topography of peptide-immunoreactive neurons and nerve fibers in the gut. ENK, enkephalin; SP, substance P; VIP, vasoactive intestinal polypeptide; NPY, neuropeptide Y; SS, somatostatin; BOM, bombesin; SG, sympathetic ganglion; m, mucosa and submucosa; sp, submucosal plexus; cm, circular muscle; mp, myenteric plexus; lm, longitudinal muscle.

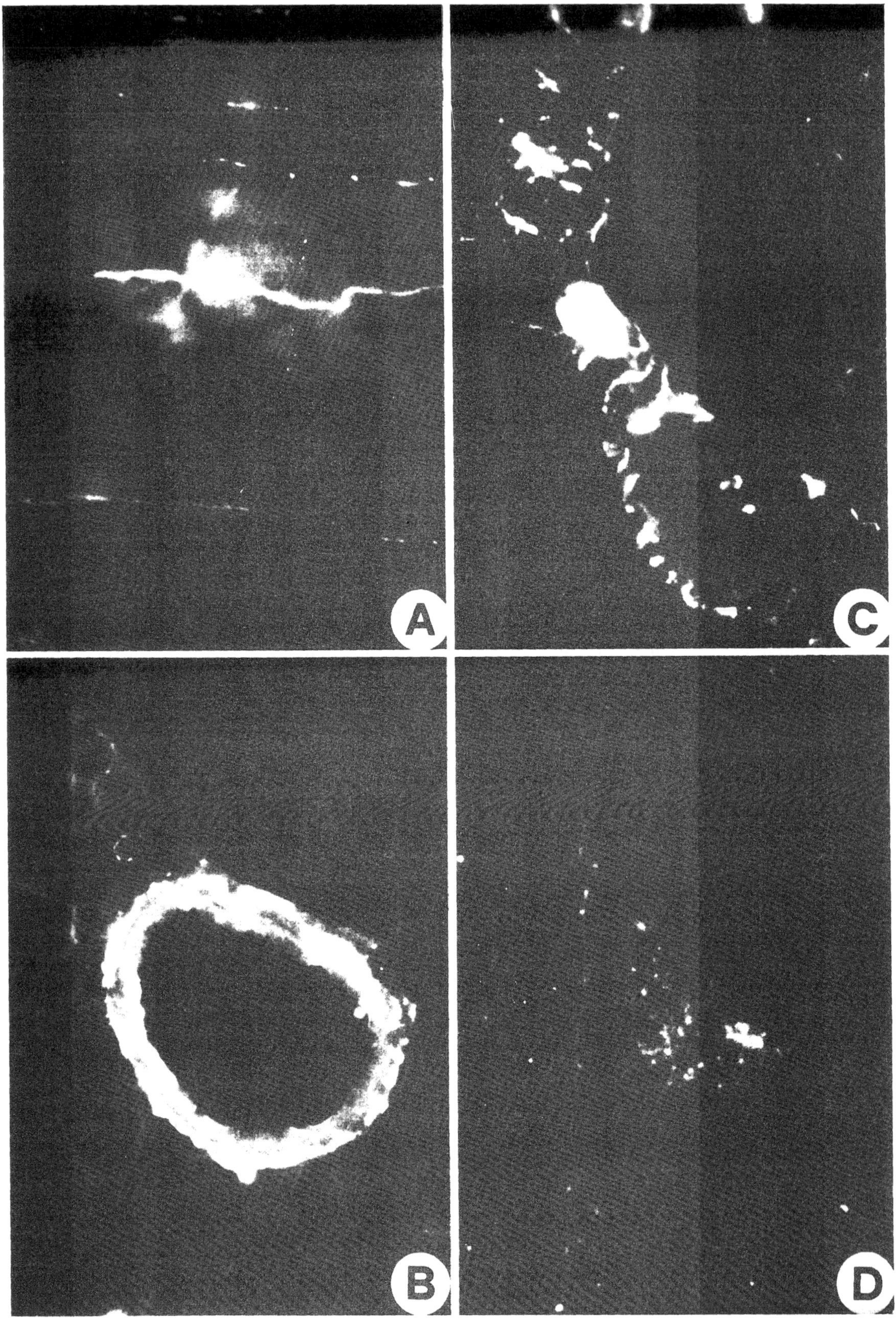
A
C
B
D

that these fibers innervated only small areas of the circular muscle layer. Subsequent investigations discovered nonadrenergic, noncholinergic nerves which inhibited gastrointestinal tract smooth muscle (39). These neurons propel material through the gut, open the various sphincter muscles, and cause relaxation of the stomach as well as "descending inhibition" during peristalsis, while the adrenergic nerves modulate gastrointestinal function at the ganglionic level (40).

Once nonadrenergic, noncholinergic nerves were discovered, attention was concentrated on discovering what agents were the actual neurotransmitters. To be considered as a neurotransmitter, an agent had to meet each of the following five criteria (38,41):

1. The agent was manufactured and stored in the nerve terminals.
2. The agent was released during depolarization.
3. The agent caused a postsynaptic response.
4. The agent was inactivated by enzymes or there was an uptake system which removed either the agent or its metabolites.
5. There were drugs that produced blocking or potentiating effects on the responses to both the agent and the nerve.

As investigation is still continuing, it is impossible to make a definitive list of all neurotransmitters affecting the alimentary canal. What follows is a compendium of the most important and most interesting candidates.

3.4.1. Noradrenergic nerves

Using Falck's fluorescence histochemical technique, we can see that postganglionic sympathetic noradrenergic neurons divide extensively within both Auerbach's and Meissner's plexuses. These nerves also supply the circular layer in some intestinal regions, especially the sphincter areas (Fig. 10-4A). Noradrenergic nerves supply arterioles (Figs. 10-4B, 10-4C) as well as the mucosa. The function of these nerves at the ganglionic level is an inhibitory one; however, with high levels of stimulation they also affect the smooth muscle.

The ultrastructural morphology is still unclear. There have been reports that SFV and LGV contain noradrenaline, as well as the more usual adrenaline. This confusion may possibly be due to the presence of "false transmitters" (42).

3.4.2. Cholinergic nerves

There are cholinergic excitatory neurons within the wall of the alimentary canal. Profiles containing AGV are very similar to the cholinergic neurons found in motor nerves of striated muscle, sympathetic ganglia, and the adrenal medulla. However, AGV is not definitive for cholinergic neurons as certain 5-hydroxytryptamine-containing and some enkephalin-containing neurons of the central nervous system also contain AGV. These neurons readily stain for acetylcholinesterase (AChE), but this reaction is not cholinergic-specific as AChE acts not only on acetylcholine but on substance P as well. A histochemical method specific for choline acetyltransferase has been developed but this method proved to be disappointing, as the activity of this enzyme in autonomic nerves is too low to permit adequate staining (43).

3.4.3. 5-Hydroxytryptamine (5-HT)-containing nerves

While demonstration of 5-HT in gastrointestinal neurons using the normal histofluorescence methods is difficult, the neurotransmitter is definitely present. The exact role these neurons play is still questionable; however, some investigators feel that they are excitatory and cause some of the atropine-resistant responses seen in gastrointestinal smooth muscle (44). 5-HT fluorescence in vivo occurs after the addition of the precursor amino acid, tryptophan; this neurotransmitter is

←

Figure 10-4. (A). Rat pylorus. Catecholamine histofluorescence. Noradrenergic nerve fibers, the majority of which carry varicosities, run parallel to the circular smooth muscle. X 400. (B). Branch of mouse mesenteric artery. Catecholamine histofluorescence. Noradrenergic nerve fibers are located in the adventitial-medial transitional zone. X 320. (C). Human stomach. Catecholamine histofluorescence. A dense plexus of noradrenergic fibers supply gastric arterioles. X 480. (D). Rat small intestine. Whole mount. Quinacrine histofluorescence. Myenteric nerve cell body. X 680.

inhibited by monoamine oxidase. 5-HT synthesis also occurs following incubation in ^{3}H-L-tryptophan. Likewise, neurons showing 5-HT and tryptophan-like activity have been described in the gut. Finally, some alimentary canal neurons are positive for radioactive labeled 5-HT, allowing their visualization by autoradiography.

3.4.4. Gamma aminobutyric acid (GABA)-containing nerves

Only 3% to 5% of the cells in Auerbach's plexus synthesize and take up GABA; none of the cells of Meissner's plexus do so. GABA is a neurotransmitter in certain invertebrate peripheral nervous systems. It is now known to be a major neurotransmitter in the vertebrate central nervous system as well. The myenteric plexus of the guinea pig, for example, possesses a small population of cells which readily take up GABA. These neurons have long extensions which penetrate deeply into the muscle layer. In vitro studies have shown that incubation of Auerbach's plexus with ^{3}H-glutamic acid forms GABA and homocarnosine, which is a dipeptide of unknown function.

3.4.5. Purinergic nerves

Other nonadrenergic, noncholinergic transmitters in the alimentary canal besides ATP have been described recently (see 45). Nerve stimulation causes strong inhibitory responses which continue to be present even after sympathectomy or adrenolytic drugs. Of further interest is the fact that this inhibition has been seen in the feline anal sphincter which is contracted by catecholamines (46). It has also been reported that intraintestinal inhibitory nerves develop in the fetal rabbit before the adrenergic nerves develop (47).

Electron microscopic study has revealed that these nonadrenergic, noncholinergic endings are in the form of LOV (93). As at least nine different morphologic types of neurons have been indentified in Auerbach's plexus of the guinea pig (36), identification of these endings is understandably difficult. Burnstock and coworkers (48) have identified these intramural neurons as containing adenosine triphosphate or a closely related purine by using the antimalarial drug quinacrine. With this method, 10% to 16% of all neurons located within the mammalian myenteric plexus stain intensely and are thus thought to be intrinsic inhibitory neurons rich in ATP (Fig. 10-4D).

3.4.6. Peptidergic nerves

The presence of many biologically active polypeptides located within the wall of the gastrointestinal tract as well as in extrinsic nerves which supply the gut can be demonstrated by immunohistochemical methods. Hokfelt, who has done much of the work on peptidergic nerves, feels that peptides may not meet all the criteria for neurotransmitters as listed above. These nerves may serve as special modulators or even as trophic factors. It is strange that while the so-called classical transmitters are synthesized in the nerve ending, the peptidergic transmitters originate within the cell body. It is also curious that the peptides lack the reuptake mechanism seen in the presynaptic nerve endings of the classical neurotransmitters (44, 49).

3.4.6.1. Substance P. Substance P is a neurotransmitter in the primary afferent nerves of the skin and spinal cord. Substance P is likewise located in the sensory neurons of the gastrointestinal tract as well. Substance P has several peripheral effects such as blood vessel dilation and intestinal smooth muscle contraction (50, 51). Its gastrointestinal effect is believed to be that of a transmitter of excitatory interneurones within Auerbach's plexus (52).

3.4.6.2. Vasoactive intestinal polypeptide (VIP). Large amounts of VIP have been demonstrated within both Meissner's and Auerbach's plexuses but, as might be expected, amounts vary depending on which portion of the gut is under study (53). For example, the musculature of the gastroesophageal junction, the pylorus, and the sphincter of Oddi of the rat and the guinea pig have a high VIP fiber content. Possibly, VIP causes vasodilation of gut vasculature in addition to controlling sphincter muscles.

3.4.6.3. Bombesin. By using antibodies raised against bombesin, a definite immunoreactivity was demonstrated within nerves of the guinea pig small intestine (54). These immunoreactive nerves were located in both plexuses as well as in the circular muscle and the celiac ganglion. Bombesin-immunoreactive nerve cell bodies in the myenteric ganglia may project from the myenteric plexus to other myenteric ganglia located more aborally. It is of interest to note that the bombesin-reactive neurons do not innervate directly the small intestine mucosa. This suggests that these fibers are not directly involved with mucosal function. There are, on the other hand, reactive nerve fibers in close proximity to submucosal nerve cell bodies which may in turn affect the mucosa. It is highly unlikely that bombesin nerves regulate vascular diameters as no perivascular fibers were found. These neurons always project in an anal direction. This suggests that they act as descending interneurons; since it is still unknown which specific peptide or peptides are contained and released by these nerves, it is unknown whether bombesin-immunoreactive nerves are inhibitory or stimulatory.

3.4.6.4. Somatostatin. Somatostatin is known to occur in nerves and endocrine cells of the alimentary canal in a large number of mammalian species. In the human stomach and small and large intestine, somatostatin was located within the external musculature, submucosa, and the mucosa. Somatostatin within the muscle exists only within neurons, while the neurotransmitter exists within both neuronal and endocrine cells in the submucosa and mucosa (55).

Somatostatin has both neuronal and endocrine functions in the gut. For example, in the stomach somatostatin-positive cells are endocrine in nature, indicating that the peptide probably regulates gastrin release and thereby HCl release. Somatostatin is also contained within endocrine cells in the crypt regions of the small intestine as well as in the large intestine and the rectum. Here, its endocrine function is unknown. In the small and large intestine, however, there are large numbers of somatostatin-containing nerves, and several possible functions have been proposed. One interesting idea is that somatostatin is involved in water and electrolyte transport (56). The presence of somatostatin-containing nerves immediately beneath the epithelium of the crypts and villi in the small intestine, where a large amount of water absorption occurs, and the much smaller number of nerves in the large intestine, where less water is absorbed, supports this hypothesis.

3.4.6.5. Enkephalin. Enkephalins potently inhibit cholinergic nerve-mediated smooth muscle contractions of the guinea pig ileum. They, along with morphine, likewise inhibit neuronal discharge within Auerbach's plexus. Enkephalin-containing cell bodies are located mainly within the myenteric plexus (57), while they are, for the most part, absent in the submucosal plexus. There are enkephalin-positive fibers within the myenteric plexus and the circular muscle layer, but they are rare in the submucosa and mucosa. Enkephalins function by causing hypopolarization of the neurons.

3.4.6.6. Other peptides. Continuing investigation has elicited various peptides which act as possible neurotransmitters. Among those of importance within the gastrointestinal tract are gastrin, neuropeptide Y (NPY), cholecystokinin (58), and angiotensin. As yet, there is no definite evidence that amino acids such as glycine or glutamate function as neurotransmitters in the gut.

3.5. Neurotransmitter receptors within the gastrointestinal tract

The effects of neurotransmitters, putative transmitters, or drugs on the gastrointestinal tract are determined by the binding of these substances (ligands) to specific receptors located in the cell surface membrane. The autoradiographic demonstration of a ligand binding site can show where a physiological phenomenon occurs, and it may help to understand biological functions by showing the sites of action of a variety of substances.

For many years the visualization of receptor binding sites has been limited by the lack of ligands that would not diffuse during histological processing. In 1979, an elegant technical procedure was introduced by Young and Kuhar that allowed the autoradiographic localization of receptor sites by the use of reversible radioligands.

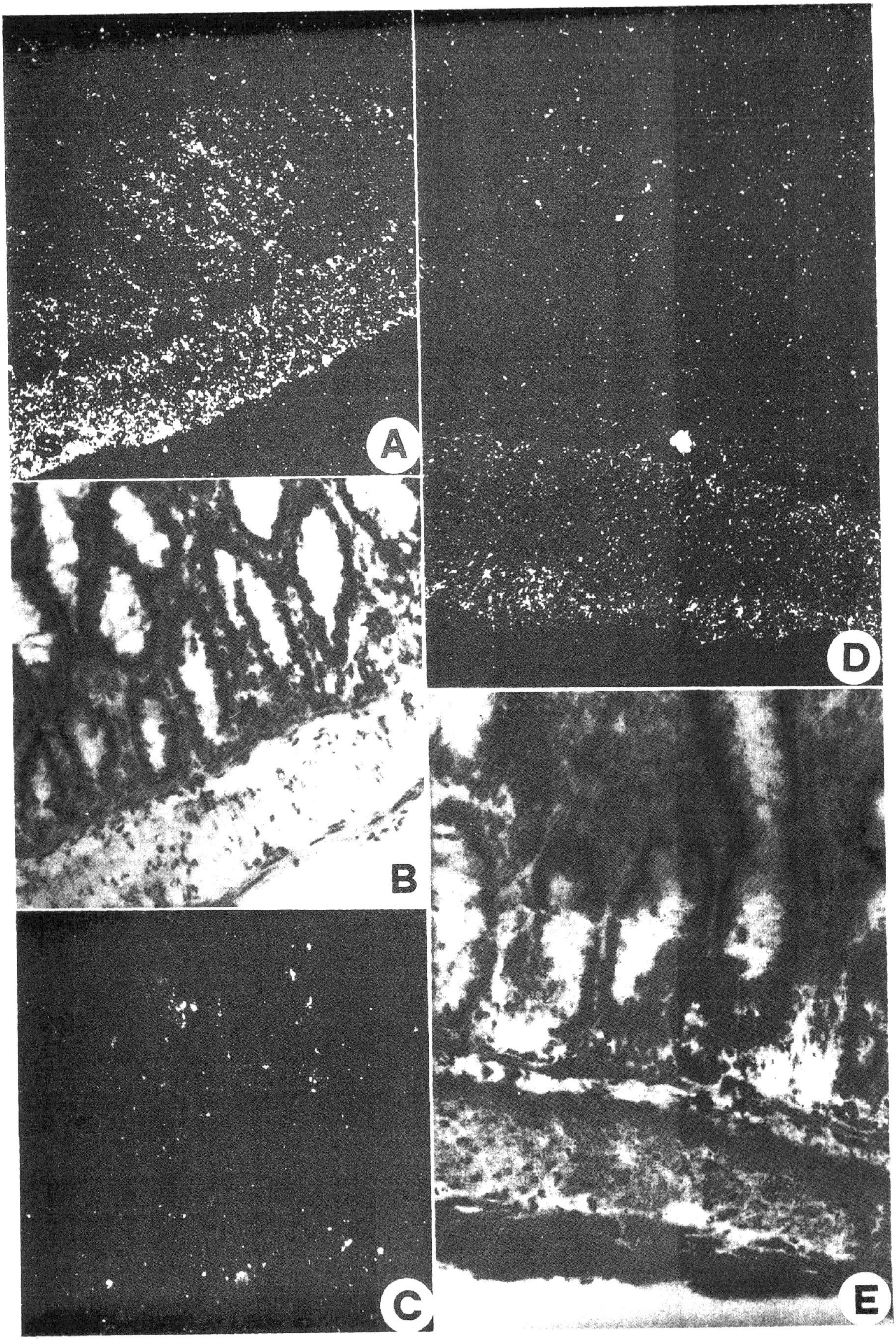
A
B
C
D
E

This technique, now widely used in several laboratories, allows the analysis, in adjacent cryostat sections, of the pharmacological characteristics and of the distribution of a given receptor binding site.

By the use of the above-mentioned method some neurotransmitter receptors have been localized in the gastrointestinal tract. Because these techniques hold promise for solving the many unanswered questions concerning gastrointestinal physiology and pharmacology, there is hope that the evaluation of the distribution and pattern of neurotransmitter receptors will be accomplished in a short time.

Cholinergic muscarinic receptors were visualized in sections of guinea pig intestine mainly using [^{3}H]-quinuclidinyl benzilate (^{3}H-QNB) as a label (59). Muscarinic receptors were found to be distributed over the myenteric and submucous plexuses as well as in the longitudinal and circular muscle layers. In addition, the muscularis mucosa of the colon was labeled by ^{3}H-QNB.

Enteric receptors for 5-hydroxydryptamine (5-HT) were demonstrated in rabbit, guinea pig, and mouse smal intestine using ^{3}H-5HT as a label (for a review see 60). High affinity 5-HT receptors were found primarily over the ganglia of the myenteric plexus as well as in the mucosal-submucosal interface. No ^{3}H-5HT binding sites were observed in the serosa, tunica muscularis, submucosa, or villus epithelium (60).

GABA-receptors were demonstrated in the rat gastrointestinal tract using ^{3}H-muscimol as a label (61). Radioreceptor binding studies carried out to evaluate the characteristics and the kinetics of ^{3}H-muscimol binding to stomach, duodenum, jejunum, and ileum sections established the existence of high affinity $GABA_A$ binding sites in the portions of gut examined. ^{3}H-muscimol was bound chiefly to the tunica muscularis of the alimentary tract portions examined. The muscularis mucosa, on the contrary, did not show significant presence of ^{3}H-muscimol binding sites. Of the enteric plexi the myenteric one appeared to be labelled by ^{3}H-muscimol. The pattern of the radioactive compound distribution was uniform over the entire ganglionic area without clusters in any particular portion of the ganglion (Fig. 10-5).

Alpha-adrenergic receptors were analyzed in a series of preliminary experiments carried out in our laboratory using ^{3}H-dihydroergocryptine as a label and sections of rat stomach, jejenum, and ileum. In each of these three portions of the gut, silver grains were located within smooth muscle as well as in the glandular tissue. The jejunum was the gastrointestinal area richest in ^{3}H-dihydroergocryptine binding sites followed, in descending order, by the stomach and the ileum (Fig. 10-5). No important effect on the distribution and pattern of gastrointestinal ^{3}H-dihydroergocryptine binding sites was found after chemical sympathectomy with the neurotoxin 6-hydroxydopamine, suggesting the postsynaptic nature of the visualized binding sites.

3.6. Gastrointestinal disorders caused by impaired innervation

Esophageal achalasia, intestinal neuronal dysplasia, and Hirschsprung's disease are the most important gastrointestinal disorders characterized by impaired innervation.

Esophageal achalasia is a neuromuscular disorder of the esophagus characterized by several abnormalities including absence of peristalsis in the esophageal body and impaired lower sphincter relaxation in response to swallowing. From a histopathological point of view, the disorder is

←

Figure 10-5. Left. Sections of rat jejunum incubated with ^{3}H-muscimol for the autoradiographic visualization of GABA-A binding sites. (A). Dark field picture showing that silver grains were accumulated primarily in the smooth muscle(s) and in smaller amounts in correspondence to glandular elements. X 200. (B). Bright field picture of Figure A stained with toluidine blue to verify microanatomical details. X 200. (C). Section adjacent to those shown in Figure A incubated with ^{3}H-muscimol in the presence of excess of unlabelled GABA (1 mM) to evaluate nonspecific binding. Dark field picture. X 200.
Right. Section of rat jejunum incubated with ^{3}H-dihydroergocryptine for the autoradiographic visualization of alpha-adrenergic receptors. (D). Dark field picture showing that alpha-adrenergic binding sites were located within smooth muscle coats. X 320. (E). Bright field picture of Figure D stained with toluidine blue to verify microanatomical details. X 320.

characterized by a significant loss of myenteric ganglion cells in the affected part of the esophagus (for a review see 62). Recent immunohistochemical findings showed a specific loss of vasoactive intestinal polypeptide-containing nerve fibers as well as a significant reduction of the polypeptide levels in tissue specimens taken from patients with achalasia (63). In view of these data it has been suggested that the cause, or one of the most important causes, of incomplete relaxation or increased resting tone of the achalatic esophagus may be due to the reduced number of vasoactive intestinal polypeptide fibers (see 63).

Intestinal neuronal dysplasia (termed also colonic neuronal dysplasia) (64) is a condition characterized by enterocolitis or colonic obstruction symptoms. Histopathologically this disorder is characterized by the presence of scattered ganglion cells in the colonic lamina propria as well as by an apparent excess of submucosal ganglion cells. Moreover, an increase of fine acetylcholinesterase-positive nerve fibers in the lamina propria, which is not accompanied by an increase in the muscularis mucosae, has been shown (64). Since an acetylcholinesterase activity may be present in cholinergic, substance-P, and VIP-containing neurons (65), further studies are necessary to establish the putative neurotransmitter(s) involved in the pathogenesis of this disease.

Hirschsprung's disease is a congenital disease characterized by an impressive dilation of the large intestine (colon or sigma). The absence of ganglion cells in the portions of the intestine affected is considered as the primary cause of the illness. It has been demonstrated that distal zones of aganglionic bowel tend to contain more acetylcholinesterase-positive nerve fibers in the inner layer of colonic musculature than normal. Ascending the bowel there was a reduction in number of nerves. The colonic area most proximal to the aganglionic zone contained fewer acetylcholinesterase-positive nerve fibers than normal. The bowel area immediately above the aganglionic portion (normally called the "transitional zone") usually had fewer nerves in its muscle layer than normal (66). Adrenergic nerve fibers, on the contrary, showed no important changes in their distribution and pattern (66).

More recently, it has been demonstrated that VIP-positive nerves are also affected in Hirschsprung's disease. VIP-immunoreactive nerve cell bodies, nerve fibers, and nerve endings were distributed throughout the ganglionic and oligoganglionic segments. On the contrary, aganglionic bowel portions contained no VIP-immunoreactive nerve cell bodies and a very limited number of nerve endings and nerve fibers (67).

3.7. Conclusions

In conclusion, several additional comments can be made. It is interesting to note the progression in which our understanding of the alimentary canal was made. Initial studies were done using light microscopy; knowledge was increased when the electron microscope became available. However, we really began to understand the musculature and innervation of the gut only when advanced techniques such as immunohistofluorescence and autoradiography were widely applied. This is not surprising; at the level of the electron photomicrograph it is possible to distinguish vesicles, but only sophisticated techniques will tell what neurotransmitter is contained within the vesicle. The use of these new analytical techniques has allowed us to discover and investigate vast numbers of neurotransmitters that were quite literally undreamt of ten years ago.

Having discovered new neurotransmitters we can now begin to consider their role in sickness and in health. What happens when a certain transmitter no longer functions because it is no longer manufactured in the aged or because an autoimmune disorder destroys a synapse? In certain cases we know the answer, in others we are just beginning to experiment. Thus, we who are doing research on the musculature and innervation of the alimentary canal can be enthusiastic about the near future. We now have the research tools to enable us to deepen our understanding of an interesting system of the body, and we can be assured that the results gathered will have definite clinical applications.

Acknowledgments

The present study was supported by grants from the Ministry of Education and from the University "La Sapienza," Rome (Progetti di Ateneo e di Facoltá).

The authors are greatly indebted to Drs. W.L. Collier, M. De Rossi and F. Ferrante for help in the preparation of this chapter. The excellent secretarial work of Mrs. C. Collier is also gratefully acknowledged.

References

1. Bloom W, Fawcett D: A Textbook of Histology, Philadelphia: WB Saunders Company, p 598–599, 1975.
2. Di Fiore M: Atlas of Human Histology. Philadelphia: Lea & Febiger, p 122–152, 1977.
3. Gabella G: Structure of muscles and nerves in the gastrointestinal tract. In: Physiology of the Gastrointestinal Tract. LR Johnson (ed), New York: Raven Press, p 197–241, 1981.
4. Guyton A: Textbook of Medical Physiology, Philadelphia: WB Saunders Company, p 852–853, 1976.
5. Gabella G: Smooth muscle cell junctions and structural aspects of contraction. *Br Med Bull* 35: 213–218, 1979.
6. Taxi J: Contribution à L'étude des connexions des neurones moteurs du systéme nérveux autonome. *Ann Sci Nat Zool*, 7: 413–674, 1965.
7. Gabella G: Innervation of the gastrointestinal tract. *Int Rev Cytol* 59: 129–193, 1979.
8. Yamamoto M: Electron microscopic studies on the innervation of the smooth muscle and the interstitial cell of Cajal in the small intestine of the mouse and bat. *Arch Hist Jpn* 40: 171–201, 1977.
9. Taylor A, Kreulen D, Prosser C: Electron microscopy of the connective tissue between longitudinal and circular muscle of the small intestine of cat. *Am J Anat* 150: 427–442, 1977.
10. Goodford P: Ion movements in smooth muscle. In: Membranes and Ion Transport, E Bittar (ed), New York: John Wiley & Sons, p 33–74, 1970.
11. Muggli R, Baumgartner H: Patterns of membrane invaginations at the surface of smooth muscle cells of rabbit arteries. *Experientia* 28: 1212–1214, 1972.
12. Orci L, Perrelet A: Membrane-associated particles: Increase at sites of pinocytosis demonstrated by freeze-etching. *Science* 181: 868–869, 1973.
13. Prosser C, Burnstock G, Kahn J: Conduction in smooth muscle: Comparative structural properties. *Am J Physiol* 199: 545–552, 1960.
14. Cooke P: A filamentous cytoskeleton in vertebrate smooth muscle fibers. *J Cell Biol* 68; 539–556, 1976.
15. Debbas G, Hoffman L, Landon E, Hurwitz L: Electron microscopic localization of calcium in vascular smooth muscle. *Anat Rec* 182: 447–472, 1975.
16. Heumann, H-G: The subcellular localization of calcium in vertebrate smooth muscle: Calcium-containing and calcium-accumulating structures in muscle cells of mouse intestine. *Cell Tissue Res* 169: 221–231, 1976.
17. Jonas L, Zelch U: The subcellular calcium distribution in the smooth muscle cells of the pig coronary artery. *Exp Cell Res* 89: 352–358, 1974.
18. Popescu L, Diculescu I: Calcium in smooth muscle sarcoplasmic reticulum in situ. *J Cell Biol* 67: 911–918, 1975.
19. Ashton F, Somlyo AV, Somlyo AP: The contractile apparatus of vascular smooth muscle: Intermediate high voltage stereo electron microscopy. *J Mol Biol* 98: 17–29, 1975.
20. Somlyo AV: Ultrastructure of vascular smooth muscle. In: Handbook of Physiology, Section 2, Vol II. Washington, DC: American Physiology Society, p 33–67, 1980.
21. Nonomura Y, Ebashi S: Calcium regulatory mechanism in vertebrate smooth muscle. *Biomed Res* 1: 1–14, 1980.
22. Small J: Studies on isolated smooth muscle cells: The contractile apparatus. *J Cell Sci* 24: 327–349, 1977.
23. Groschel-Stewart U: Comparative studies of human smooth and striated muscle myosins. *Biochim Biophys Acta* 229: 322–334, 1971.
24. Pollard T, Weihing R: Actin and myosin and cell movement. *CRC Crit Rev Biochem* 2: 1–65, 1974.
25. Somlyo AP, Somlyo AV: Ultrastructure of smooth muscle. In: Methods in Pharmacology. E Daniel, D Paton, (eds), New York: Plenum, p 3–43, 1975.
26. Rosenbluth J: Obliquely striated muscle. III. Contraction mechanism of *Ascaris* body muscle. *J Cell Biol* 34: 15–33, 1967.
27. Urwin P, Zampighi G: Structure of the junction between communicating cells. *Nature* (Lond) 283: 545–549, 1980.
28. Peracchia C: Structural correlates of gap junction permeation. *Int Rev Cytol* 66: 81–146, 1980.
29. Agostoni E, Chinnock J, Daly M, Murray J: Functional and histological studies of the vagus nerve and its branches to the heart, lungs and abdominal viscera in the cat. *J Physiol* (Lond) 135: 182–205, 1975.
30. Ahlman B, Lundberg J, Dahlstrom A, Larsson I, Pettersson G, Kewenter J, Nyhus L: Evidence for innervation of the small intestine from cervical sympathetic ganglia. *J Surg Res* 24: 142–149, 1978.
31. Gabella G: Neuron size and number in the myenteric plexus of the newborn and adult rat. *J Anat* 109: 81–95, 1971.
32. Gabella G: Structure of the Autonomic Nervous System. London: Chapman & Hall, 1976.
33. Gerhon M, Bursztajn S: Properties of the enteric nervous system: Limitation of access of intravascular macromolecules to the myenteric plexus and muscularis externa. *J Comp Neurol* 180: 467–487, 1978.
34. Geffen L, Livett B: Synaptic vesicles in sympathetic neurons. *Physiol Rev* 51: 98–157, 1971.
35. Baumgarten H, Holstein A-F, Owman C: Auerbach's plexus of mammals and man: Electron microscopic identification of three different types of neuronal processes in myenteric ganglia of the large intestine from rhesus monkeys, guinea pigs and man. *Z Zellforsch* 106: 376–397, 1970.

36. Cook R, Burnstock G: The ultrastructure of Auerbach's plexus in the guinea pig. I. Neuronal elements. *J Neurocytol* 5: 171–194, 1976.
37. Campbell G, Gibbins I: Nonadrenergic, noncholinergic transmission in the autonomic nervous system: Purinergic nerves. In: Trends in Autonomic Pharmacology, Vol. I. S Kalsner (ed), Baltimore and Munich: Urban & Schwarzenberg, p 103–144, 1979.
38. Burnstock G: Cytochemical studies in the enteric nervous system. In: Cytochemical Methods in Neuroanatomy. New York: Alan R Liss, Inc p 129–149, 1982.
39. Burnstock G: Evolution of the autonomic innervation of visceral and cardiovascular systems in vertebrates. *Pharmacol Rev* 21: 247–324, 1969.
40. Burnstock G, Costa M: Inhibitory innervation of the gut. *Gastroenterology* 64: 141–144, 1973.
41. Burnstock G: Purinergic nerves. *Pharmacol Rev* 24: 509–581, 1972.
42. Llewellyn-Smith I, Wilson A, Furness J, Costa M, Rush R: Ultrastructural identification of nonadrenergic axons in the enteric plexuses of the guinea pig ileum. *Neurosci Abstr* 6: 274, 1980.
43. Chubb I, Hodgson A, White G: Acetylcholinesterase hydrolyzes substance-P. *Neuroscience* 5: 2065–2072, 1980.
44. Furness J, Costa M: Types of nerves in the enteric nervous system. *Neuroscience* 5: 1–20, 1980.
45. Bennett M, Burnstock G, Holman M: Transmission from intramural inhibitory nerves in the smooth muscle of the guinea pig taenia coli. *J Physiol* (Lond) 182: 541–558, 1966.
46. Garrett J, Howard E: Neural control of the internal anal sphincter of cats after chemical sympathectomy with 6-hydroxydopamine. *J Physiol* (Lond) 247: 25–27P, 1975.
47. Gershon M, Thompson E: The maturation of neuromuscular function in a multiply innervated structure: Development of the longitudinal smooth muscle of the foetal mammalian gut and its cholinergic excitatory, adrenergic inhibitory innervation. *J Physiol* (Lond) 234: 257–277, 1973.
48. Burnstock G, Campbell G, Satchell D, Smythe A: Evidence that adenosine triphosphate or a related nucleotide is the transmitter substance released by nonadrenergic inhibitory nerves in the gut. *Br J Pharmacol* 40: 668–688, 1970.
49. Sundler F, Håkanson R, Leander S: Peptidergic nervous systems in the gut. *Clin Gastroenterol* 9; 517–543, 1980.
50. Franco R, Costa M, Furness J: Evidence for the release of endogenous substance-P from intestinal nerves. *Naunyn-Schmiedeberg's Arch Pharmacol* 306: 195–201, 1979.
51. Franco R, Costa M, Furness J: Evidence that axons containing substance-P in the guinea pig ileum are of intrinsic origin. *Naunyn-Schmiedeberg's Arch Pharmacol* 307: 58–63, 1979.
52. Katayama Y, North R, Williams J: The action of substance-P on the neurones of the myenteric plexus of the guinea pig intestine *Proc R Soc B* 206: 191–208, 1979.
53. Larsson L-I: Ultrastructural localization of a new neuronal peptide (VIP). *Histochemistry* 5: 173–176, 1977.
54. Costa M, Furness J, Yanaihara N, Yanaihara C, Moody T: Distribution and projections of neurons with immunoreactivity for both gastrin-releasing peptide and bombesin in the guinea pig small intestine. *Cell Tissue Res* 235: 285–293, 1984.
55. Keast J, Furness J, Costa M: Somatostatin in human enteric nerves. *Cell Tissue Res* 237: 299–308, 1984.
56. Mitchenere P, Adrian T, Kirk R, Bloom S: Effect of gut regulatory peptides on intestinal luminal fluid in the rat. *Life Sci* 29: 1563–1570, 1981.
57. Alumets J, Hakanson R, Sundler F, Chang K-J: Leu-enkephalin-like material in nerves and enterochromaffin cells in the gut. *Histochemistry* 56: 187–196, 1978.
58. Larsson L-I, Rehfeld J: Localization and molecular heterogeneity of cholecystokinin in the central and peripheral nervous system. *Brain Res* 165: 201–218, 1979.
59. Buckley N, Burnstock G: Autoradiographic localization of muscarinic receptors in guinea pig intestine: Distribution of high and low affinity agonist binding sites. *Brain Res* 294: 15–22, 1984.
60. Gershon M, Takaki M, Tamir H, Branchek T: The enteric neural receptor for 5-hydroxytryptamine. *Experientia* 41: 863–868, 1985.
61. Amenta F: Autoradiographic localization of GABA receptor sites in peripheral tissues. In: GABAergic Mechanisms in Mammalian Periphery. N Bowery, S Erdo (eds), New York: Raven Press, p 99–115, 1986.
62. Fisher R, Cohen S: Disorders of the lower esophageal sphincter. *Ann Rev Med* 26: 373–390, 1975.
63. Aggestrup S, Uddman R, Sundler F, Fahrenkrug J, Hakanson R, Sorensen H, Hambraeus G: Lack of vasoactive intestinal polypeptide nerves in esophageal achalasia. *Gastroenterology* 84: 924–927, 1983.
64. Sharli A, Meir-Ruge W: Localized and disseminated forms of neuronal intestinal dysplasia mimicking Hirschsprung's disease. *J Pediatric Surg* 16: 164–169, 1983.
65. Burnstock G, Griffith S: Neurohumoral control of the vasculature. In: Biology and Pathology of the Vessel Wall. N Woolf (ed), New York: Praeger, 1983.
66. Garrett J, Howard E, Nixon H: Autonomic nerves in rectum and colon in Hirschsprung's disease: A cholinesterase and catecholamine histochemical study. *Arch Dis Childh* 44: 406–417, 1969.
67. Tsuto T, Okamura H, Fukui K, Obata H, Terubayshi H, Iwai N, Majima S, Yanaihara N, Ibata Y: An immunohistochemical investigation of vasoactive intestinal polypeptide in the colon of patients with Hirschsprung's disease. *Neuroscience Letters* 34: 57–62, 1982.

Author's address:
Francesco Amenta
Dipartimento di Scienze Neurologiche
Via A. Borelli, 50, L
00161, Roma, Italy

CHAPTER 11

The lymphoid system and immunologic defense of the digestive tract

JACQUES PAPPO & ROBERT L. OWEN

1. Lymphoid system of the digestive tract

Lymphocytes and macrophages are distributed in the intestinal tract as isolated cell populations in the intestinal epithelium and lamina propria and are present in focal aggregates forming lymphoid follicles. Ultrastructural investigation has played a major role in understanding the cellular interactions and cytoarchitectural details critical to function of the mucosal immune system (1–3). The lymphoid tissue of the intestinal tract, which includes that tissue present in the oral cavity, pharynx, Peyer's patches, appendix, and isolated follicles in the small and large intestines, represents a substantial fraction of the lymphoid elements which make up the immune system. Collectively, these immunologic structures, termed gut–associated lymphoid tissues (GALT), present a large surface area to a variety of potential antigens and respond in a highly coordinated fashion during the recognition of foreign molecules and pathogens. The distribution of lymphoid cells reflects varying antigen load in different regions of the intestine. Lymphoid tissue is prominent in the oropharynx, minimal in the esophagus and stomach, where rapid transit time and acid secretions reduce the antigenic burden, and becomes abundant again in the distal small intestine and the large intestine, where stasis promotes bacterial proliferation.

1.1. Salivary glands

Previously salivary glands were not known to have a capacity for immunologic response but were viewed as target organs accumulating antibody producing cells arising in the intestine after direct antigenic stimulation. The functional capacity of these organs for initiating antibody responses has now been shown by studies in which specific IgA antibodies (4) and antibody-producing cells (5) are generated after local antigen challenge of salivary glands. Structural observations have demonstrated the presence of organized lymphoid follicles in close association with the glandular duct system (6). Oral antigens may gain access into the salivary glands by a retrograde route (7), be translocated by duct cell endocytosis (8), and be deposited in the gland interstitium where they activate cells of the immune system via antigen presenting cells (9). The existence of an immunologic defense mechanism in the oral cavity has particular advantages for dealing locally with continuous challenge by soluble and particulate oral bacterial antigens. In vivo, this local defense pathway may overlap substantially with antibody-producing cells which home from the intestine after being stimulated at that site.

1.2. Tonsils

The aggregates of submucosal lymphoid tissue circling the entrance to the digestive and respiratory tracts include the lingual, palatine, and pharyngeal tonsils which form Waldeyer's lymphoid ring (10). Due to their strategic anatomical location, these lymphoid tissues encounter dietary, environmental, and microbial antigens first. The epithelium lining lymphoid follicles

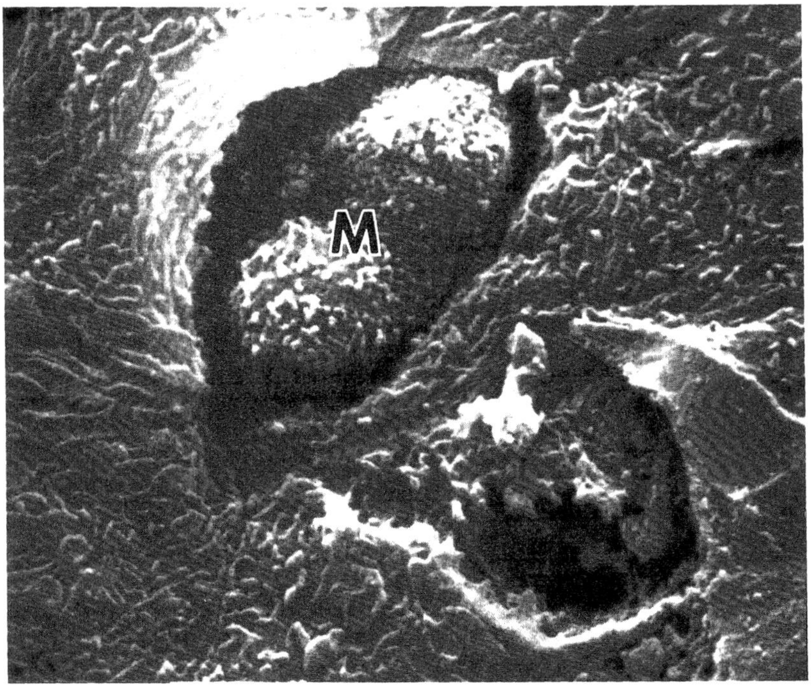

Figure 11-1. Human tonsil crypt. An open passage into a follicle and an adjacent passage occluded by an M cell (M) among stratified squamous epithelial cells covering a lymphoid follicle. SEM X 8,750. (Reproduced from ref 2 by permission.)

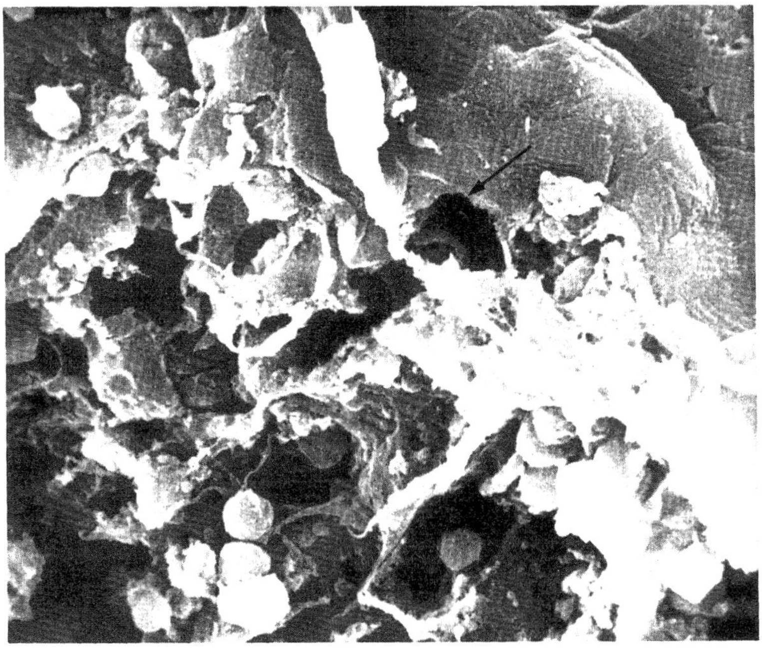

Figure 11-2. Human tonsil. Fracture through a lymphoid follicle dome with an open pore (arrow) leading into passageways occupied by lymphoid cells. SEM X 1,600. (Reproduced from ref 11 by permission.)

samples antigens through specialized antigen-uptake M cells (Fig. 11-1) and allows passage of antigens through surface pores (Fig. 11-2) (11). Tonsillar tissues appear to respond to immunologic stimuli (12) and contain populations of immunocompetent T cells surrounding B cell-dependent lymphoid follicles. Tonsils are capable of generating antibody producing cells (13, 14) but their full contribution to intestinal defense has not been approached experimentally.

1.3. Esophagus and stomach

The esophagus and stomach have not been systematically studied as regions where immunologic events might occur. The esophageal surface is populated by epithelial Langerhan's cells, which process and present antigen, and by scattered T cells and B cells in the lamina propria (15). In the stomach, both isolated lymphoid cells and lymphoid follicles can be found in the lamina propria and muscularis mucosae in the antrum (16). These lymphoid cell populations comprise part of the GALT, but their role in local immune responses and connections with other intestinal immunocompetent cells are not known.

1.4. Peyer's patches

Peyer's patches are collections of lymphoid follicles (Fig. 11-3) distributed within the lamina propria and submucosa of the small intestine (17). Their number varies from 10 to 20 in rodents (18) to several hundred in humans (19). Peyer's patches contain as few as two to three follicles or may be constituted by nearly 1,000 nodules. These immunologic structures are formed early in development by focal accumulations of reticular/dendritic cells which are later infiltrated by T cells (20). In rodents, germinal center formation occurs several weeks after birth and is dependent on antigenic stimulation (21), but appreciable numbers of germinal centers are found before birth in sheep and humans.

Peyer's patch lymphoid follicles harbor an enriched population of IgA B cells (Fig. 11-4) (22, 23). Activation of lymphoid follicle cells by antigen results in the generation of IgA producing

Figure 11-3. Human Peyer's patch. A rounded follicle surrounded by finger-shaped villi in the terminal ileum. Goblet cells which appear as white dots on the villi are absent from the corrugated follicle surface. SEM X 63. (Reproduced from ref 41 by permission.)

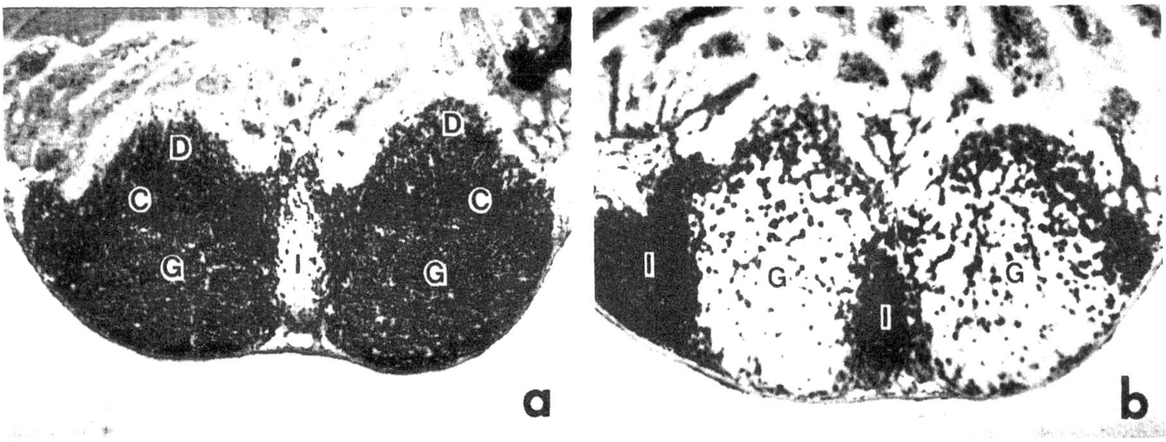

Figure 11-4. Mouse Peyer's patches. Cross section through two follicles: a) Labeling with anti-B220 shows B cells in the germinal centers (G), corona (C), dome (D), and scattered in the interfollicular area (I). LM X 50. b) Labelling with anti-Thy-1.2 shows T cells in the interfollicular area (I) and scattered in the coronas and germinal centers. LM X 60. (Reproduced from ref 22 by permission.)

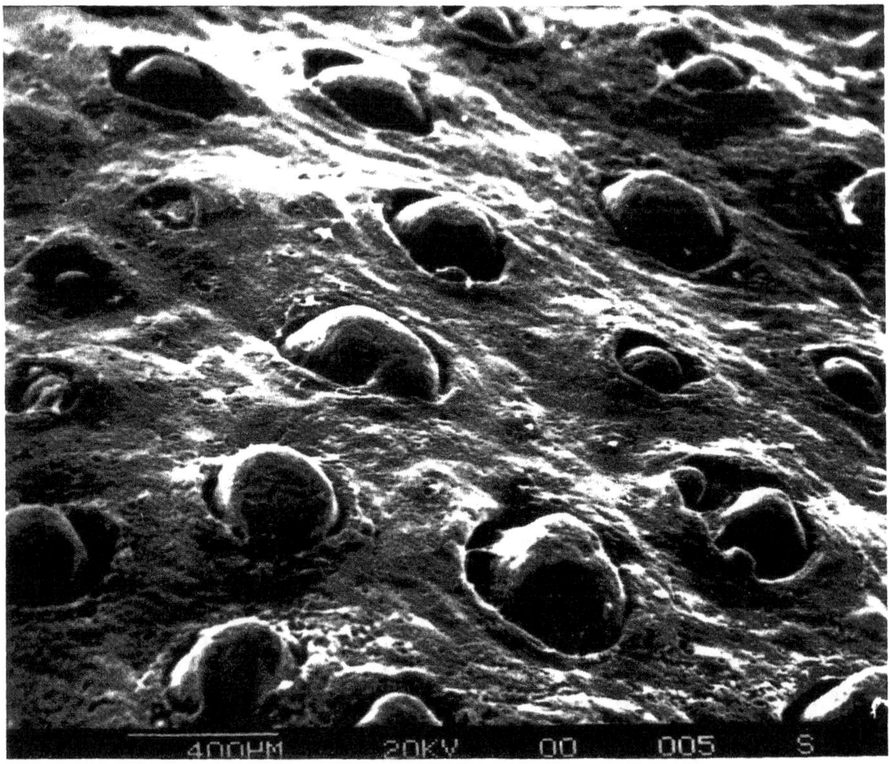

Figure 11-5. Rabbit appendix. Surrounded by crevices, distinct particle-trapping follicle domes protrude from the colonic mucosa. SEM X 50.

cells which migrate out of the patch and populate distant mucosal sites (24). This antigen activated, IgA committed population synthesizes IgA antibody into secretions which bathe mucosal surfaces and is central to defense processes in mucosae.

1.5. *Appendix*

The appendix is a diverticulum found in the ceca of humans, rabbits, and some other mammals. The wall of the appendix is composed predominantly of lymphoid follicles and intervening colonic mucosa. In the appendix, as in Peyer's patches, large numbers of B cells are arranged in lymphoid follicles surrounded by interfollicular T cell regions (25, 26). In rabbits, individual follicles are separated from the surrounding mucosa by an encircling crevice (Fig. 11-5). In humans, individual follicles are less clearly demarcated (Fig. 11-6). The appendix performs functions of antigen uptake, presentation, and lymphocyte proliferation. The epithelium over the follicle dome contains M cells (Fig. 11-7) beneath which intraepithelial lymphoid cells in discrete pockets (27) are in continuous transit into the follicle, to mesenteric nodes, and into the lumen (28). The follicle associated epithelium (FAE) transports tracer markers (29). Luminal bacteria (30) which enter appendix follicle domes are phagocytosed by macrophages.

1.6. *Colon and rectum*

In mice, rats, and other mammals which lack an appendix, organized patches of lymphoid tissue are found in the cecum (Figs. 11-8, 11-9). In humans and in other animals, organized lymphoid tissue in the form of intramucosal and submucosal follicles occur at intervals throughout the colon and rectum (31, 32). Colonic follicles, sometimes called lymphoglandular complexes, are spherical aggregates of lymphocytes, which penetrate the muscularis mucosae and are surrounded by lymphatics (33). Most lymphoglandular complexes are penetrated by branched and splayed tubular or pitlike invaginations, but some, particularly in the rectum, bulge into the lumen between crypts,

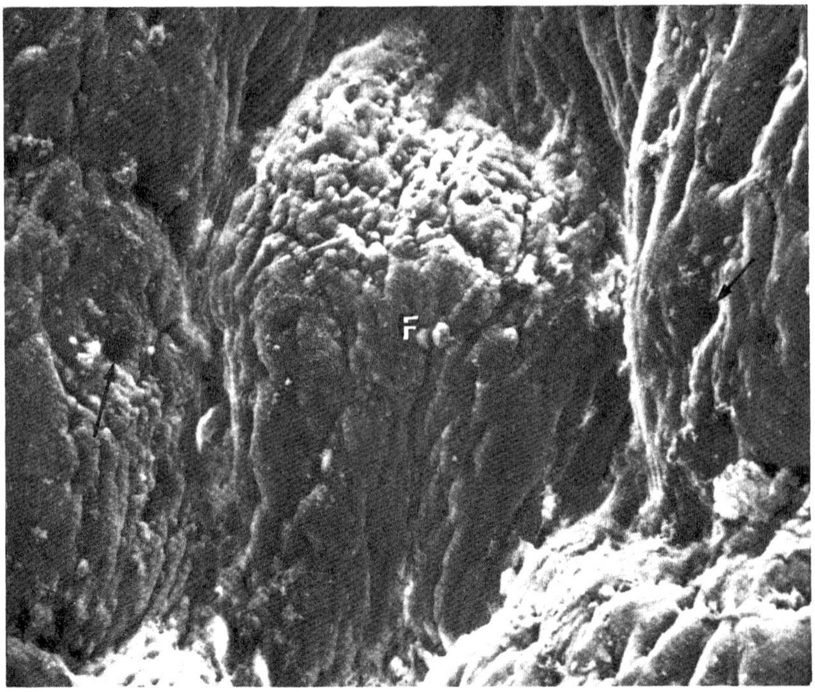

Figure 11-6. Human appendix. A lymphoid follicle (F) without a discrete encircling crevice bulges from the colonic mucosa pierced by crypt mouths (arrows). SEM X 300. (Reproduced from ref 2 by permission.)

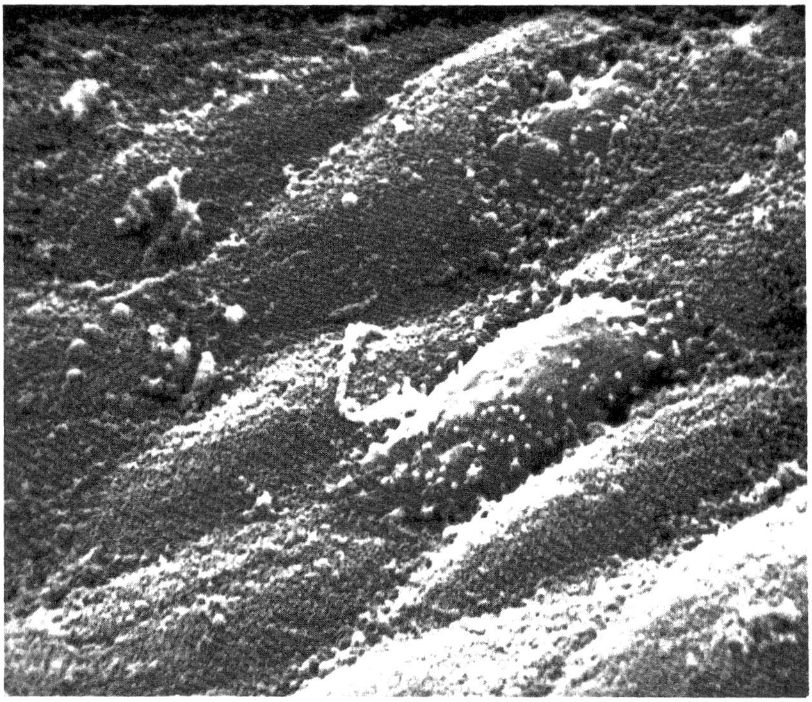

Figure 11-7. Human appendix. M cells with short dispersed microvilli lie among enterocytes with closely packed microvilli. SEM X 5,900. (Reproduced from ref 2 by permission.)

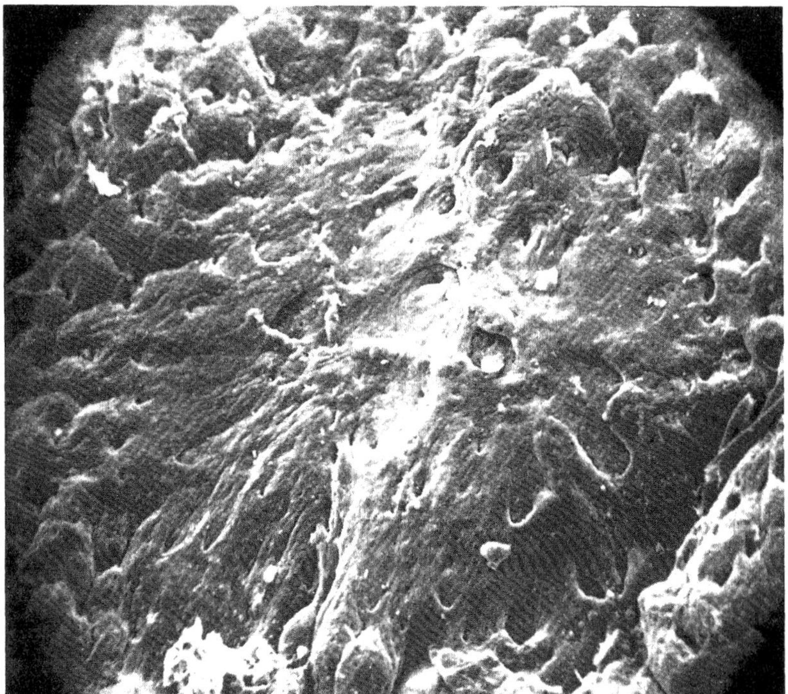

Figure 11-8. Mouse cecal patch. Lymphoid follicles in a position analagous to that of the human appendix. Crypt mouths in the surrounding colonic mucosa are pulled toward the edges of the patch. SEM X 118. (Reproduced from ref 2 by permission.)

are covered by attenuated FAE, and resemble Peyer's patch domes. Colonic FAE cells exhibit structural characteristics of M cells (32, 33) and their surfaces adsorb tracer markers (32), suggesting a role for antigen sampling. In the normal human colon, lymphoglandular complexes increase in number threefold from the cecum to the rectum. In neoplastic and inflammatory diseases, the distribution of these colonic follicles increases sharply (33).

2. Role of mucosal lymphoid tissue in intestinal defense

2.1. Antigen trapping at mucosal surfaces

The intestinal epithelium excludes antigenic substances over most of its surface, samples luminal contents in localized areas, and functions as an organ for intestinal immunologic responses. The epithelial surfaces over villi and in GALT have distinct structural characteristics which facilitate their diverse roles. Villus epithelial cells function as a barrier which restricts the passage of macromolecules from the intestinal environment, while GALT epithelium samples and translocates antigens to the mucosal immune system. Passage of tracer macromolecules through villus epithelium has been demonstrated following intestinal infusion of tracer molecules (34) and their subsequent identification in lymph and portal blood. Tracer markers, taken up by villus epithelium in pinocytic vesicles, are shuttled to the lysosomal compartment for degradation and sequestration or are discharged through the basolateral membranes into the extracellular space. Because trauma can induce disruption of epithelial cell integrity (35), the physiologic significance of experimental demonstrations of antigen transport through villus epithelium remains inconclusive. Transport studies using cholera toxin (36), which is handled at the cell surface by receptor-mediated endocytosis, provide more convincing evidence for molecular absorption through mature intestinal epithelia. Antigen sequestration after absorption in villus epithelium contrasts sharply

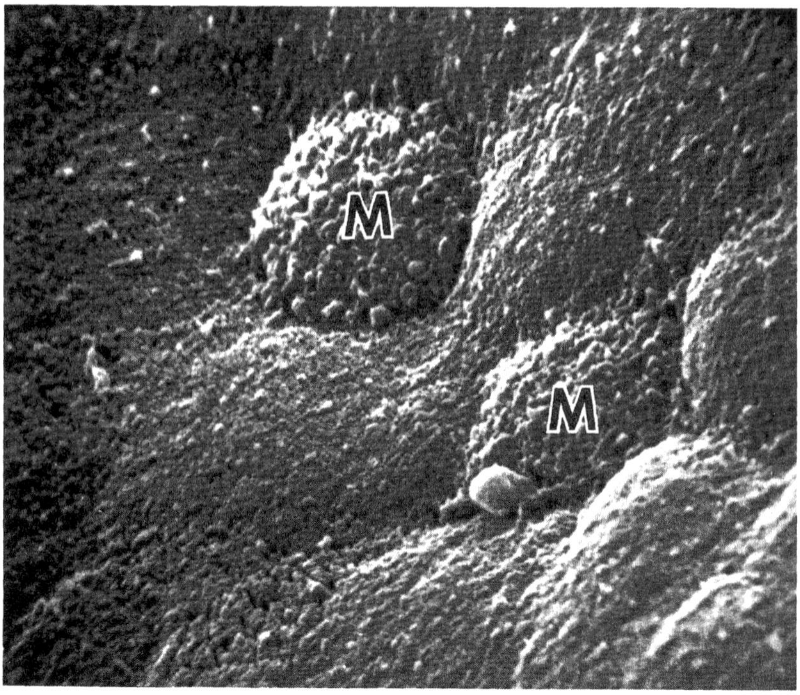

Figure 11-9. Mouse cecal patch. Irregular microvilli over M cells (M) contrast with uniform, closely packed microvilli on surrounding enterocytes. SEM X 7,400. (Reproduced from ref 2 by permission.)

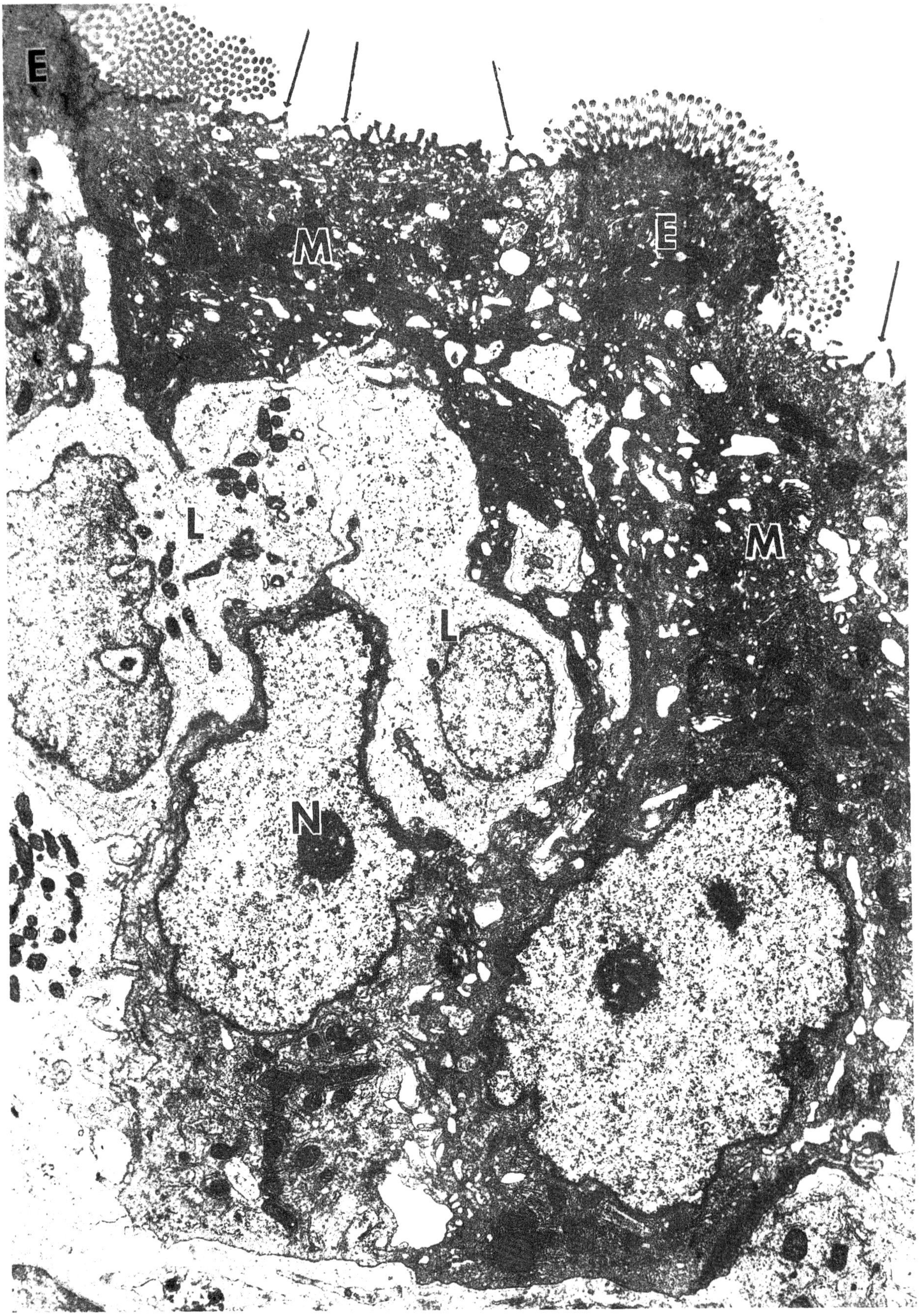
E
M
E
L
M
L
N

with the antigen handling properties of GALT epithelium. This epithelium is structurally adapted to bind, take up, and transport antigens in close proximity to cells of the immune system and will be described in Section 2.2.1.

2.2. Functions of cellular constituents in antigen processing

The GALT provides an initial site for interaction between immunoreactive cells and intestinal luminal antigens. These interactions occur in intestinal lymphoid tissues and depend on discrete cell communication events. The amount of antigen which eventually traverses the basal lamina is restricted in the absence of disease. Due to the amplifying nature of the mucosal immune apparatus, a quantitatively small fraction of immunogenic molecules can nonetheless stimulate antigen presenting cells which are responsible for activating immunoregulatory T cells and antibody producing B cells (37).

2.2.1. M cells

The (FAE) constitutes the interface between complex and heterogeneous antigens in the lumen and immunologic cells in the intestinal wall (38). The FAE is replenished by proliferation of stem cells in crypts encircling each follicle (39, 40) and is populated by cuboidal or low columnar absorptive cells and specialized M cells. Goblet cells are largely excluded from the flow of crypt cells onto the follicle surface, reducing local mucus production and enhancing the opportunities for antigen contact with FAE cells. Structurally, M cells form a thin cytoplasmic membrane-like separation between the intestinal lumen and underlying cells of the immune system (Fig. 11-10). M cells lack closely packed microvilli, well developed surface glycocalyx, and apical terminal webs found in enterocytes. Instead, they have irregular surfaces with folds (Fig. 11-11), surface indentations, and multiplecoated apical vesicles which allow efficient contact with luminal antigens and transport into the epithelium where lymphoid cells invaginate M cell basolateral membranes (Fig. 11-12) (41).

The close association of lymphocytes with FAE basolateral surfaces may trigger the maturation of M cells (42). Uptake and transport occur at much faster rates in intestinal epithelium containing Peyer's patches (43), yet antigen catabolism appears to be substantially lower at these anatomical regions. The reduced lysosomal content in M cells (44) may help transport essentially undegraded molecules to antigen-presenting cells. M cells take up macromolecules by pinocytosis from pits in the cell surface, then transport these in vesicles for subsequent release into the extracellular space (45,46). The mechanism of uptake by M cells resembles uptake in neonatal intestine. Whether M cells represent focal retention of immature epithelial uptake or have evolved as specific defense elements in the gut is not established. The structural features of M cells enable them to continuously sample luminal antigens and function as a specific antigen delivery pathway for intestinal immune responses.

More recently, M cells have been found to be specialized sites for the attachment and phagocytosis of viruses (Fig. 11-13) (47), bacteria (Fig. 11-14) (48), and protozoa (49), which are transported in vesicles and discharged to subjacent lymphocytes and macrophages. The binding and internalization by M cells of type I reovirus has also been demonstrated to represent a route for systemic dissemination of virus particles which are not trapped by Peyer's patch cells.

M cells exhibit a deficiency in alkaline phosphatase activity (Fig. 11-15) (50), which may reflect cell differentiation or may be related to the ability of M cells to interact with antigenic macromolecules and microorganisms, but M cells also display surface glycoconjugates similar to surrounding absorptive cells. The surface recognition molecules expressed on M cells as well as the factors which regulate antigen uptake are currently unknown. Although it has become increasingly clear that M cells are focal points of

←

Figure 11-10. Human Peyer's patch. Endocytosis by stubby microvilli (arrows) forms prominent vesicles in M cells (M), which are shorter than adjacent enterocytes (E). Two lymphoid cells (L) invaginate the M cell on the left between its nucleus (N) and the lumen. TEM X 8,400.

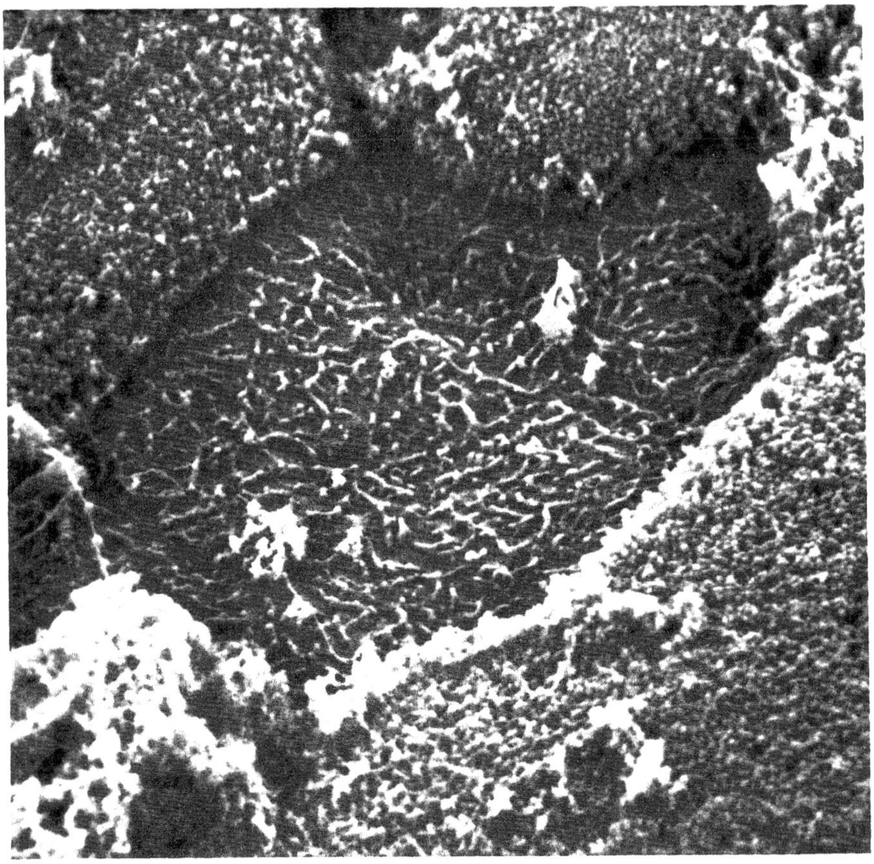

Figure 11-11. Human Peyer's patch. Microfolds converge and diverge over the surface on an M cell surrounded by enterocytes. SEM X 6,500. (Reproduced from ref 41 by permission.)

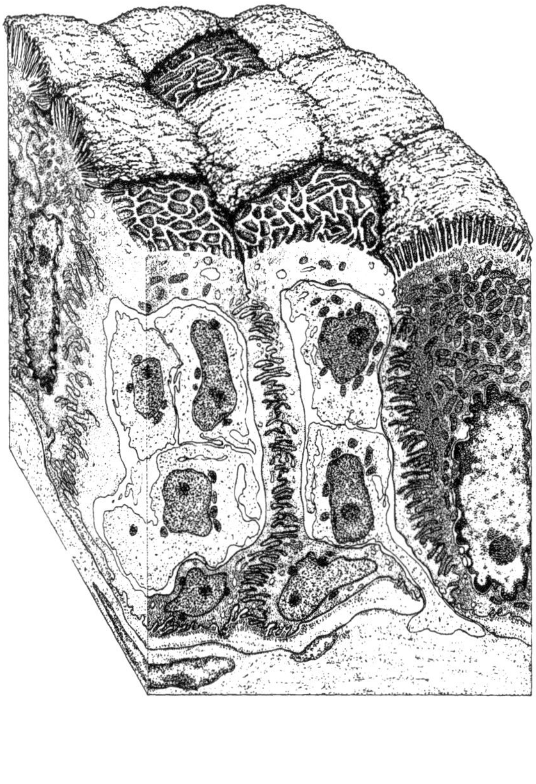

contact between antigens and immune cell surfaces, the nature of interactions and the extent to which M cells might be required to trigger immune responses remain largely speculative at present. Intestinal immunologic responses occur even after extirpation of all Peyer's patches, however M cells are found in the epithelium over isolated lymphoid nodules as well as in Peyer's patches (51).

2.2.2. *Macrophages*

Cells of the mononuclear phagocyte system are widely distributed in the intestine, form a network for containment of antigen, and perform complex

Figure 11-12. Diagram of M cell showing specialized surfaces, lateral interdigitation with enterocytes, invagination by migrating lymphoid cells and orientation along the basal lamina. (Reproduced from ref 2 by permission.)

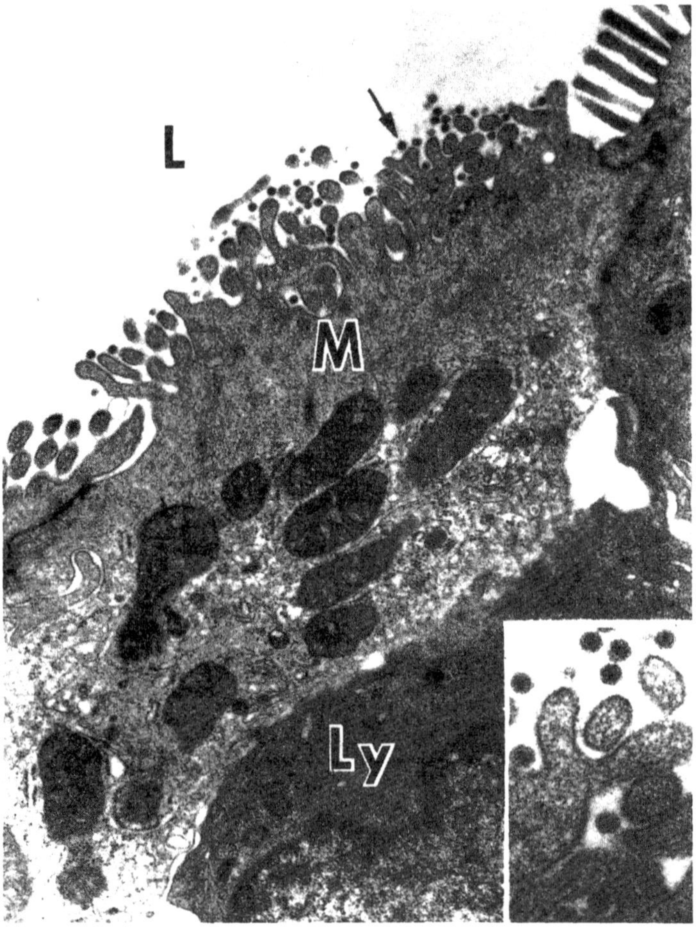

Figure 11-13. Mouse Peyer's patch. Reovirus type 1 particles (arrow) in the lumen (L) adhere to the surface of the M cell but not to microvilli of adjacent enterocytes. A lymphocyte (Ly) lies beneath the M cell. TEM X 5,250. Inset: Detail of viral attachment to microvilli. (Reproduced from ref 47 by permission.)

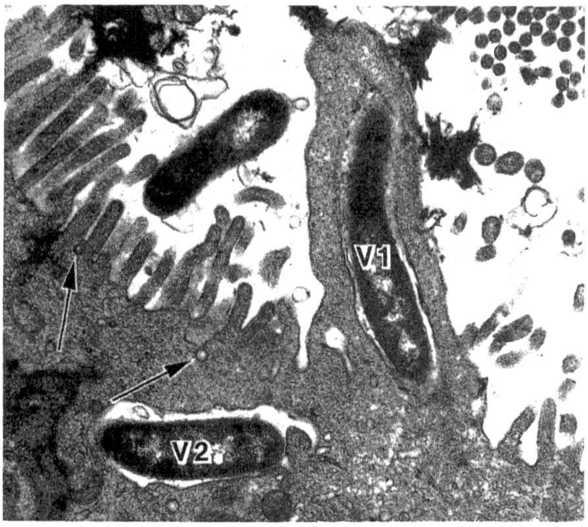

functions of catabolism and antigen presentation to T cells (52). Macrophages can be identified by ultrastructural, histochemical, immunohistochemical, or functional uptake assays. In villi, macrophages are commonly found in apical regions where they reside in the lamina propria interspersed with plasma cells. These macrophages contain an abundant lysosomal apparatus (53) and are actively phagocytic, engulfing microorganisms and soluble antigens.

Figure 11-14. Rabbit Peyer's patch. Cholera vibrios being phagocytosed (V1) and transported (V2) by an M cell. Small coated pinocytotic vesicles (arrows) are also found among apical microvilli. TEM X 14,400. (Reproduced from ref 48 by permission.)

Populations of macrophage-like cells seed the intestinal lamina propria during fetal development (20) and appear prior to germinal center formation. Intestinal macrophages reside in the tissue for approximately two to three months (54), but their kinetics of migration into the lumen and within GALT are just being recognized.

In Peyer's patches, macrophages residing between FAE cells trap *Giardia* (55), and as particles move into tissue domains, subepithelial macrophages contain the antigen flow. In germinal centers, large tingible body macrophages containing phagocytosed lymphoid cells predominate (53).

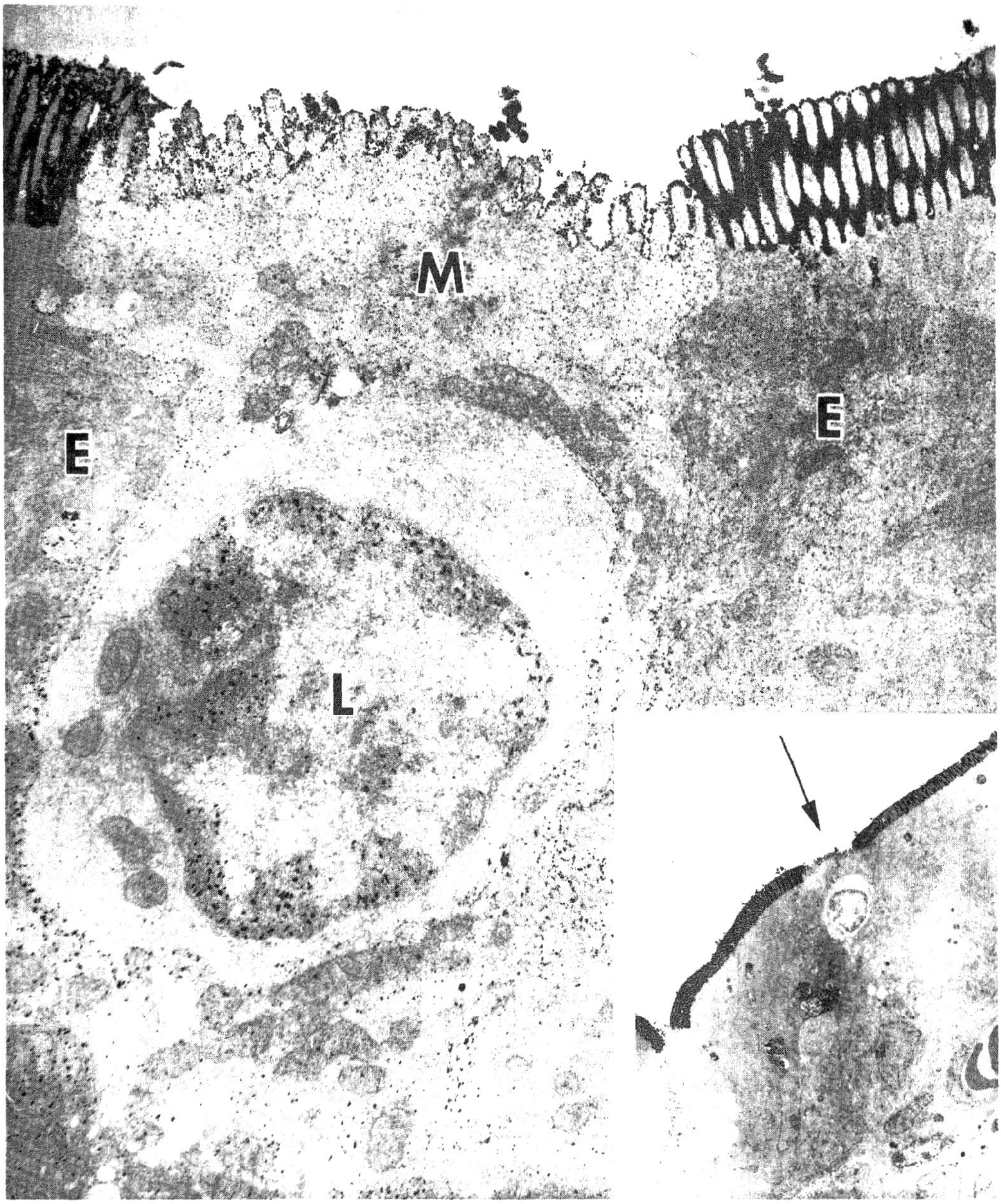

Figure 11-15. Rat Peyer's patch. Alkaline phosphatase which is prominent over microvilli of enterocytes (E) is much reduced over the luminal surface of the M cell. TEM X 15,800. Inset: At lower magnifications M cells form apparent gaps in the alkaline phosphatase reaction product. TEM X 1,450. (Reproduced from ref 50 by permission.)

2.2.3. Lymphocytes and antibody producing cells

The GALT supplies antigen sensitized cells to mucosal tissues. Peyer's patches contain an enriched population of IgA B cell precursors which populate the gastrointestinal lamina propria (56). At this site, the numbers of IgA producing cells can reach 2×10^5 cells/mm^3 of tissue (57). Stimulation of GALT by antigen results in proliferation of germinal center IgA B cells (58), migration of cells to mucosal and glandular sites, and production of IgA antibody in secretions (59).

Peyer's patches also contain heterogeneous subpopulations of T cells which control cellular interactions. Specific T helper cells augment IgA responses but suppress IgM and IgG production (60). Specialized switch T lymphocytes mediate the transition of B cells from surface IgM into surface IgA bearing lymphocytes (61). Suppressor T cells migrate from GALT to spleen where they limit systemic immune responses to orally ingested molecules (62).

2.3. Effector immune responses. Mechanisms of defense

IgA is quantitatively the major immunoglobulin in secretions bathing mucosal surfaces. The functional significance of secretory IgA is based on its ability to interfere with the adsorption or attachment of potential pathogens (63), thus providing a mechanism of antigen disposal on mucosal surfaces. By limiting adherence, IgA retards the association of antigen with mucosal cell surfaces and prevents the penetration of bacterial antigens into mucosae. It is clear that stimulation of IgA antibody confers protection against bacterial disease (64). In addition to a role as a blocking antibody, IgA can also mediate antibody-dependent cellular cytotoxicity against bacteria (65) and can specifically enhance the antibacterial activity of mouse Peyer's patch cells (66).

3. Implications of animal studies for human pathology

Animal studies of the mucosal immune system provide a conceptual framework for understanding pathology of digestive tract lymphoid organs which are subject to infectious, inflammatory, and malignant processes. Despite the morbidity and mortality associated with these organs, the extreme rarity of congenital absence of the appendix (67) and other intestinal lymphoid organs suggests in evolutionary terms that aggregation of lymphoid cells in follicles must be beneficial.

In many instances, pathological processes are extensions of the same mechanisms by which microorganisms are taken up and processed for antigen presentation and by which macrophages and other cellular immune constituents are marshalled for containment of transported microorganisms. If microbial load exceeds containment capacities, systemic infection ensues. After an immune response has been elaborated, if uptake of target antigens occurs, intramural inflammatory responses may develop. Natural selection may favor development of receptors for adherence to and transport through follicle surfaces on those microorganisms which enter the body through the intestinal tract but have their major effects in tissues remote from mucosal surfaces. Even though entrance to the body through mucosal lymphoid organs may initiate mucosal immune defenses, once microorganisms have breached the intestinal wall, such defenses would have little effect in containing systemic infection. Uptake of molecules and particles by lymphoid organs may play a role in malignant, toxic, hypersensitive, and pharmacologic responses to transported materials. Ultrastructural and immunologic techniques provide opportunities for examining hypotheses of pathogenesis, although limitation of investigational material to biopsies, surgical resection specimens, and autopsies necessarily restricts the studies which can be carried out in humans.

3.1. Tonsillitis

Removal of tonsils and adenoids reduces nasopharyngeal response to poliovirus (68), suggesting that tonsils perform a strategic role in immune surveillance at the entrance to the digestive and pulmonary tracts. Prior to development of this hypothesis, tonsillectomy was one of the most frequently performed surgical procedures. Now it is realized that inflammation may reflect physiologic as well as pathophysiologic processes within tonsils, and tonsillectomy has become quite controversial. Crypts penetrating tonsil tissue are lined by modified squamous epithelium containing cells analagous to M cells in intestinal lymphoid nodules. Tonsillar M cells cap tubulovesicular passageways down into tonsil lymphoid tissue (Fig. 11-1). Macrophages, plasma cells, lymphocytes, lymphoblasts, and granulocytes within these passageways (Fig. 11-2) reflect continuing antigenic stimulation by ingested materials and initial contact with infectious agents. Now that this cellular infiltration is recognized as a physiologic process, histologic criteria for diagnosis of chronic tonsillitis have come into question. In ultrastructural studies, tonsillectomy specimens removed from patients with recurrent tonsillitis show atrophy of germinal centers, fibrosis, crypts lined by stratified or keratinized epithelium, greatly reduced numbers of M cells, and reduction of the tubulovesicular transport system of M cells (69–71). If this loss of M cells is accompanied by reduced antigen uptake and diminished activation of lymphocytes, reduced defense against subsequent infection might contribute to recurrent tonsillitis. Whether a reduced tonsillar capacity for antigen processing is the cause or result of recurrent tonsillar infection remains an area of active investigation.

3.2. Effects of prostaglandins on intestinal lymphoid nodules

It has been postulated that prostaglandins may protect animals from inflammatory responses to normally occurring antigens. If so, then prostaglandin inhibitors might be expected to compromise this normal modulating influence. In dogs given prostaglandin inhibitors by mouth or intravenously, Peyer's patches ulcerate giving the appearance of ulcerative enterocolitis (72). Administration of the prostaglandin inhibitor, indomethacin, also results in gastric lesions, but these follow enlargement of lymphoid nodules in the stomach (16), suggesting that the effects of prostaglandin inhibitors on gastrointestinal lymphoid nodules may be more complex than originally suspected.

3.3. Appendicitis

Appendectomy like tonsillectomy has decreased in the past 30 years as knowledge of the immunologic response of the appendix has developed. Inflammation of the appendix accompanied by fever and leukocytosis is still considered an indication for removal, because of the possibility of perforation and peritoneal soilage by intestinal contents. Incidental appendectomy during the course of abdominal operations has, however, decreased markedly with the realization that the appendix is immunologically active and not a vestigial organ.

3.4. Diseases of Peyer's patches

The largest Peyer's patch in the intestines of humans, sheep, pigs, and many other mammals is located just above the ileocecal valve at the narrowest portion of the small intestine. At this location transit of liquid ileal contents is retarded, prolonging contact of luminal contents with Peyer's patch follicles. Regurgitation of liquid cecal contents into the terminal ileum also increases the antigenic load to these lymphoid tissues and may contribute to pathologic processes. It has been known for at least 50 years that ingested mycobacteria are taken up by Peyer's patches, but only recently has it been shown that M cells take up and transport mycobacteria into lymphoid follicles (73). Although such uptake may eventuate in protective immunologic responses, containment of transported mycobacteria by macrophages and other lymphoid elements is also associated with granuloma formation, which may contribute to mucosal breakdown, ulceration, and narrowing of the terminal ileum, sometimes resulting in intestinal obstruction.

In regional enteritis (Crohn's disease), the radiologic and pathologic appearance of the terminal ileum produced by ulceration, granuloma formation, and narrowing of the terminal ileum resembles intestinal tuberculosis. These similarities suggest that an infectious agent may also contribute to the pathogenesis of Crohn's disease, even though none has been specifically identified. Limited or mild lesions considered "early" have been found in association with lymphoid nodules, leading to the hypothesis that the unknown agent or agents are transported through follicle surfaces from the lumen, initiating characteristic granuloma formation, ulceration, and scarring (74).

3.4.1. *Salmonella typhi infection*

In experimental infections, salmonella are preferentially taken up from the intestinal lumen by Peyer's patches of the distal ileum and cecum. In contrast, cultures of homogenates from intestine adjacent to Peyer's patches contain very few viable salmonellae, demonstrating the specificity of the uptake process through Peyer's patch surfaces (75). Local cellular imune responses appear inadequate to sequester transported salmonella, as evidenced by their rapid movement to mesenteric nodes and subsequent systemic distribution. In humans with *Salmonella typhi* infection, systemic symptoms follow hematogenous dissemination. Proliferation of *S. typhi* in Peyer's patches, as immune responses are developing, is associated with a marked inflammatory reaction, which may lead to perforation of the intestinal wall through Peyer's patch follicles. Intestinal perforation is a complication of major significance in those underdeveloped countries in which intestinal typhoid infection continues to occur (76).

In horses *Corynebacterium equi* infection, which presents as a suppurative bronchopneumonia, has been found to invade through Peyer's patches (77). Failure of phagocytes to eliminate the organism is followed by a pyogranulomatous inflammatory response, intestinal lymphadenitis, and mucosal ulceration.

3.4.2. *Peyer's patches and development of intestinal malignancies*

The demonstrated capacity of Peyer's patch follicles for uptake of viruses from the intestinal lumen carries with it the potential for uptake of cancer-associated viruses. In patients with nonendemic Burkitt's lymphoma, selective involvement of small bowel and germinal centers in adjacent Peyer's patches has been noted (78). This may reflect uptake of tumor-associated virus or the risk that malignant degeneration may be associated with rapid proliferation of B cells in Peyer's patch germinal centers.

Peyer's patches may have an important role in absorption and retention of ingested luminal carcinogens and may be target organs of carcinogenesis (79). Studies of experimental administration of carcinogens by the oral route have demonstrated that 3-methylcholanthrene is taken up most avidly by Peyer's patches in which it is retained and metabolized. Interestingly, in a case of partially-obstructing cecal carcinoma, we found greater prominence of M cells in the adjacent ileal Peyer's patch (Fig. 11-16) than we had observed in humans previously. Although this may have reflected antigenic stimulation by bacterial overgrowth, it is also possible that M cell proliferation increases susceptibility to luminal carcinogens. Carcinogens injected directly into Peyer's patches, lymph nodes, or spleen, result in B and T cell lymphoproliferative disorders in Peyer's patches but not in other lymphoid tissue (80).

In the colon, solitary lymphoid nodules are distributed more diffusely than in the small intestine. Groups of lymphoid follicles, which have been referred to as colonic patches, are also

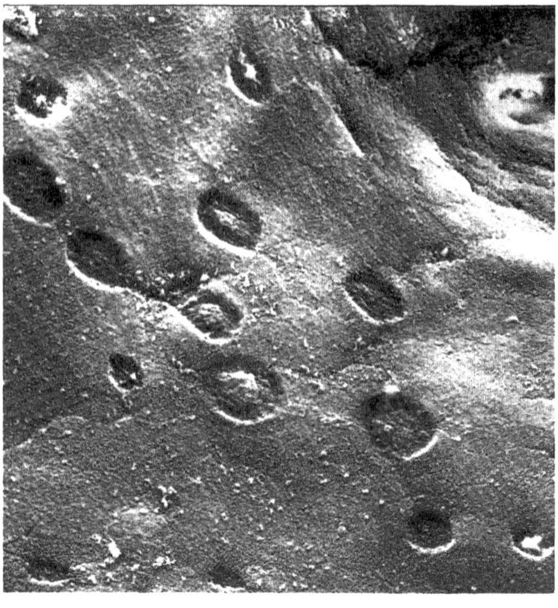

Figure 11-16. Human Peyer's patch. Prominent M cells over a follicle in the terminal ileum of a man with partially obstructing cecal carcinoma. SEM X 950.

observed in experimental animals and in humans but have been studied much less than Peyer's patches. A diffuse and uniform increase in the number of lymphoglandular complexes (lymphoid nodules) is found in colon carcinoma regardless of location of the tumor (33). Overall, only 1% of lymphoglandular complexes have germinal center formation, but in Crohn's disease, ulcerative colitis, and carcinoma, 10% of complexes have germinal centers, suggesting either a primary or secondary relationship to the underlying disease processes. In patients without familiar polyposis, 56% of microscopic adenomas originated in association with lymphoid follicles (81), suggesting that in the colon as in the small intestine, lymphoid follicles may play a role in carcinogenesis.

4. Endoscopic study of Peyer's patches

Endoscopic methods for routinely examining the terminal ileum by traversing the ileocecal valve during colonoscopy, now permit direct assessment of Peyer's patches in vivo. Follicles can be recognized through the endoscope and biopsied (82). In a baseline study of 100 cases at endoscopy and 48 from autopsy, it was possible to recognize Peyer's patches by their appearance or by interruption of Kerckring's folds (82). The incidence and type did not vary by sex but did vary by age. Flat follicles were frequent in newborns, raised follicles were most common in teenagers, and then, with increasing age, raised follicles became less frequent and flat follicles again more common (83). Application of in vivo methodologies for correlation of Peyer's patch structure and function with disease states will await further investigation.

Acknowledgments

This work was supported by the Veterans Administration and by the National Institutes of Health DE07117, AM36563, and DK38550.

References

1. Oláh I, Röhlich P, Törö I: Ultrastructure of Lymphoid Organs — An Electron Microscopic Atlas. Philadelphia and Toronto: JB Lippincott, Budapest: Akademiai Kiado, 1975.
2. Owen RL, Nemanic P: Antigen processing structures of the mammalian intestinal tract: An SEM study of lymphoepithelial organs. *Scan Electron Microsc* II: 367–378, 1978.
3. Owen RL, Bhalla DK: Lympho-epithelial organs and lymph nodes. In: Biomedical Research Applications of Scanning Electron Microscopy Vol 3, GM Hodges, KE Carr, (eds), London: Academic Press, p 79–169, 1983,
4. Emmings FG, Evans RT, Genco RJ: Antibody response in the parotid fluid and serum of Irus monkeys (*Macaca fascicularis*) after local immunization with *Streptococcus mutans*. *Infect Immun* 12: 281–292, 1975.
5. Nair PNR, Schroeder HE: Local immune response to repeated topical antigen application in the simian labial mucosa. *Infect Immun* 41: 399–409, 1983.
6. Schroeder HE, Dorig-Schwarzenbach A: Structure and composition of the oral mucous membrane on the lips and cheeks of the monkey, *Macaca fascicularis*. *Cell Tissue Res* 224: 89–104, 1982.
7. Nair PNR, Schroedr HE: Retrograde access of antigens to the minor salivary glands in the monkey *Macaca fascicularis*. *Arch Oral Biol* 28: 145–152, 1983.
8. Coleman R, Hand AR: Endocytosis of native and cationized ferritin by rat parotid duct cells. *J Dent Res* (special issue) 64: 223 (abstract), 1985.
9. Pappo J, Ebersole JL, Taubman MA: Immunol, in press. Resident salivary gland macrophages function as accessory cells in antigen dependent T cell proliferation. *Immunol*, in press.
10. Oláh I: Structure of the tonsils. In: Tonsils — Structure, Immunology and Biochemistry, F Antoni, M Straub, (eds), Budapest: Akademiai Kiado, p 5–49, 1978.
11. Howie AJ: Scanning and transmission electron microscopy on the epithelium of human palatine tonsils. *J Pathol* 130: 91–98, 1980.
12. Anderson JC: The response of the tonsil and associated lymph nodes of gnotobiotic piglets to the presence of bacterial antigen in the oral cavity. *J Anat* 117: 191–198, 1974.
13. Watanabe T, Yoshizaki K, Yagura T, Yamamura Y: In vitro antibody formation by human tonsil lymphocytes. *J Immunol* 113: 608–616, 1974.
14. Hoffman MK, Schmidt D, Oettgen HF: Production of antibody to sheep red cells by human tonsil cells in vitro. *Nature* 243: 408–410, 1973.
15. Geboes K, De Woolf-Peeters C, Rutgeerts P, Janssens J, Vantrappen G, Desmet V: Lymphocytes and Langerhans cells in the human oesophageal epithelium. *Virchows Arch* [A] 401: 45–55, 1983.

16. Tabata K, Ohtsuki H, Okabe S: Role of lymphoid nodules in pathogenesis of indomethacin-induced gastric lesions in dogs. *Dig Dis Sci* 29: 346–352, 1984.
17. Carlson JR, Owen RL: Structure and functional role of Peyer's patches. In: Immunopathology of the Small Intestine, MN Marsh, (ed), Chichester: John Wiley and Sons Ltd, p 21–40, 1986.
18. Abe K, Ito T: A qualitative and quantitative morphologic study of Peyer's patches of the mouse. *Arch Histol Jpn* 40: 407–420, 1977.
19. Cornes JS: Number, size and distribution of Peyer's patches in the human small intestine. I. The development of Peyer's patches. *Gut* 6: 225–229, 1965.
20. Mayrhofer G, Pugh CW, Barclay AN: The distribution, ontogeny and origin in the rat of Ia positive cells with dendritic morphology and of Ia antigen in epithelia, with special reference to the intestine. *Eur J Immunol* 13: 112–122, 1983.
21. Pollard M, Sharon N: Response of the Peyer's patches in germ-free mice to antigenic stimulation. *Infect Immun* 2: 96–100, 1970.
22. Ermak TH, Owen RL: Differential distribution of lymphocytes and accessory cells in mouse Peyer's patches. *Anat Rec* 215: 144–152, 1986.
23. Spencer J, Finn T, Isaacson PG: Human Peyer's patches: An immunohistochemical study. *Gut* 27: 405–410, 1986.
24. Weisz-Carrington P, Roux ME, McWilliams M, Phillips-Quagliata JM, Lamm ME: Organ and isotype distribution of plasma cells producing specific antibody after oral immunization: Evidence for a generalized secretory immune system. *J Immunol* 123: 1705–1708, 1979.
25. Bockman DE: Functional histology of appendix. *Arch Histol Jpn* 46: 271–292, 1983.
26. Gorgollón P: The normal human appendix: A light and electron microscopic study. *J Anat* 126: 87–101, 1978.
27. Ogawa K, Miyoshi M: Intercellular spaces in the lymph nodule-associated epithelium of the rabbit Peyer's patch and appendix. *Arch Histol Jpn* 48: 53–67, 1985.
28. Heatley RV, Bienenstock J: Luminal lymphoid cells in the rabbit intestine. *Gastroenterology* 82: 268–275, 1982.
29. Bockman DE, Cooper MD: Early lymphoepithelial relationships in human appendix: A combined light and electron microscopic study. *Gastroenterology* 68: 1160–1168, 1975.
30. Bockman DE, Boydston WR: Participation of follicle associated epithelium (FAE), macrophages, and plasma cells in the function of the appendix. *Scan Electron Microsc* III: 1341–1350, 1982.
31. Laufer I, deSa D: Lymphoid follicular pattern: A normal feature of the pediatric colon. *Am J Roentgenol* 130: 51–55, 1978.
32. Bland PW, Britton DC: Morphological study of antigen-sampling structures in the rat large intestine. *Infect Immun* 43: 693–699, 1984.
33. O'Leary AD, Sweeney EC: Lymphoglandular complexes of the colon: Structure and distribution. *Histopathol* 10: 267–283, 1986.
34. Warshaw AL, Walker WA, Cornell R, Isselbacher KJ: Small intestinal permeability to macromolecules: Transmission of horseradish peroxidase into mesenteric lymph and portal blood. *Lab Invest* 25: 675–684, 1971.
35. Rhodes RS, Karnovsky MJ: Loss of macromolecular barrier function associated with surgical trauma to the intestine. *Lab Invest* 25: 220–229, 1971.
36. Peterson JW, Verwey WF: Radiolabeled toxin for studying binding of cholera toxin and toxoids to intestinal mucosal receptor sites. *Proc Soc Exp Med Biol* 145: 1187–1191, 1974.
37. Waksman BH, Ozer H: Specialized amplification elements in the immune system: The role of nodular lymphoid organs in the mucous membranes. *Prog Allergy* 21: 1–113, 1976.
38. Wolf JL, Bye WA: The membranous epithelial (M) cell and the mucosal immune system. *Annu Rev Med* 35: 95–112, 1984.
39. Bhalla DK, Owen RL: Cell renewal and migration in lymphoid follicles of Peyer's patches and cecum — an autoradiographic study in mice. *Gastroenterology* 82: 232–242, 1982.
40. Bye WA, Allan CH, Trier JS: Structure, distribution and origin of M Cells in Peyer's patches of mouse ileum. *Gastroenterology* 86: 789–801, 1984.
41. Owen RL, Jones AL: Epithelial cell specialization within human Peyer's patches: An ultrastructural study of intestinal lymphoid follicles. *Gastroenterology* 66: 189–203, 1974.
42. Smith MW, Peacock MA: M cell distribution in follicle associated epithelium of mouse Peyer's patches. *Am J Anat* 519: 167–175, 1980.
43. Keljo DJ, Hamilton JR: Quantitative determination of macromolecular transport rate across intestinal Peyer's patches. *Am J Physiol* 244: 637–644, 1983.
44. Owen RL, Apple RT, Bhalla DK: Morphometric and cytochemical analysis of lysosomes in rat Peyer's patch follicle epithelium: Their reduction in volume fraction and acid phosphatase content in M cells compared to adjacent enterocytes. *Anat Rec* 216: 521–527, 1986.
45. Owen RL: Sequential uptake of horseradish peroxidase by lymphoid follicle epithelium of Peyer's patches in the normal unobstructed mouse intestine: An ultrastructural study. *Gastroenterology* 72: 440–451, 1977.
46. Neutra MR, Hall TL, Mayer EL, Fishkind DJ: Transport of membrane-bound macromolecules by M cells in follicle-associated epithelium of rabbit Peyer's patch. *Cell Tissue Res*, 247: 537–546, 1987.
47. Wolf JL, Rubin DH, Finberg R, Kauffman RS, Sharpe AH, Trier JS, Fields BN: Intestinal M Cells: A pathway for entry of reovirus into the host. *Science* 212: 471–472, 1981.
48. Owen RL, Pierce NF, Apple RT, Cray WC: M cell transport of *Vibrio cholerae* from the intestinal lumen into Peyer's patches: A mechanism for antigen sampling and for microbial transepithelial migration. *J Infect Dis* 153: 1108–1118, 1986.
49. Marcial MA, Madara JL: *Cryptosporidium*: Cellular localization, structural analysis of absorptive cell-parasite membrane-membrane interactions in guinea pigs, and suggestion of protozoan transport by M cells. *Gastro-*

enterology 90: 583–594, 1986.

50. Owen RL, Bhalla DK: Cytochemical analysis of alkaline phosphatase and esterase activities and of lectin-binding and anionic sites in rat and mouse Peyer's patch M cells. *Am J Anat* 168: 199–212, 1983.
51. Rosner AJ, Keren DF: Demonstration of M cells in the specialized follicle-associated epithelium overlying isolated lymphoid follicles in the gut. *J Leukocyte Biol* 35: 397–404, 1984.
52. Pappo J: Macrophages of the secretory immunologic system: Identification and functional interactions with T lymphocytes. Doctor of Medical Sciences Thesis, Harvard University, Cambridge, Massachusetts, 1984.
53. Sobhon P: The light and electron microscopic studies of Peyer's patches in non germ-free adult mice. *J Morphol* 135: 457–482, 1971.
54. Joel DD, Laissue JA, LeFevre ME: Distribution and fate of ingested carbon particles in mice. *J Reticuloendothel Soc* 24: 477–487, 1978.
55. Owen RL, Allen CL, Stevens DP: Phagocytosis of *Giardia muris* by macrophages in Peyer's patch epithelium in mice. *Infect Immun* 33: 591–601, 1981.
56. Vecchi M, Berti E, Primignani M, Monti M, Agape D, Torgano G, Arosio E, de Franchis R: In situ identification of immune competent cells in gastrointestinal mucosa: An evaluation by immunoelectronmicroscopy. *Virchows Archiv* [A] 406: 407–415, 1985.
57. Crabbe PA, Carbonara AO, Heremans JF: The normal human intestinal mucosa as a major source of plasma cells containing gamma A-immunoglobulin. *Lab Invest* 14: 235–248, 1965.
58. Carlson JR, Heyworth MF, Owen RL: Response of Peyer's patch lymphocyte subsets to *Giardia muris* infection in BALB/c mice. II. B-cell subsets. Enteric antigen exposure is associated with immunoglobulin switching by Peyer's patch B cells. *Cell Immunol* 97: 51–58, 1986.
59. Mestecky J, McGhee JR, Arnold RR, Michalek SM, Prince SM, Babb JC: Selective induction of an immune response in human external secretions by ingestion of bacterial antigen. *J Clin Invest* 61: 731–737, 1978.
60. Richman LK, Graeff AS, Yarchoan R, Strober W: Simultaneous induction of antigen-specific IgA helper T cells and IgG suppressor T cells in the murine Peyer's patch after protein feeding. *J Immunol* 126: 2079–2083, 1981.
61. Kawanishi H, Saltzman LE, Strober W: Mechanisms regulating IgA-class specific immunoglobulin production in murine gut-associated lymphoid tissue. I. T cells derived from Peyer's patches that switch sIgM B cells to sIgA B cells in vitro. *J Exp Med* 157: 433–450, 1983.
62. Mattingly JA, Waksman BH: Immunologic suppression after oral administration of antigen. I. Specific suppressor cells formed in rat Peyer's patches after oral administration of sheep erythrocytes and their systemic migration. *J Immunol* 121: 1878–1883, 1978.
63. Williams R, Gibbons R: Inhibition of bacterial adherence by secretory immunoglobulin A: A mechanism of antigen disposal. *Science* 177: 697–699, 1972.
64. Taubman MA, Smith DJ: Effects of local immunization with *Streptococcus mutans* on induction of salivary immunoglobulin A antibody and experimental dental caries in rats. *Infect Immun* 9: 1079–1091, 1974.
65. Lowell GH, Smith LF, Artenstein MS, Nash GS, MacDermott RP Jr: Antibody-dependent cell-mediated antibacterial activity of human mononuclear cells. I. K lymphocytes and moncytes are effective against meningococci in cooperation with human immune sera. *J Exp Med* 150: 127–137, 1979.
66. Tagliabue A, Nencioni L, Villa L, Keren DF, Lowell GH, Boraschi D: Antibody-dependent cell-mediated antibacterial activity of intestinal lymphocytes with secretory IgA. *Nature* 306: 184–186, 1983.
67. Robinson JO: Congenital absence of vermiform appendix. *Br J Surg* 39: 344–345, 1952.
68. Ogra PL: Effect of tonsillectomy and adenoidectomy on nasopharyngeal antibody response to poliovirus. *N Engl J Med* 284: 59–64, 1971.
69. Surján L Jr: Reduced lymphocyte activation in repeatedly inflamed human tonsils. *Acta Otolaryngol* (Stockh) 89: 187–194, 1980.
70. Maeda S, Mogi G: Functional morphology of tonsillar crypts in recurrent tonsillitis. *Acta Otolaryngol* (Stockh) (suppl) 416: 7–19, 1984.
71. Karchev T, Kabakchiev P: Electron microscope observations on the tonsillar epithelium in children with recurrent tonsillitis. *Int J Pediatr Otorhinolaryngol* 4: 149–156, 1982.
72. Stewart THM, Hetenyi C, Rowsell H, Orizaga M: Ulcerative enterocolitis in dogs induced by drugs. *J Pathol* 131: 363–378, 1980.
73. Fujimura Y: Functional morphology of microfold cells (M cells) in Peyer's patches — phagocytosis and transport of BCG by M cells into rabbit Peyer's patches. *Gastroenterol Jpn* 21: 325–335, 1986.
74. Rickert RR, Carter HW: The "early" lesion of Crohn's disease: Correlative light and scanning electron microscopic studies. *J Clin Gastroenterol* 2: 11–19, 1980.
75. Carter PB, Collins FM: The route of enteric infection in normal mice. *J Exp Med* 139: 1189–1203, 1974.
76. Bitar R, G Tarpley J: Intestinal perforation in typhoid fever: A historical and state-of-the-art review. *Rev Infect Dis* 7: 257–271, 1985.
77. Johnson JA, Prescott JF, Markham JF: The pathology of experimental *Corynebacterium equi* infection in foals following intragastric challenge. *Vet Pathol* 20: 450–459, 1983.
78. Mann RB, Jaffe ES, Braylan RC, Nanba K, Frank MM, Ziegler JL, Berard CW: Nonendemic Burkitt's lymphoma: A B-cell tumor related to germinal centers. *N Engl J Med* 295: 685–691, 1976.
79. Bost KL, Cuchens MA: The role of the Peyer's patch in carcinogenesis. I. The adsorption from the gut and retention of 3-methylcholanthrene by Peyer's patches. *Carcinogenesis* 7: 1251–1256, 1986.
80. Bost KL, Garrett LR, Cuchens MA: The role of the Peyer's patch in carcinogenesis. II. 3-methylcholanthrene-induced lymphoid malignancies in rat Peyer's patches. *Carcinogenesis* 7: 1257–1265, 1986.
81. Oohara T, Ogino A, Tohma H: Microscopic adenoma in nonpolyposis coli: Incidence and relation to basal cells and lymphoid follicles. *Dis Colon Rectum* 24: 120–126, 1981.

82. Fujikura S: A study of Peyer's patch of the terminal ileum. Part 1: A study of the mucosal surface appearance and histological finding in endoscopic cases and autoptic cases. *Gastroenterol Endosc* (Japan) 26: 1246–1261, 1984.
83. Fujikura S: A study of Peyer's patch of the terminal ileum. Part 2: A clinico-statistical study. *Gastroenterol Endosc* (Japan) 27: 326–336, 1985.

Author's address:
Jacques Pappo, DDS, DMSc
Cell Biology and Aging Section (151E)
Veterans Administration Medical Center
4150 Clement Street
San Francisco, CA 94121, USA
Phone number (415) 750-2101

CHAPTER 12

Microvascularization of the alimentary canal as studied by scanning electron microscopy of corrosion casts

AIJI OHTSUKA, OSAMU OHTANI, & TAKURO MURAKAMI

1. Introduction

Scanning electron microscopy (SEM) of corrosion casts allows a clearer three-dimensional view of fine vascular architecture than conventional light or transmission electron microscopy of sectioned or injected tissue specimens (1–8). Positional relations of each vessel to the tissue elements cannot be examined in the cast samples, since the cast itself is observed in isolation from the surrounding tissues. However, SEM of vascular corrosion casts has already shown itself to be valuable and useful in the study of microvascular architecture in various organs and tissues, including the gastrointestinal canal (9–24).

This chapter reviews the findings on the vascular architecture of the alimentary canal and associated lymphoid tissue as observed by the corrosion casting/SEM method.

2. Preparation and scanning electron microscopy of vascular casts

The corrosion casting/SEM method consists of vascular casting, microdissection, and scanning electron microscopy. Many casting media are available. Among them, low viscosity methacrylate media, including commercially available Mercox and Batson's plastic, are usually used because of ease in handling and the good casting they produce of minute vessels (1–4, 6–8). A monomeric methyl and hydroxypropyl methacrylate mixture is very fluid and useful for reproduction of bile canaliculi (25).

Fresh stomach, small intestine, and large intestine are thoroughly perfused through their afferent arteries with Ringer or other physiological solutions. Then the casting medium is injected through the arteries into the perfused organs until the efferent veins are filled with the injected resin. After resin injection, the organs are immersed in a hot water bath (60°C), corroded in a 10% to 20% NaOH solution, and washed in running tap water (1, 3, 7). For preparation of lymphatic casts, the casting medium (Mercox diluted with monomeric methyl methacrylate, 1.5–2.0 centistokes) is injected intraparenchymally into the intestinal submucosa (24).

The blood vascular or lymphatic corrosion casts thus prepared are cut into blocks of appropriate sizes and air dried. They are mounted on metal stubs, coated with gold or platinum, and observed with a scanning electron microscope.

3. Blood vasculature

3.1. Stomach

In the stomach, the arteries in the subserosa penetrate the tunica muscularis and reach the submucosa where they form the submucosal arterial plexus (Figs 12-1 to 12-4). In the submucosal layer, the veins also form the submucosal venous plexus. This venous plexus is drained into the veins in the subserosa. The blood vascular beds of the muscular sheets are fairly independent and mainly supplied by the subserosal arteries and veins.

The microvascular pattern of the gastric mu-

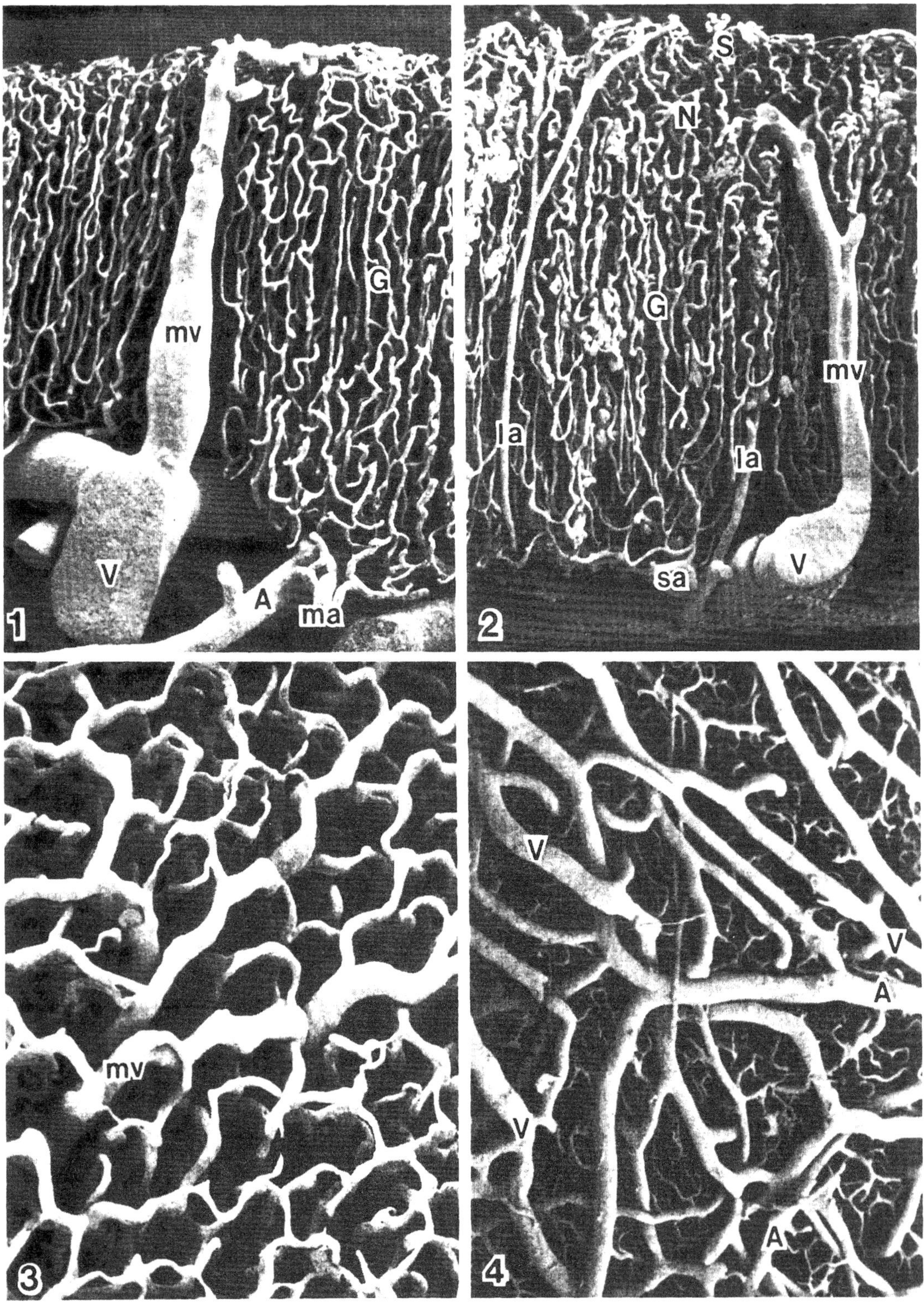
mv
G
V
A
ma
1
S
N
G
mv
la
la
sa
V
2
mv
3
V
V
A
V
A
4

cosa has some species-specific differences. In the rat (Fig. 12-1) (17,20), the submucosal arterial plexus gives off numerous terminal arterioles, the mucosal arterioles, which ascend for short distances toward the glandular mucosa. Near the base of the fundic glands, the mucosal arterioles break up into capillaries. These capillaries ascend toward the luminal surface and form the glandular capillary networks around the fundic glands. When observed from the luminal surface, the networks have a honeycomb appearance; each cell corresponds to one foveola (Fig. 12-3). The mucosal venules originate just beneath the surface epithelium and descend perpendicularly to empty into the submucosal venous plexus (Figs. 12–1, 12–3). The mucosal venules are fairly independent and receive few capillaries in their descending course. The capillaries around the lower half and bottom of the fundic glands are drained into short and small venules which in turn empty into the submucosal venous plexus. In the rabbit (Fig. 12-2), the submucosal arterial plexus gives off two types of arterioles, short and long, to the gastric mucosa (23). The short arterioles give rise to capillaries at the base of the fundic glands. The long arterioles ascend along the gland and supply the capillary network beneath the surface epithelium. This superficial capillary network is denser than the deep capillary network around the fundic glands. In the glandular neck region, the superficial and deep networks connect with each other and converge into common venules which descend along the glands to empty into the submucosal veins. The microvascular architecture of the gastric mucosa in the human is essentially equivalent to that in the rat; the mucosal capillary networks of the human stomach receive arterial supply at the base of the glands and drain beneath the surface epithelium into venules (10,22).

Arteriovenous anastomoses in the submucosal plexus of the human (26) or rat (27,28) stomach have been described by some light microscopists, though denied by many others (29–31). Recent corrosion casting/SEM studies have reported neither a short pathway nor a wide channel between the arterial and venous vessels in the stomachs of the human (22), rat (17,20), and rabbit (23).

The corrosion casting/SEM findings, together with light microscopy of living tissues (17,29), give evidence that the blood in the glandular mucosal capillaries flows upward from the glandular bottom to the surface epithelium. This gastric microcirculation may have two physiological significances in protecting the gastric mucosa from acid injury: (a) secretion of H^+ by the parietal cells is accompanied by the release of HCO_3^- into the adjacent interstitial space, where deep segments of the mucosal capillaries exist. Released HCO_3^- may be transported to the surface epithelium by the mucosal blood flow. Thus, the capacity of the surface mucous cells to secrete HCO_3^--rich mucus or to neutralize back-diffusing H^+ would be increased (17). (b) The upward mucosal blood flow may also transport some substances activating surface mucous cells (23), such as serotonin (32) or somatostatin (33) secreted by basal granulated (endocrine) cells which are located in close proximity to capillaries in the deep segment of the gastric gland (34).

3.2. Small intestine

The subserosal arteries penetrate the muscular layer and form the submucosal arterial plexus. The veins also form the submucosal venous plexus, which in turn empty into the subserosal veins. The mucosa and the muscular sheet in the small intestine are separately supplied by the submucosal vessels (Fig. 12-5) (9).

The capillary bed of the tunica muscularis is divided into two layers, each of which consists of

Figures 12-1–12-4. Scanning electron micrographs of blood vascular corrosion casts of the rat glandular stomach (1, lateral view; 3, luminal view; 4, submucosal aspect) and rabbit gastric mucosa (2, lateral view). (1, 3, 4) In the rat, the mucosal arteriole (ma) arising from the submucosal artery (A) gives rise to the glandular capillaries (G) near the bottom of the gland. The mucosal venule (mv) collects the glandular capillaries immediately beneath the surface epithelium and descends through the glands to empty into the submucosal vein (V). (2) In the rabbit, the superficial capillary network (S) beneath the surface epithelium is supplied by long arterioles (la) ascending through the glandular region, and the deep capillary network surrounding the gland is supplied at the glandular bottom by short arterioles (sa). These two networks drain into the mucosal venule in the glandular neck region (N). (1, X 75; 2, X 100; 3, X 180; 4, X 75.)

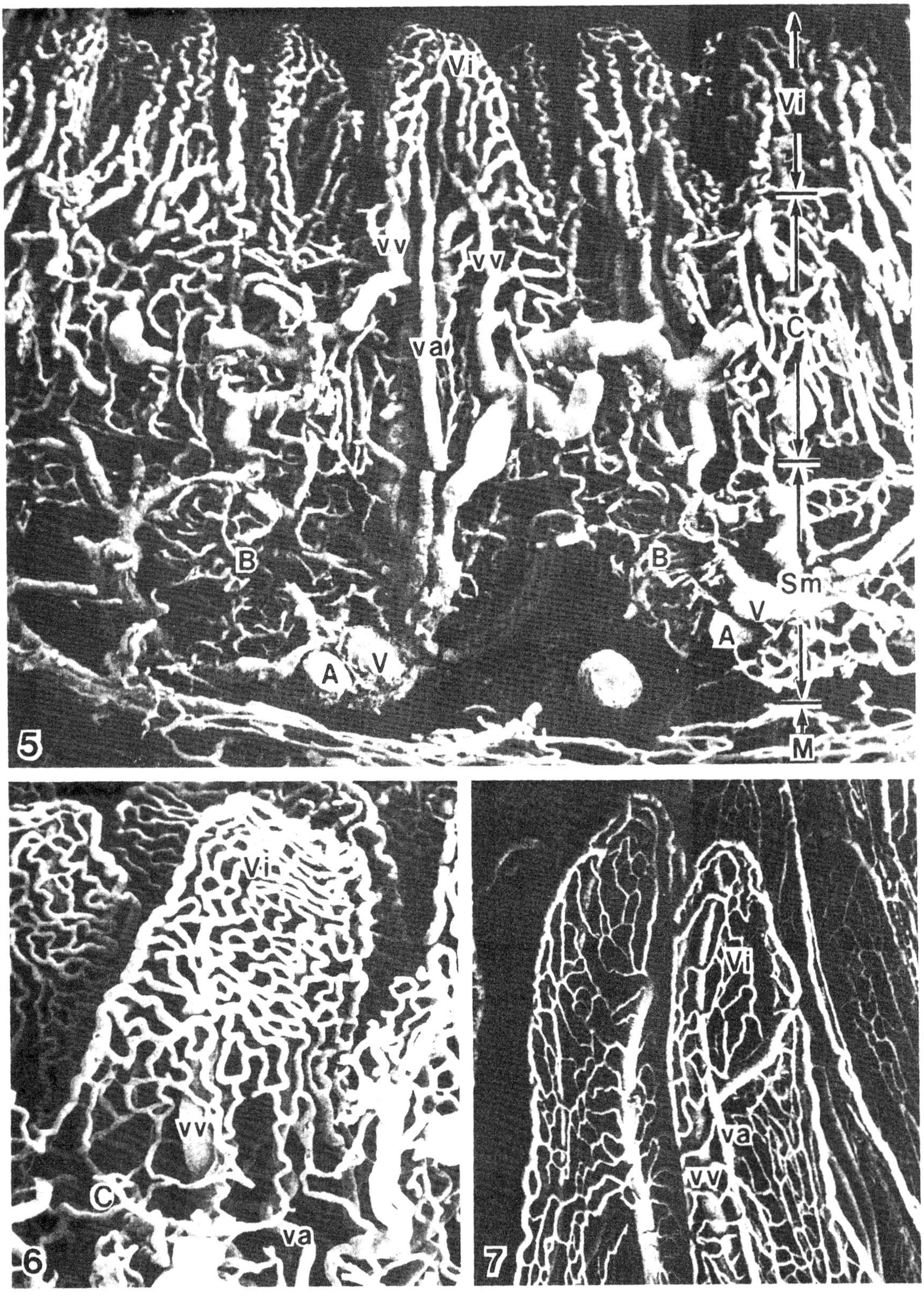
Vi
vv
vv
va
Vi
C
B
B
Sm
V
A
A
V
M
5
Vi
vv
C
va
6
Vi
va
vv
7

capillaries running parallel to the muscle fibers (Fig. 12-9). These capillary networks of the tunica muscularis receive arterial supply from the submucosal arterial plexus and drain into the submucosal venous plexus.

The submucosal arterial plexus gives off two types of mucosal twigs, long or villous arterioles and short or cryptal ones. The villus arteriole ascends in the villus and breaks up into the capillary network beneath the villus epithelium. This villous capillary network drains into villous venules, which descend in the villus and empty into the submucosal venous plexus. The cryptal arterioles reach the bottom of the intestinal crypts and give rise to the capillary plexuses surrounding the crypts. The cryptal capillary plexuses and the villous capillary networks are connected with each other at the villous base (Figs. 12-5, 12-6, 12-8). The cryptal plexus has thin or faint venous drainages (cryptal venules) into the submucosal venous plexus (9). The capillary networks surrounding the Brunner's glands in the duodenum receive a distinct or isolated arterial supply from the submucosal arterial plexus and drain into the submucosal venous plexus (Fig. 12-5) (35).

Some interspecies differences are observed, which may reflect differences in the shapes of the intestinal villi. In the cylindrical or conical villus of the cat (Fig. 12-5), the arteriole ascends in the villus and branches near the tip to supply the villus plexus (16). The villus capillary plexus converges, at or near the villus base, into the villous venules. In the flattened villus of the rat (Fig. 12-6), the villus arteriole ascends centrally through the villus and connects at the villus tip with the capillary network (9, 18, 20). The villous network converges, at the middle segment of the villus, into villous venules. There are usually two venules in a villus in the upper part of the intestine and often three in the lower part (9). In the flattened conical villus of the rabbit (Fig. 12-7), the villus arteriole branches near the tip to supply the villous capillary network. In addition to this main villus arteriole, each villus possesses several thin arterioles which break up into the villous subepithelial capillaries at the upper segment of the villus (24). A twig of the villus venule usually runs transversely at or near the middle segment of the villus and collects the capillaries of the lower half of the villus as well as those of the upper half (16, 24). In the tongue-shaped villus of the human, the arteriole connects at or near the villus tip with the capillary network which, in turn, drains at the middle segment of the villus into venules (14, 16).

The findings obtained by SEM of vascular casts confirm those obtained by previous light microscopic or radiographic studies in various animals, including man and monkey (36–50). All of the SEM studies of vascular casts have particularly revealed that the villus microvasculature principally shows a fountain pattern (35, 36).

3.3. Lymphoid tissue of small intestine

In the small intestine, lymphoid follicles occur both singly and in groups (Peyer's patches). The microvascular architecture of the Peyer's patch has been studied with conventional light microscopy of injected samples in the rat (51), mouse (52), guinea pig (53), rabbit (51), and cat (51) and also with corrosion casting/SEM method in the rat (15) and mouse (21).

The submucosal arteries and veins supplying the Peyer's patch run, as the interfollicular arteries and veins, around or between the lymphoid follicles (Fig. 12-9). The interfollicular arteries give off the follicular arterioles, which enter the follicle. Some of the follicular arterioles directly supply a domelike capillary network beneath the follicular epithelium (Figs. 12-8, 12-10). The domelike network drains at its marginal region into the venules, which in turn empty into the interfollicular veins. Thus, the microcirculation of the follicular dome appears to be a modification of the Mall's fountain pattern seen in

←

Figures 12-5–12-7. Blood vascular casts of the cat duodenum (5), rat jejunum (6), and rabbit ileum (7). (5) The mucosa and muscular layers (M) are isolatedly supplied by the submucosal artery (A) and drain into the submucosal vein (V). The capillary network surrounding the Brunner's glands (B) in the submucosa (Sm) is supplied by the submucosal arteries and drain into the submucosal veins. (5–7) The villus arteriole (va) connects at the villus tip with the villus capillary network (Vi) which drains into the villus venule (vv). C, cryptal vascular network. (5, X 110; 6, X 170; 7, X 75.)

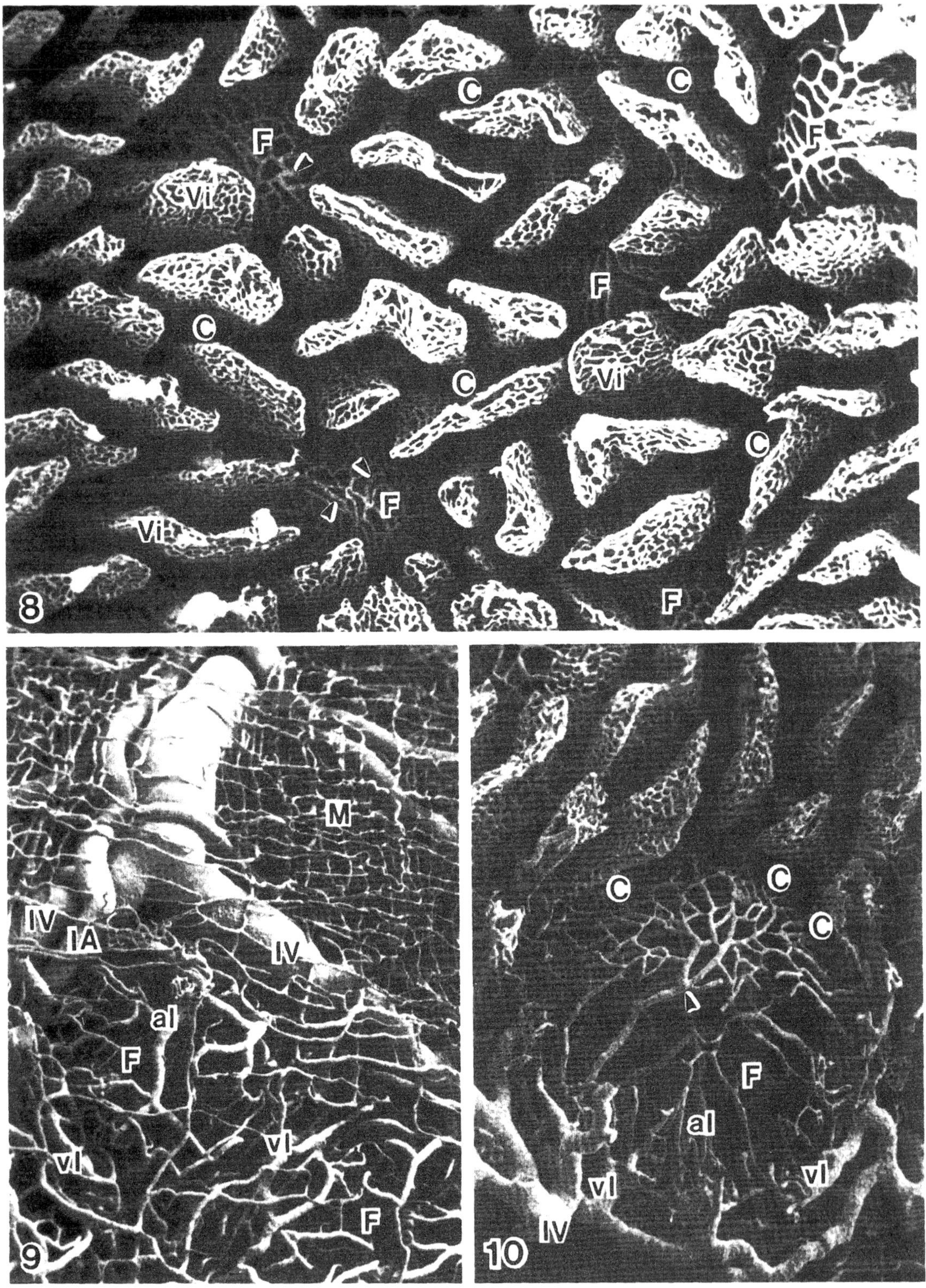
C
C
F
F
Vi
F
C
C
Vi
C
F
Vi
F
8
M
IV
IA
IV
al
F
vl
vl
F
9
C
C
C
F
al
vl
vl
IV
10

villi (15). Some marginal capillaries of the dome-like network continue into the capillary plexus of the crypts surrounding the follicle. These connections may provide a route for the migration of humoral factors from the follicle apex to the crypts, which may regulate the migration of proliferating cells from the crypts to the follicle surface (15).

Capillaries within the follicle are thin and form a coarse network which consists of hairpin shaped or inverted tuning fork-like capillaries. The intrafollicular capillary plexus drains into the collecting venules in the basal area or periphery (thymus-dependent area) of the follicle (Figs. 12-9, 12-10) (15). These venules reach the interfollicular space, where they empty into the interfollicular veins. The collecting venules in the thymus-dependent area are high endothelial (15, 21) and regarded as the sites where lymphocytes migrate from the vascular lumen to the lymphoid follicle (54, 55).

3.4. *Large intestine*

The arteries and veins enter the submucosa where they form the submucosal arterial and venous plexus, respectively. As in the small intestine, the tunica muscularis and the mucosa are separately supplied by the submucosal vessels.

The submucosal arterial plexus gives off short mucosal arterioles which reach the base of the glands, where they break up into capillaries. The capillaries ascend along the glands and reach the surface epithelium, where they surround the openings of the glands in a huneycomb pattern (Fig. 12-11) (12). The mucosal venules collect the capillaries just beneath the surface epithelium, descend through the glandular mucosa, and empty into the submucosal venous plexus (Fig. 12-12) (56). This pattern is quite similar to that seen in the gastric mucosa and described above. Light microscopic observation of the India ink injected human colon (57) has shown that the mucosal capillary plexus receives its arterial supply at the base of the glands and drains beneath the surface epithelium into venules, which agrees with SEM findings of the corrosion casts of the rat large intestine illustrated here.

3.5. *Lymphoid tissue of large intestine*

The rat cecum has a cluster of lymphoid follicles in the apex, the cecal patch (Fig. 12-11). This patch seems to correspond to the appendix vermiformis of other animals including the rabbit and man (58). The blood vascular architecture of the rat cecal patch shows a pattern which is basically similar to that of the rat Peyer's patch illustrated above.

Blood vasculature of the appendix has been studied only by a light microscope (51). It has been reported that the vascular architecture of the appendicular lymphoid follicle in the cat consists of the dome and intrafollicular plexuses resembling that seen in the Peyer's patch.

4. Lymphatic architecture

4.1. *Small intestine*

The lymphatics begin with blind ended or sacculated vessels, the initial lymphatics. In the small intestine, the initial lymphatics, as the central lacteals, are located in the villi (13, 24). The central lacteals descend through the mucosa and empty into the submucosal lymphatic plexus (Fig. 12-13). The submucosal lymphatic plexus consists of thin and thick lymphatics; the thin lymphatics form a network which fills in the coarse mesh of thick lymphatics. The mucosal lymphatics or central lacteals are connected with both thin and thick lymphatics in the submucosa (24). The thick lymphatics give off the efferent lymphatics which

←

Figures 12-8–12-10. Blood vascular casts of the rat Peyer's patch (8, luminal view; 9, serosal view; 10, partially dissected cast). (8) The dome capillaries of the lymphoid follicles (F) are exposed among the villus capillary networks. (9, 10) The intrafollicular capillaries are supplied by the follicular arterioles (al) and drain into the collecting venules (vl) at the base or periphery of the follicle. (8, 10) The arterioles (arrowheads) break up into subepithelial capillaries at the top of the dome. C, crypts; IA, IV, interfollicuar artery and vein; M, muscular layer. (8, X 75; 9, X 65; 10, X 65.)

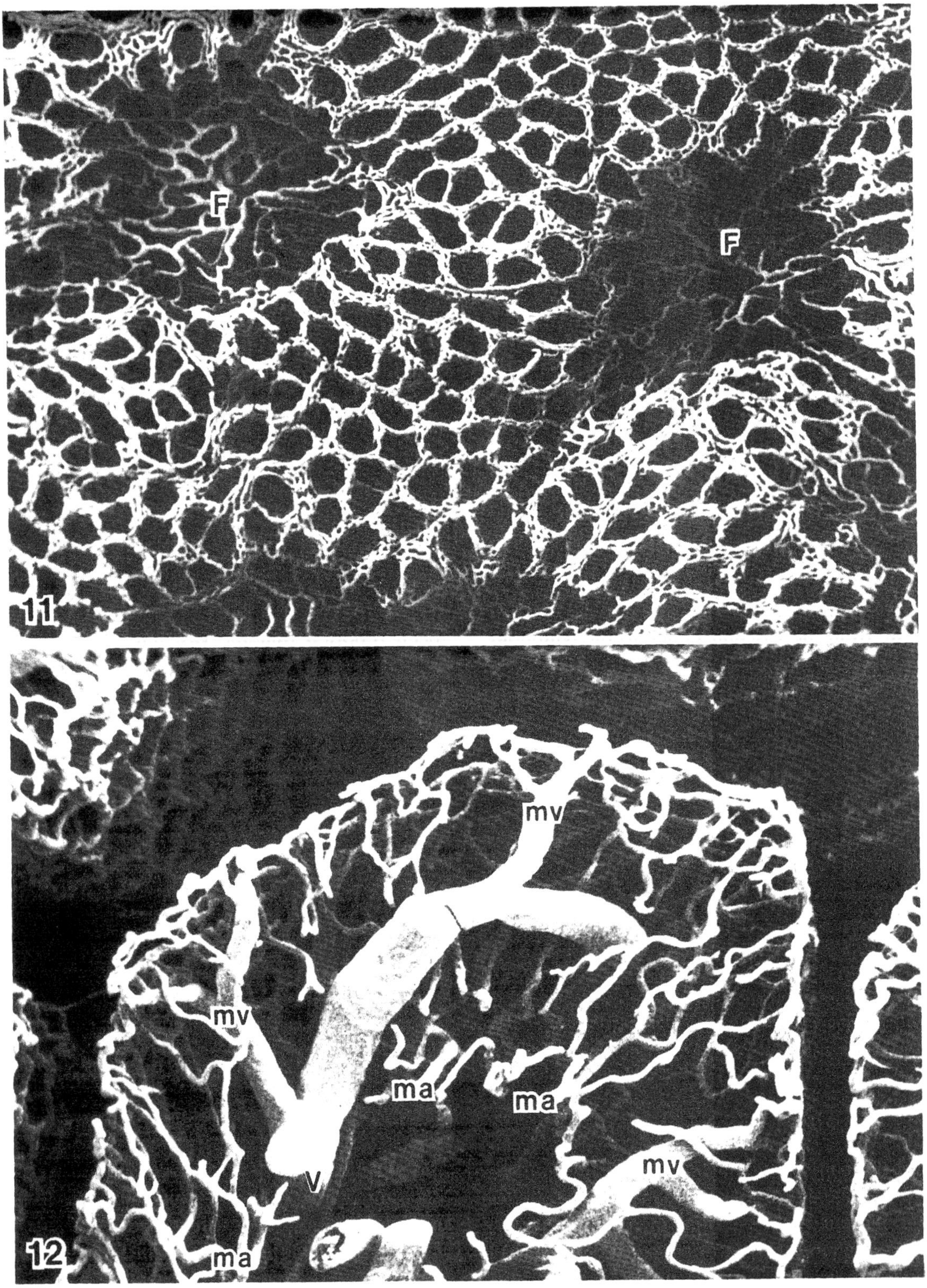

Figures 12-11–12-12. Blood vascular casts of the rat cecum (11) and rectum (12). (11) The luminal aspect of the glandular capillary network shows a honeycomb pattern. The capillary networks of two lymphoid folliclular domes (F) are seen in the glandular capillary network. (12) The mucosal arterioles (ma) give rise to the glandular capillaries at the bottom of the gland. The mucosal venules (mv) collect the glandular capillaries immediately beneath the surface epithelium. V, submucosal vein. (11, X 80; 12, X 180.)

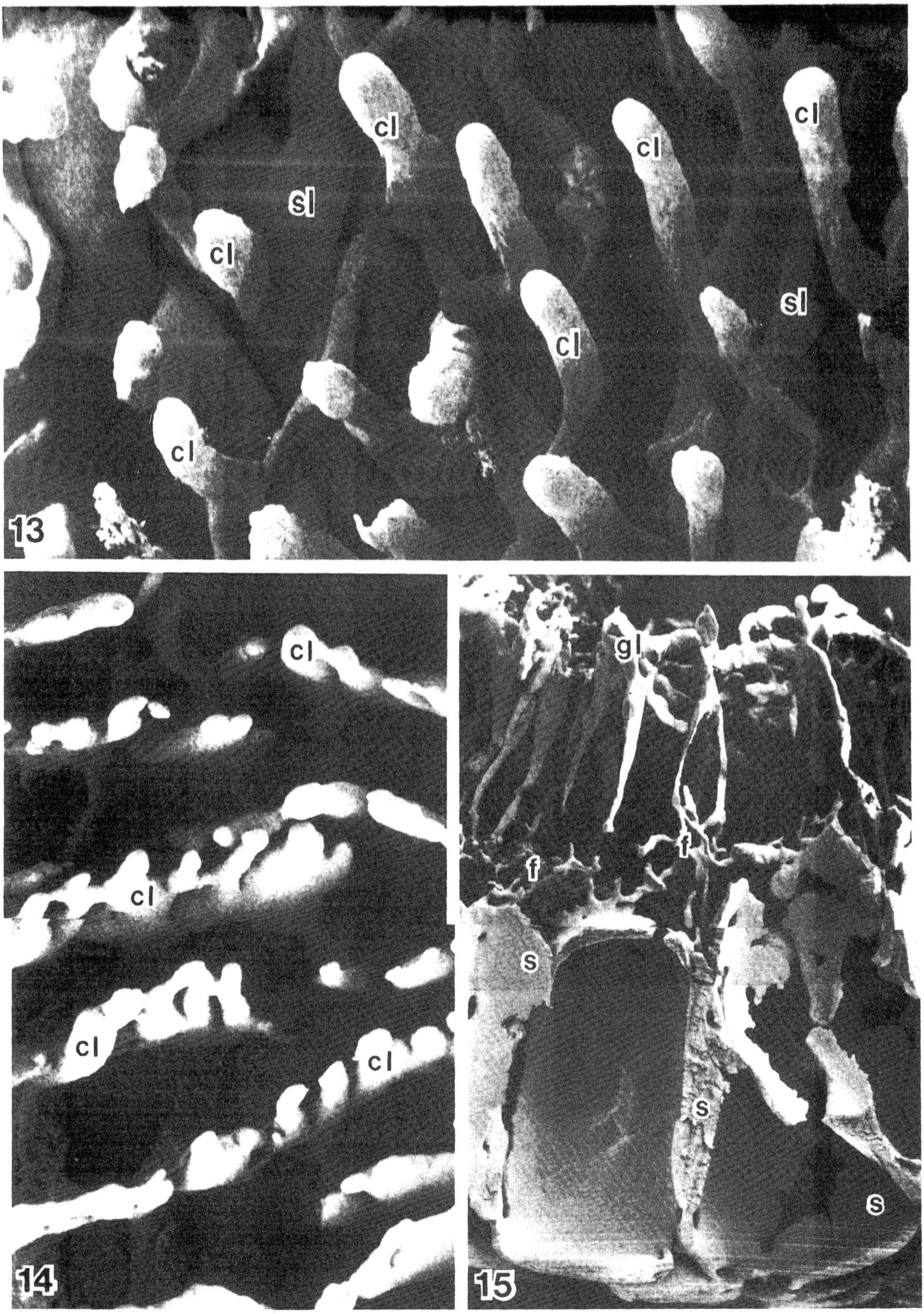

Figures 12-13–12-15. Lymphatic casts of the lower part of the small intestine in the rabbit (13) and rat (14) and of the rabbit appendix (15). (13, 14) The central lacteals (cl) beginning with blind ends are observed as fingerlike protrusions. The central lacteals descend to lead into the submucosal lymphatics (sl). (15) The lateral side of the lymphoid follicle is surrounded by well-developed perifollicular lymphatic sinus (s). The upper segment of the follicle is surrounded by fine lymphatic network (f). The glandular lymphatics (gl) connect with the perifollicular sinus. No lymphatics are seen within the follicle. (13, X 160; 14, X 120; 15, X 30.)

obliquely pierce the tunica muscularis and drain into the lymphatics in the mesentery.

The number and shape of the lacteals varies in each villus (24, 59). The cylindrical villus, such as in the rabbit ileum, usually contains one lacteal (Fig. 12-13). The flattened conical villus in the upper part of the rabbit small intestine usually contains two to five lacteals with interconnections among them. In such cases, a sinus is formed near the villus base. The leaflike villus of rat small intestine always contains five or more central lacteals (Fig. 12-14) (60).

4.2. Appendix vermiformis

A lymphatic sinus (perifollicular sinus) surrounds the lateral side of the appendicular lymphoid follicle (Fig. 12-15) (19, 61). The sinus receives a network of fine lymphatics which are distributed in the thymus-dependent area and drains into the submucosal lymphatics (61). The lymphatics surrounding the intestinal glands in the appendix continue into the sinus. No lymphatic vessels are reproduced within the lymphoid follicles.

5. Conclusion

Scanning electron microscopy of corrosion casts clearly visualizes the blood vascular beds and the lymphatic channels.

The gastrointestinal canal is richly supplied by blood vessels. The arteries and veins form arterial and venous plexuses, respectively, in the submucosa. The mucosa and the tunica muscularis are supplied separately. The mucosal glands or crypts receive arterial supplies at their bases and emit efferent veins in their necks or opening regions. The villous arterioles form a dense capillary network beneath the villus epithelium. The villous venules usually arise at the lower half of the villus.

Lymphatics in the small intestine begin with the central lacteals in the villi. They penetrate the mucosa and empty into the submucosal lymphatic plexus. In the appendix vermiformis, the lymphoid follicle is surrounded by the lymphatic sinus into which fine lymphatics located in the thymus-dependent area drain. No lymphatic vessel is reproduced within the lymphoid follicle.

References

1. Murakami T: Application of the scanning electron microscope to the study of the fine distribution of the blood vessels. *Arch Histol Jpn* 32: 445–454, 1971.
2. Nowell JA, Lohse CL: Injection replication of the microvasculature for SEM. In: Scanning Electron Microscopy/1974-I, O Johari, I Corvin (eds), Chicago: IIT Research Institute, p 267–274, 1974.
3. Murakami, T: Methyl methacrylate injection replica method. In: Principles and Techniques of Scanning Electron Microscopy, vol. 6. MA Hayat (ed), New York: Van Nostrand Reinhold Co., p 159–169, 1978.
4. Gannon, BJ: Vascular casting. In: Principles and Techniques of Scanning Electron Microscopy, vol 6. MA Hayat (ed), New York: Van Nostrand Reinhold Co., p 170–193, 1978.
5. Frasca JM, Carter HW, Schaffer WA: An improved latex injection media for replicating the pulmonary microvasculature. In: Scanning Electron Microscopy/1978/II. O Johari (ed), Chicago: AMF O'Hare, SEM Inc, p 485–489, 1978.
6. Hodde KC, Nowell JA: SEM of microcorrosion casts. In: Scanning Electron Microscopy/1980/II. RP Becker, O Johari (eds), Chicago: AMF O'Hare, SEM Inc, p 88–106, 1980.
7. Murakami T, Ohtani O, Ohtsuka A, Kikuta A: Injection replication and scanning electron microscopy of blood vessels. In: Biomedical Research Applications of Scanning Electron Microscopy, Vol 3. GM Hodges, KE Carr (eds), London: Academic Press, p 1–30, 1983.
8. Lametschwandtner A, Lametschwandtner U, Weiger T: Scanning electron microscopy of vascular corrosion casts — technique and applications. In: Scanning Electron Microscopy/1984/II. O Johari (ed), Chicago: AMF O'Hare, SEM Inc, p 663–695, 1984.
9. Ohashi Y, Kita S, Murakami T: Microcirculation of the rat small intestine as studied by the injection replica scanning electron microscope method. *Arch Histol Jpn* 39: 271–282, 1976.
10. Kobayashi A, Takeda S, Hattori S, Kohli Y, Hattori T, Kawai K: Vascular architecture of human gastric mucosa. *Jpn J Gastroenterol* 73: 169–178, 1976 (in Japanese).
11. Takao, M: Microcirculation of gastric ulcer. A scanning electron microscopic study of vascular casts. *Okayama Igakkai Z* 90: 915–948, 1978 (in Japanese).
12. Nopanitaya W, Aghajanian JG, Gray LD: An improved plastic mixture for corrosion casting of the gastrointestinal microvascular system. In: Scanning Electron Microscopy/1979/III. RP Becker, O Johari (eds), Chicago: AMF O'Hare, SEM Inc, p 751–755, 1979.
13. Tamura A: Scanning electron microscopic studies on lymphatic casts of the rat stomach and the rabbit small intestine. *Okayama Igakkai Z* 91: 1075–1099, 1979 (in Japanese).

14. Gannon BJ, Rogers PAW, O'Brien PE: Two capillary plexuses in human intestinal villi. *Micron* 11: 447–448, 1980.
15. Bhalla DK, Murakami T, Owen RL: Microcirculation of intestinal lymphoid follicles in rat Peyer's patches. *Gastroenterol* 81: 481–491, 1981.
16. Gannon BJ, Gore RW, Rogers PAW: Is there an anatomical basis for a vascular countercurrent mechanism in rabbit and human intestinal villi? *Biomed Res* 2 (suppl): 235–241, 1981.
17. Gannon B, Browning J, O'Brien P: The microvascular architecture of the glandular mucosa of rat stomach. *J Anat* 135: 667–683, 1982.
18. Métry JM, Neff M, Knoblauch M: The microcirculatory system of the intestinal mucosa of the rat. An injection cast and scanning electron microscopy study. *Scand J Gastroenterol* 17 (suppl 71): 159–162, 1982.
19. Bockman DE: Functional histology of appendix. *Arch Histol Jpn* 46: 271–292, 1983.
20. Ohtani O, Kikuta A, Ohtsuka A, Taguchi T, Murakami T: Microvasculature as studied by the microvascular corrosion casting/scanning electron microscope method. I. Endocrine and digestive system. *Arch Histol* Jpn 46: 1–42, 1983.
21. Yamaguchi K, Schoefl GI: Blood vessels of the Peyer's patch in the mouse: I. Topographic studies. *Anat Rec* 206: 391–401. 1983.
22. Gannon B, Browning J, O'Brien P, Rogers P: Mucosal microvascular architecture of the fundus and body of human stomach. *Gastroenterol* 86: 866–875, 1984.
23. Ohtsuka A, Ohtani O: The microvascular architecture of the rabbit stomach corpus in vascular corrosion casts. In: Scanning Electron Microscopy/1984/IV. O Johari (ed), Chicago: AMF O'Hare, SEM Inc, p 1951–1956, 1984.
24. Ohtani O, Ohtsuka A: Three-dimensional organization of lymphatics and their relationship to blood vessels in rabbit small intestine. A scanning electron microscopic study of corrosion casts. *Arch Histol Jpn* 48: 255–268, 1985.
25. Murakami T, Itoshima T, Hitomi K, Ohtsuka A, Jones AL: A monomeric methyl and hydroxypropyl methacrylate injection medium and its utility in casting blood capillaries and bile canaliculi for scanning electron microscopy. *Arch Histol Jpn* 47: 223–237, 1984.
26. Barlow TE, Bentley FH, Walder DN: Arteries, veins, and arteriovenous anastomoses in the human stomach. *Surg Gynecol Obstet* 93: 657–671, 1951.
27. Oka S: Studies on the microcirculation of gastrointestinal mucosa. *Saishin Igaku* 25: 1705–1713, 1970 (in Japanese).
28. Hase T, Moss BJ: Microvascular changes of gastric mucosa in the development of stress ulcer in rats. *Gastroenterology* 65: 224–234, 1973.
29. Guth PH, Rosenberg A: In vivo microscopy of the gastric microcirculation. *Am J Digest Dis* 17: 391–398, 1972.
30. Piasecki C: Blood supply to the human gastroduodenal mucosa with special reference to the ulcer-bearing areas. *J Anat* 118: 295–335, 1974.
31. Piasecki C: Observations on the submucous plexus and mucosal arteries of the dog's stomach and first part of the duodenum. *J Anat* 119: 133–148, 1975.
32. Fujita H, Fujita T: Textbook of histology, part 2, 2nd ed. Tokyo: Igaku-Shoin, 1984 (in Japanese).
33. Johansson C, Aly A: Stimulation of gastric mucus output by somatostatin in man. *Eur J Clin Invest* 12: 37–39, 1982.
34. Kusumoto Y, Iwanaga T, Ito S, Fujita T: Juxtaposition of somatostatin cell and parietal cell in the dog stomach. *Arch Histol Jpn* 42: 459–465, 1979.
35. Browning J, Gannon B: The microvascular architecture of rat proximal duodenum, with particular reference to Brunner's glands. *Biomed Res* 5: 245–258, 1984.
36. Mall JP: Die Blut- und Lymphwege im Dünndarm des hundes. *Abh Mat-Phys I König Sachs Gesell Wiss* 14: 153–189, 1887.
37. Spanner R: Neue Befunde über die Blutwege der Darmwand und ihre funktionelle Bedeutung. *Morphol Jahrb* 69: 394–454, 1932.
38. Jacobson LF, Noer RJ: The vascular pattern of the intestinal villi in various laboratory animals and man. *Anat Rec* 114: 85–101, 1952.
39. Nylander G, Olerud S: The vascular pattern of an isolated jejunal loop. A microangiographic study in the rat. *Acta Chir Scand* 121: 39–46, 1961.
40. Mohiuddin A: Blood and lymph vessels in the jejunal villi of the white rat. *Anat Rec* 156: 83–90, 1966.
41. Reynolds DG, Brim J, Sheehy TW: The vascular architecture of the small intestinal mucosa of the monkey (*Macaca mulatta*). *Anat Rec* 159: 211–218, 1967.
42. Miller DS, Rahman MA, Tanner R, Mathan VI, Baker SJ: The vascular architecture of the different forms of small intestinal villi in the rat (*Rattus norvegicus*). *Scand J Gastroenterol* 4: 477–482, 1969.
43. Reynolds DG, Swan KG: Intestinal microvascular architecture in endotoxic shock. *Gastroenterol* 63: 601–610, 1972.
44. Bellamy JEC, Latshaw WK, Nielsen NO: The vascular architecture of the porcine small intestine. *Can J Comp Med* 37: 56–62, 1973.
45. Svanvik J: Mucosal blood circulation and its influence on passive absorption in the small intestine. An experimental study in the cat. *Acta Physiol Scand* (suppl) 385: 1–44, 1973.
46. Winne D: The vasculature of the jejunal villus. *Excerpta Medica International Congress Series* 391: 56–57, 1973.
47. Szyika AS: Systématisation de l'angioarchitectonie du jéjunum et de l'iléon chez l'homme adulte. *Arch Anat Histol Embryol Norm Exp* 59: 79–142, 1976.
48. Wolfram-Gabel R: Systématisation de l'angioarchitectonie du jéjunum et de l'iléon chez l'homme en période périnatale. *Arch Anat Histol Embryol Norm Exp* 63: 39–142, 1980.
49. Wolfram-Gabel R, Maillot Cl: La vascularisation des tuniques sous-muqueuse et muqueuse des première et deuxième portions du duodenum chez l'homme adulte. *Arch Anat Histol Embryol Norm Exp* 64: 129–171, 1981.
50. Eriksson B: Microangiographic pattern in the small intestine of the cat after irradiaton. *Scand J Gastroenterol* 17: 887–895, 1982.
51. Dabelow A: Die Blutgefäßversorgung der lymphatischen Organe. *Verhand Anat Gesel 46 Versam* (*Anat Anz* 87 (Erg)): 179–224, 1939.
52. Abe K, Ito T: A qualitative and quantitative morphologic

study of Peyer's patches of the mouse. *Arch Histol Jpn* 40: 407–420, 1977.

53. Blau JN: A comparative study of the microcirculation in the guinea pig thymus, lymph nodes and Peyer's patches. *Clin Exp Immunol* 27: 340–347, 1977.
54. Gowans JL, Knight EJ: The route of recirculation of lymphocytes in the rat. *Proc R Soc Lond* (Biol) 159: 257–282, 1964.
55. Marchesi VT, Gowans JL: The migration of lymphocytes through the endothelium of venules in lymph nodes: An electron microscope study. *Proc R Soc Lond* (Biol) 159: 283–290, 1964.
56. Browning J, Gannon B: Mucosal microvascular organization of the rat colon. Acta Anat 126: 73–77, 1986.
57. Wolfram-Gabel R, Maillot Cl, Koritke JG: La vascularisation des tuniques sous-muqueuse et muqueuse du colon chez l'homme. *Arch Anat Histol Embryol Norm Exp* 66: 67–98, 1983.
58. Owen RL, Nemanic P: Antigen processing structures of the mammalian intestinal tract: An SEM study of lymphoepithelial organs. In: Scanning Electron Microscopy/1978/II RP Becker, O Johari (eds). Chicago: AMF O'Hare. SEM Inc, p 367–378, 1978.
59. Shimizu S: Beiträge zur Anatomy des Lymphgefässsystems der Wirbeltiere und des Menschen (Japaner). Nr. 8. Darmzotten und ihre Gefässe, insbesondere die Chylusgefässe der Säugetiere und des Menschen. *Okajimas Folia Anat Jpn* 10: 193–227, 1932.
60. Ohtani O: Three-dimensional organization of lymphatics and its relationship to blood vessels in rat small intestine. *Cell Tiss Res* 248: 365–374, 1987
61. Ohtani O, Ohtsuka A, Owen RL: Three-dimensional organization of the lymphatics in the rabbit appendix. A scanning electron and light microscopic study. *Gastroenterology* 91: 947–955, 1986.

Authors' address:
Department of Anatomy
Okayama University Medical School
2-5-1 Shikata-cho
Okayama 700, Japan

CHAPTER 13

The endocrine cell system of the digestive tract

TSUNEO FUJITA

1. Introduction

The sources of gut hormones are basal-granulated cells dispersed in the epithelium of the gastro-enteric mucosa. They are clear cells with secretory granules in the basal portion of the cell and were discovered over 100 years ago (1; for early literature see 2). Since the beginning of this century it has been noticed that these granules showed a positive chromaffin reaction, i.e., they tinge yellow after being treated with a fixative containing K_2CrO_4 (3, 4). For this reason the term "enterochromaffin cell" has long been the synonym of the basal-granulated or gut endocrine cell. The chromaffin reaction of the cells, meanwhile, was ascribed to their content of serotonin (5-HT) (5). However, modern cytological studies have revealed that there are basal-granulated cells which are negative for the chromaffin reaction. Furthermore, recent electron microscope studies have shown that the chromaffin and non-chromaffin cells possess secretory granules of markedly different structure and also that the nonchromaffin basal-granulated cells comprise several different cell types with regard to the fine structure of the granules (6–10). More recent immunocytochemical investigations have demonstrated that the different cell types contain different peptide hormones (Fig. 13-1). It is now widely accepted that about 20 cell types correspond to different peptides (or their groups), and, besides serotonin, different kinds of monoamines are contained in the cells, together with peptide hormones (12, 13).

This review does not aim to detail the cell types, nor their hormones and localizations, as these subjects have been dealt with in a review by Kobayashi and Segi in this series (14). The present review, rather, will analyze the general structure and function of the cells, as well as their ontogenetic and phylogenetic aspects.

2. Structure and function

2.1. General distribution and structure

The gut endocrine cells are dispersed in the mucosal epithelium (Fig. 13-1, 13-2). They are generally numerous in the intestinal crypts or in the neck portion of the pyloric and duodenal glands, but some are recognized in villous epithelium. In the gastric body they occur frequently near the bottom of the fundic glands. Different cell types have favorite or even specific sites of distribution (Table 13-1), but this review will concentrate on the general aspects of the gut endocrine cells.

The gut endocrine cells are generally pyramidal cells extending from the epithelial base to the luminal surface (Figs. 13-1–13-4). As they extend a slender cytoplasmic process reaching the gut lumen, they are called *open-type* cells (15, 16). It is worth nothing that the apical process is provided with an elaborate tuft of microvilli whose morphology differs among cell types (15, 16) (Fig. 13-3). The round nucleus is shifted towards the cell base, containing one or two rather conspicuous nucleoli and dispersed chromatin masses. Golgi apparatuses are well developed near the nucleus. Cisterns of the rough endoplasmic reticulum are rather well developed and occasionally form a lamellar assembly reminiscent of a Nissl body in neurons (Figs. 13-4, 13-5). A pair of centrioles are found in the supranuclear portion. Several lysosomes may be dispersed in the cytoplasm (Fig. 13-6).

In contrast to the above-described cells of the open type, there are *closed-type* endocrine cells which are restricted to the gastric body in mammals (15, 16) and to the proventricle in birds (17). In humans, all the endocrine cells in the gastric body, i.e., the area containing fundic glands, are closed in type (18). The cells of this type lie flat on the basement membrane, extending themselves horizontally. Their apical aspect is

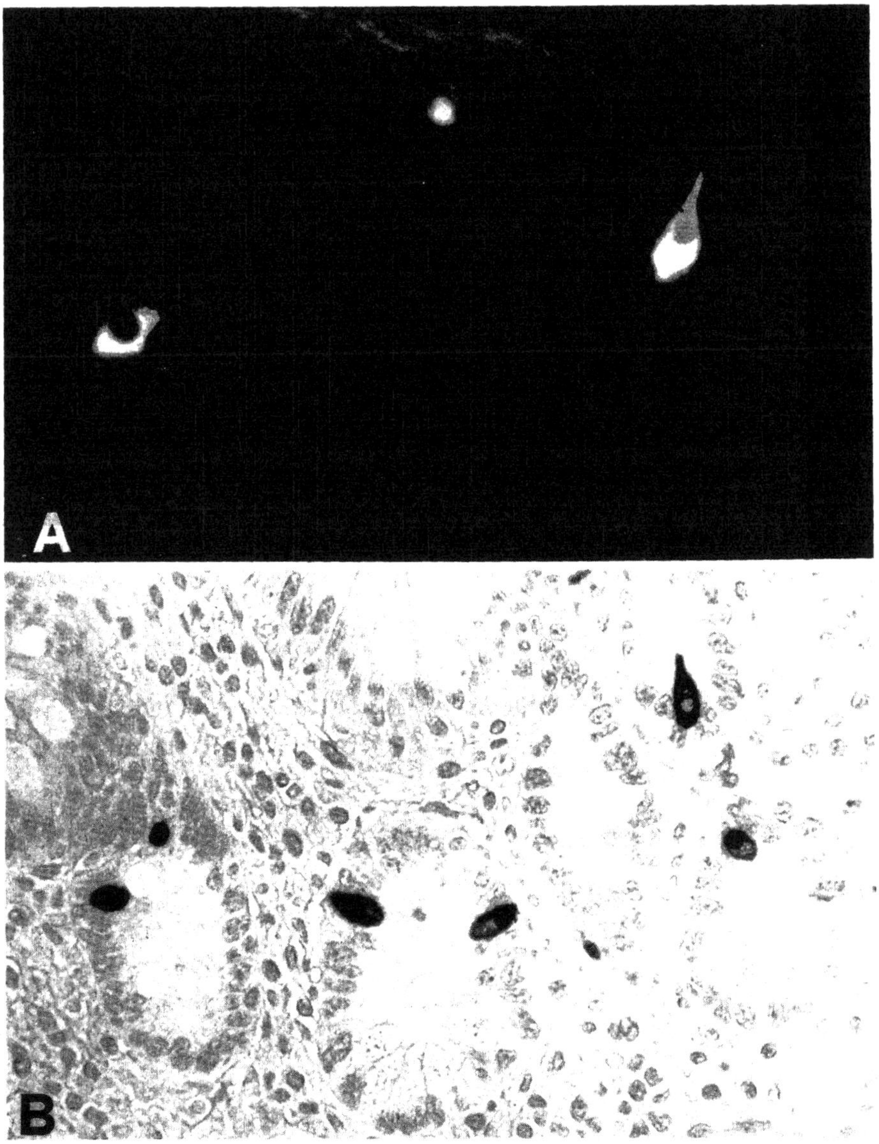

Figure 13-1. Immunohistochemical demonstration of gut endocrine cells. (A). CCK-PZ containing cells in the human duodenum shown with the immunofluorescence technique. (B). Glicentin-immunoreactive cells in the human ileum demonstrated by the peroxidase-antiperoxidase method. Micrographs by Dr. T. Iwanaga.

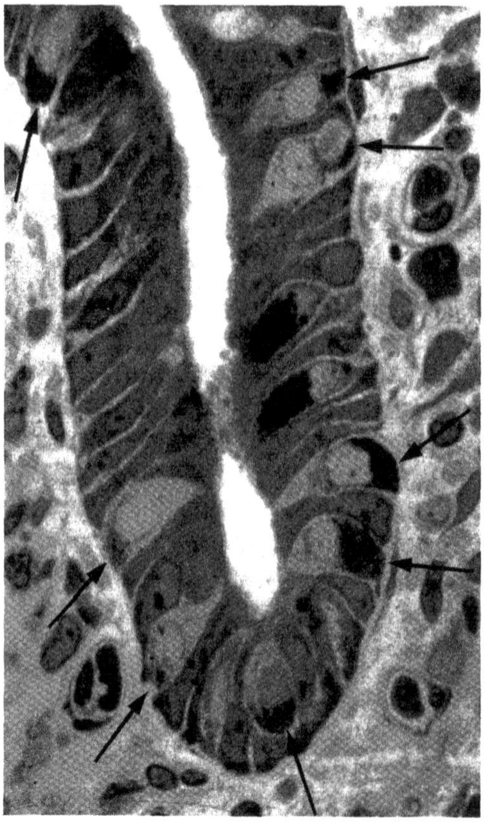

Figure 13-2. Light micrograph showing a longitudinally sectioned human duodenal crypt. Clear cells with basal granules are endocrine cells (arrows). Note also erythrocyte containing capillaries in the connective tissue which transfer the gut hormones. Toluidine blue stained upon section. Micrograph by Dr. M. Osaka.

covered by other epithelial cells, thus being separated from the gut lumen. The features of the organelles in the closed-type cells are essentially the same as those in the open-type cells, though the location of their cytoplasmic organelles is not as clearly polarized as in the open-type cells (18).

2.2. Cell types and their hormones

The main types of gastroenteric endocrine cells are shown in Table 13-1 with reference to their hormones and distribution mainly in human gut. This table also contains pancreatic endocrine elements, as they are unseparable from the overlapping with the gut endocrine cells. As reviews dealing with the cell types and their hormones are available (12, 13, 14), only brief comments on particular problems of some of the cells will be given.

2.2.1. EC cell

The EC (enterochromaffin) cell is defined by its possession of serotonin. Its secretory granules are characteristically electron dense and distorted in shape (Fig. 13-6). These granules vary markedly in size among cell populations within the same individual (19), and therefore several subtypes of EC cells have ben suggested to occur.

The occurrence of a bioactive peptide in the

Table 13-1. Cell types, distribution and messengers of gut endocrine cells.

Cell type	Stomach		Small intestine		Large intestine	Signal substance	
	Fundus	Pylorus	Upper	Lower		Peptide	Amine
D	+	+	++	+	+	Somatostatin	
D_1	+	+	+	+	+	?	
EC	+	+	++	+	+	(Enkephalins)	Serotonin
ECL	++	–	–	–	–	?	(Histamine)
G	–	++	+	–	–	Gastrin	
M or I	–	–	++	+	–	CCK-PZ	
K	–	–	+	+	–	GIP	
L	–	–	+	+	+	GLI	
Mo	–	–	+	+	–	Motilin	
N	–	–	–	+	–	Neurotensin	
PP	–	[+]	[+]	–	–	PP	
S	–	–	+	+	–	Secretin	

( ) In animals, but not identified in man
[ ] In human fetuses or newborns, rarely in adult

EC cell has been proposed (15). Motilin and substance P have been reported as the candidates but have been disputed (for reference see 20). Recently enkephalins have been demonstrated in the EC cells in the pig duodenum and pyloric antrum (20) (Fig. 13-7). The EC cells in the gastric body and colon in the same animal have been found negative for enkephalins, suggesting the occurrence of different subtypes of the EC cell (20).

2.2.2. *EC-like (ECL) cell*

This cell, characterized by vacuolar granules with a dark core, is distributed only in the gastric body. In the mouse and rat this cell possesses histamine (21), but nothing essential is known about the possible role of this monoamine upon the acid release from the parietal cells. Little is known about a peptide secreted from this cell. Intrinsic factor has been proposed to be stored in this cell (22), but this view is not widely supported.

2.2.3. *D_1 cell*

This cell contains round, gray appearing granules, smaller in diameter (150–200 nm) than those of the D cell. The granule contents are tightly surrounded by a membrane sac, leaving no clear space between them. Gastric inhibitory peptide (GIP) was found in the D_1 cell (23), but recent studies indicate the possibility that this category of cell may include different subtypes corresponding to different peptide hormones. Motilin secreting cells, today called Mo cells, were previously included in the D_1 cell category (24).

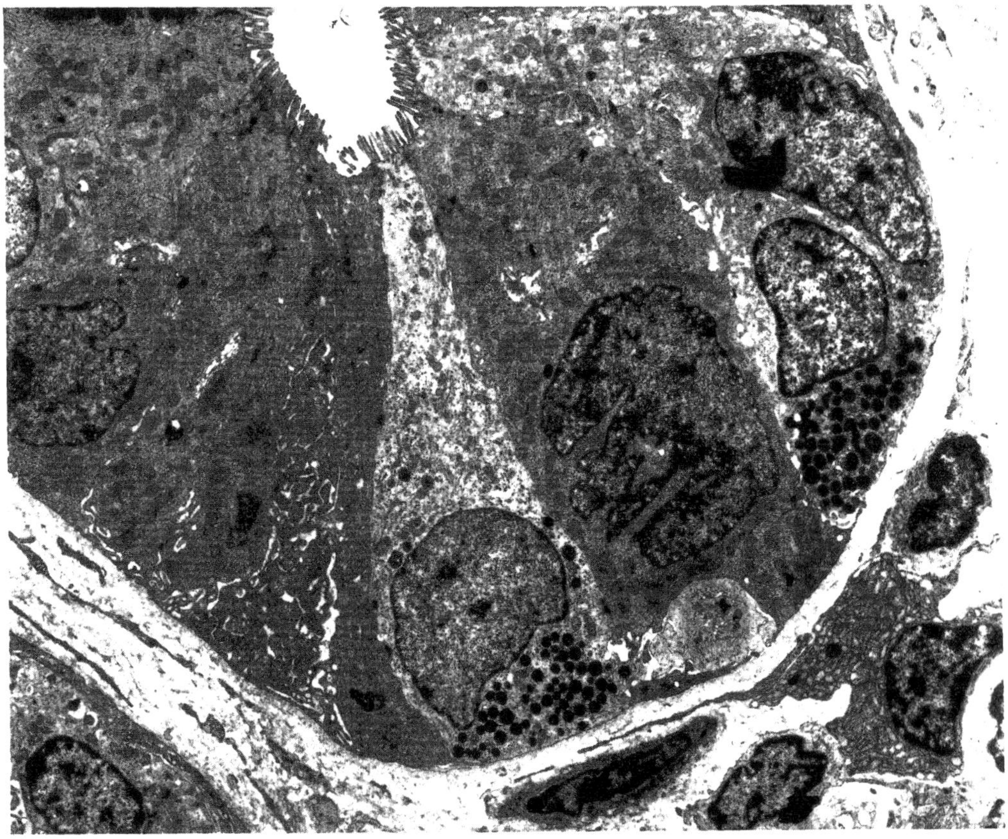

Figure 13-3. Low power electron micrograph showing a transverse section of a duodenal crypt. Two L cells with large dark granules are seen. One of them extend a process covered by microvilli to the lumen. Micrograph by Dr. M. Osaka. X 3,500.

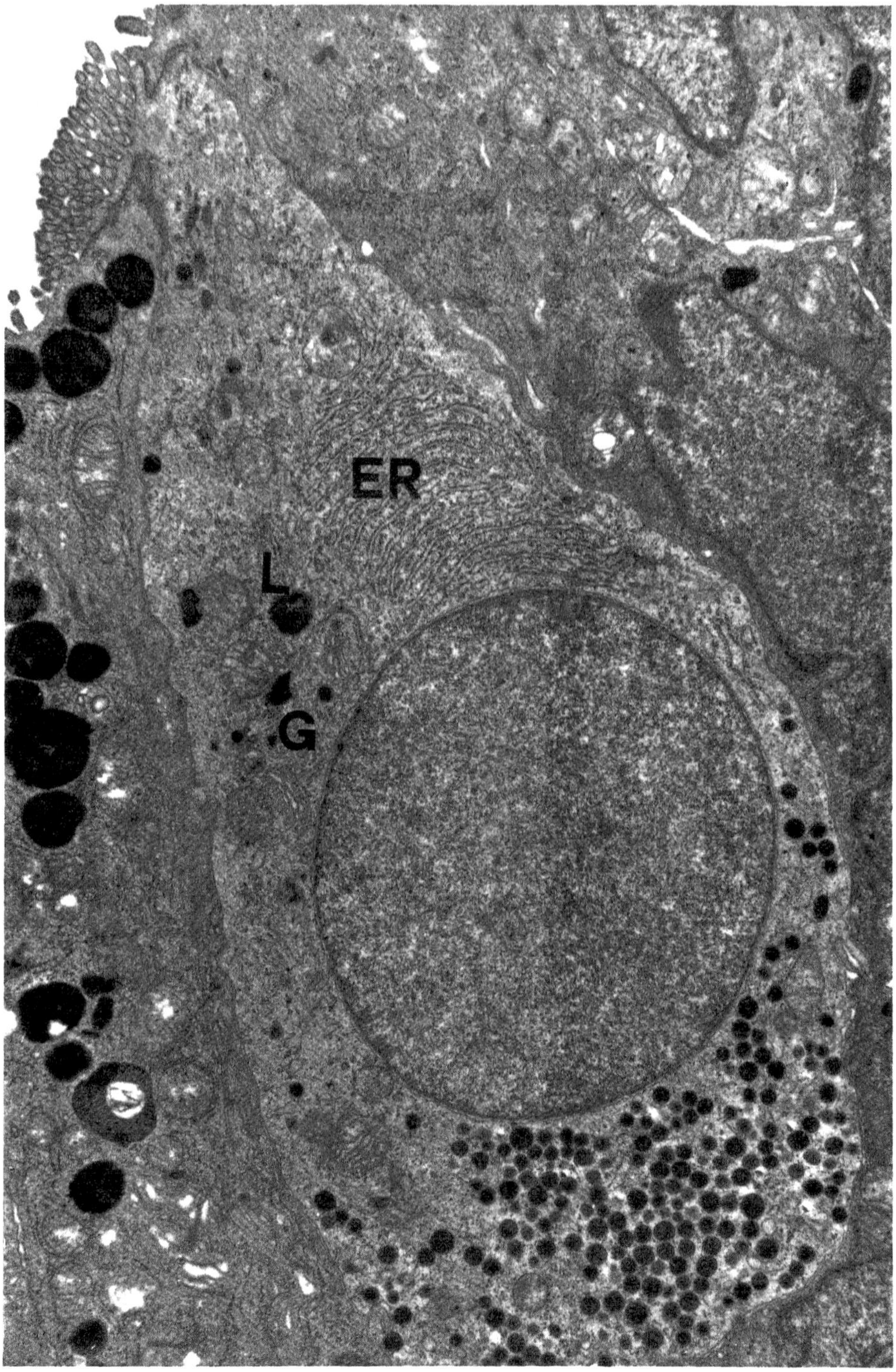

Figure 13-4. M cell characterized by medium-sized round granules and especially elaborate microvilli. Note a prominent mass of the rough endoplasmic reticulum (ER), reminiscent of the Nissl body. The Golgi complex (G) is also well-developed. L, lysosomes. Reproduced from Sasagawa et al. (Ref. 11). X 14,000.

2.3. Receptive function

The first evidence that the basal-granulated cells are really endocrine in nature was obtained by detecting the exocytotic or Ω-shape release of their granules under the electron microscope after stimulating the cells from the gut lumen with different kinds of secretagogues (25,26). Thus, luminal administration of 0.1 N hydrochloric acid in the dog and human caused granule release of D, S, L, and M cells, while $NaHCO_3$ or meat extracts caused that of G cells (25–28). A parallelism between the exocytotic incidences and blood gastrin levels has been demonstrated (29). When amino acids (phenylalanine, etc.) or an $MgSO_4$ solution are introduced into the human duodenum, the gall bladder contracts and pancreatic enzyme output is increased, indicating the release into circulation of cholecystokinin-pancreozymin (CCK-PZ). At this timepoint exo-

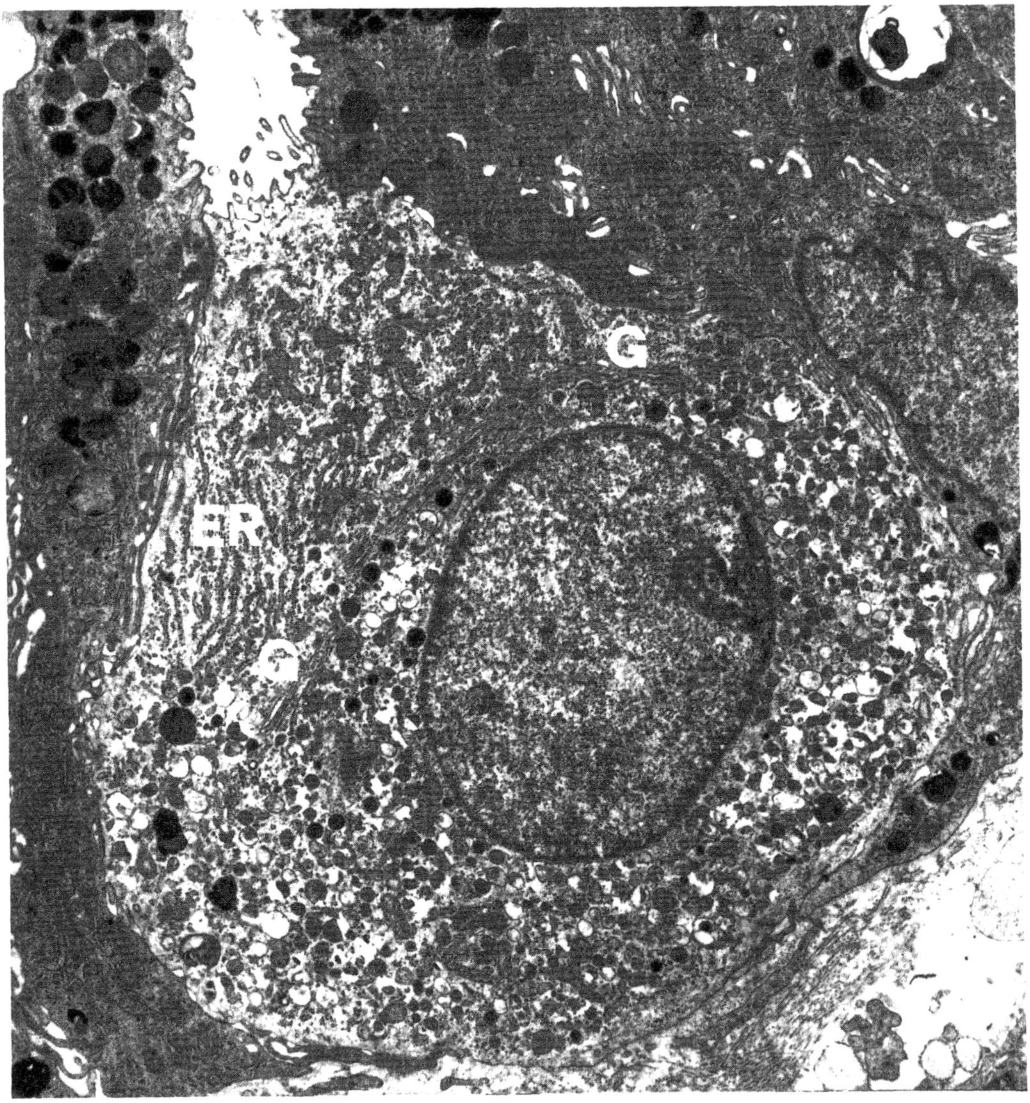

Figure 13-5. G cell in human pyloric antrum. With its characteristic stout shape, the cell also reaches the lumen with its microvillous apex. Note the well-developed endoplasmic reticulum (ER) and Golgi complexes (G). The secretory granules are partly dense but partly loose or vacant-looking in contents after the routine fixation procedures. Reproduced from Sasagawa et al. (Ref. 11). X 9,000.

cytotic figures have been detected by electron microscopy in certain endocrine cells resembling M cells (30). EC cells, on the other hand, are sensitive to agents causing diarrhea. In the dog, introduction of 50% glucose into the duodenum causes severe degranulation in the EC cells (26). In juvenile rabbits, intraluminally administered cholera toxin induces the same reaction of the

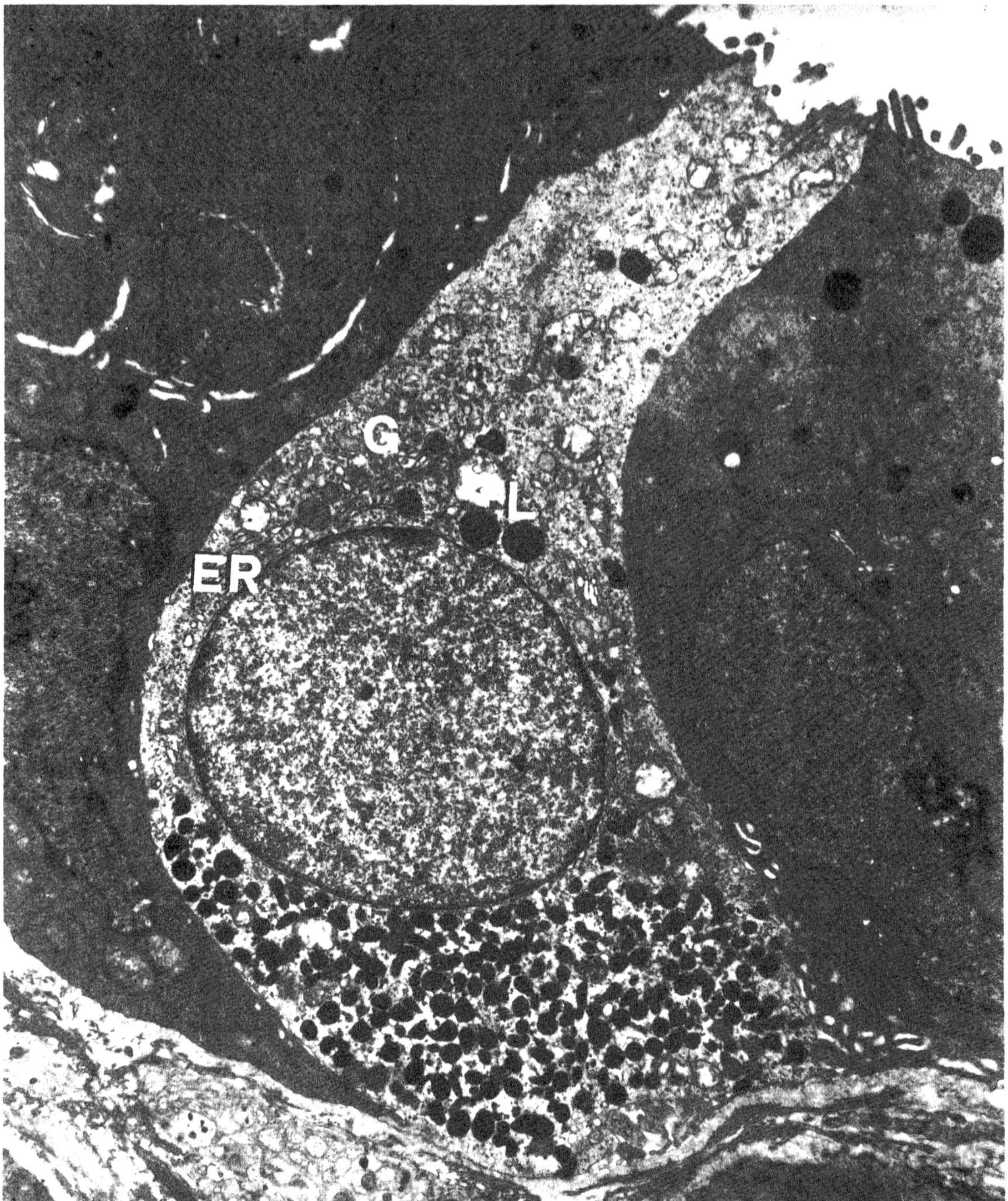

Figure 13-6. EC cell in human duodenal crypt. The granules of this cell are very dark and distorted in shape. The rough endoplasmic reticulum (ER) and Golgi complex (G) are clearly seen. L, lysosomes. Reproduced from Sasagawa et al. (Ref. 11). X 10,000.

cells (31, 32). It has been suggested that serotonin and associated peptide(s) released from the EC cells might mediate diarrhea by causing hyperexcretion of water and mucus and elevating the movement of the gut muscles (31, 32).

The above-mentioned experiments are based on the hypothesis that the luminal chemical and physicochemical stimuli are received by the apical microvilli of the bipolar endocrine cell which then responds to the stimuli by releasing its secretions from the cell base (15, 25, 26). This hypothesis is the first that could account for the mechanism of the release of gut hormones in response to the chemical information of the food.

This hypothesis is contrasted against the previous one saying that a nerve terminal reaching the mucosal surface recognizes the luminal stimuli, and this nerve transmits its excitation to an endocrine cell over some synaptic relays belonging to the vagal nerve (33).

Our morphological observations, either by silver impregnation or by electron microscopy, have never revealed the occurrence of a nerve terminal in or on the gut epithelium, nor have we recognized a nerve synapse to a gut endocrine cell. Furthermore, we performed pharmacological experiments to examine the possible intervention of nerves between the luminal stimuli and hormone release. In anesthetized dogs, the CCK-PZ release from the duodenum induced by luminally administered amino acids was measured in terms of protein output from the pancreas. Neither atropine (1 mg/kg, IV) nor tetrodotoxin (5 μg/kg, IV) inhibited the release of CCK-PZ by duodenal amino acids (34). These experiments deny the intervention of nervous elements between the luminal stimuli and hormone release.

Local anesthetics administered intraluminally have long been known to suppress the hormone release induced by luminal chemical stimuli (for reference see 35). This previously has been thought to be evidence to support the hypothesis of the occurrence of nervous elements that recognize and transmit luminal stimuli. According to our view, however, this is the evidence that the open-type endocrine cell in the gut, especially its apical receptor portion, is sensitive to local anesthetics (35).

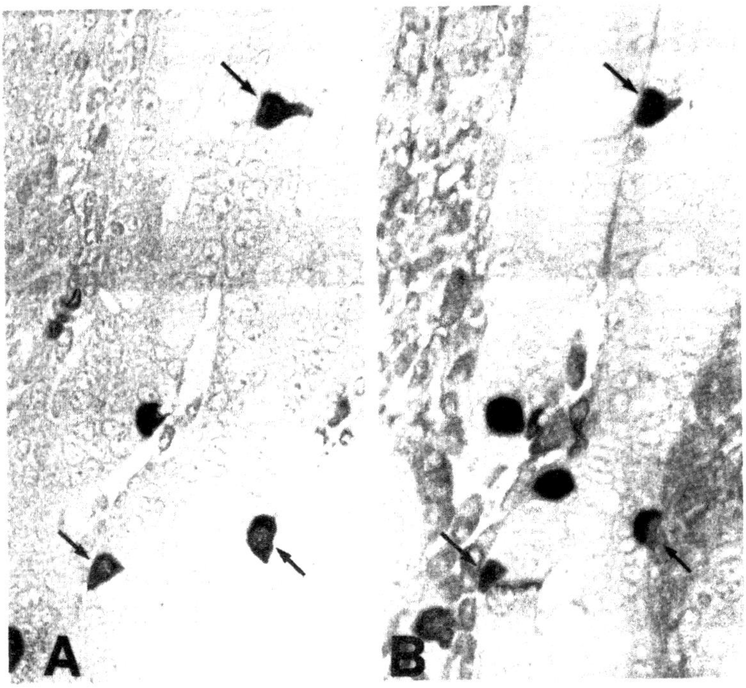

Figure 13-7. Two consequent paraffin sections are stained alternately with an antiserum against Met-enkephalin-Arg-Gly-Leu (A) and with an antiserotonin antiserum (B). It is clear that the same cells (arrows) contain both enkephalins and serotonin.

The closed-type endocrine cells, which are restricted to the gastric body, do not respond to the luminal chemical stimuli. It is postulated that mechanical stimuli, especially extension and pressure to the mucous membrane, may cause deformation of the endocrine cells, bringing about granule release (15). Although the closed-type endocrine cells in mammals are of a rather simple lens shape, they extend long dendritic processes along the basal aspect of the epithelium in the avian proventricle (17). This structure seems to be an adaptation to subtly recognize the delicate stress and strain in the proventricular wall.

2.4 Secretory function

2.4.1. Coexistence of secretory substances

Recent biochemical and immunocytochemical studies have indicated that secretions of multiple categories are produced in a single endocrine cell and, as they are packaged in the same granules, each secretory granule contains all these substances (37). The secretory substances coexisting in the cell and in the granule are as follows:

1. *Messenger peptides or so-called peptide hormones.* These are fragments born by the processing of a larger molecule, the preprohormone. Only part of the fragments are bioactive and therefore many inactive peptides are also contained in the secretory granule.
2. *Amines or biogenic monoamines.* Serotonin (5-hydroxy-tryptamine) is produced in the EC cell of a large variety of vertebrates (Figs. 13-7, 13-8), while histamine has been detected in the EC-like cell of rats and mice (see above). Advances in fluorescence histochemistry of amines have revealed that not only the EC and murine EC-like cell, but also other endocrine cells have the APUD (amine precursor uptake and decarboxylation) ability. This ability becomes manifest when one administers an animal with a precursor of monoamine such as L-dopa (38).
3. *ATP and other adenine nucleotides.* These are very constant constituents of the secretory granules of endocrine cells, including those in the digestive tract (16, 39, 40). Fluorescence

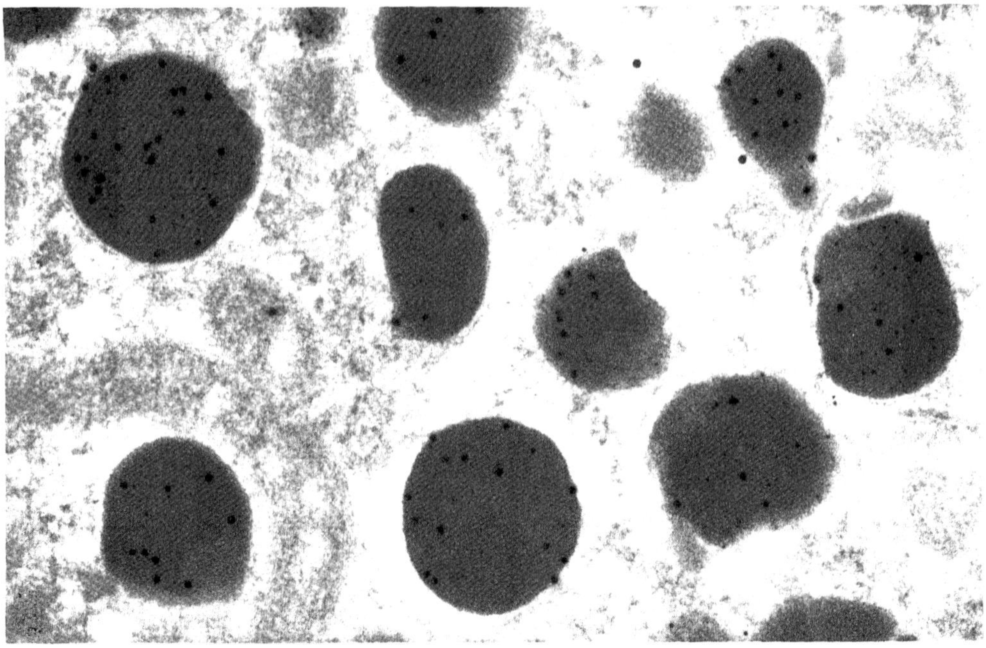

Figure 13-8. Coexistence of enkephalins and serotonin in EC cell granules shown by the protein A-gold method. The immunoreactivity for Met-enkephalin-Arg-Gly-Leu is indicated by small gold particles, while that for serotonin, by large particles. Each secretory granule contains both large and small particles. Courtesy by Dr. T. Uchida, Department of Anatomy, Yamanashi Medical College. X 56,000.

staining with quinacrine and uranaffine reaction can demonstrate the localization of ATP in the secretory granules for light and electron microscopy, respectively.
4. *Chromogranins and related acidic glycoproteins.* Chromogranin A, the representative substance of this category, obtained from the adrenal chromaffin cell granules, has recently been demonstrated by immunohistochemistry to be contained in the secretory granules of a large variety of endocrine cells including those in the digestive tract (41–43). Nonspecific and semispecific stainings of endocrine granules are believed to demonstrate these acidic carrier proteins. The metachromatic reactions of endocrine granules in toluidine blue, pH 5–6 are ascribed to this category of substances (15).

These and still other substances are combined into a large molecule to be stored in the secretory granule.

2.4.2. *Production and transport of secretory substances*

The secretory substances of proteinic and peptidic nature are synthesized in the rough endoplasmic reticulum, transported to the Golgi apparatus, and packaged in secretory granules. ATP and other adenine nucleotides enter these granules, presumably from mitochondria. Amines thereafter are taken up into the granules from the cytosol. A detailed account and references will be omitted on these intracellular events as reviews are available (16,44).

2.4.3. *Release of secretory substances*

The release of secretory substances from the gut endocrine cells occurs by an exocytotic mechanism in which granules open on the basal aspect of the cell (see above). Evidence is available that the influx of Ca^{2+} from the cell base, and not from the apical (luminal) side, is essential for the cell to exocytotically release the secretions (45).

When granule release takes place very actively, the membrane sacs of two, three, or more granules fuse together and the contents are released in tandem from the cell (28).

During or after the exocytotic opening of granules, a part of or the entire membrane sac becomes coated and appears bristlelike. These bristle-coated membrane portions invaginate into the cytoplasm, become pinched off, and are transferred to the Golgi area for the reutilization of the membrane materials (16, 26, 44).

In the gut endocrine cells, exocytotic evidence can be very rarely encountered when the digestive tract is empty. They show exocytotic images when adequate stimuli are given from the lumen (see above). During the fetal life, at least certain kinds of endocrine cells seem to be engaged in granule release continuously, i.e., apparently independently of the luminal stimuli. One of the typical cases has been shown in the D cells in human fetuses (4–7 months). Possible trophic functions of gut hormones upon developing digestive organs in fetuses have been suggested (46).

2.4.4. *Paracrine and hemocrine transport*

The messenger substances released from the gut endocrine cells are diffused through the basement membrane and connective tissue of the propria mucosae. Part of them may act upon vicinal targets, e.g., secretory cells, smooth muscles, and nerves. This short distance transfer of messengers, called paracrinia, seems very important because this is the basis of the local reactions, either secretory, contractile, or nervous, of the gut wall against luminal stimuli (15, 37).

Other messenger substances penetrate the blood capillary wall to enter the circulation. They may be transferred only a short distance by a capillary loop or network to escape soon into the connective tissue, to reach their targets such as grandular cells or smooth muscles. This is called microhemocrinia. They may also be conveyed to remote targets, i.e., first to the liver via the portal vein, and thereafter to organs and cells to be reached by the general circulation. This is macrohemocrinia (37).

It is also known that the D cell in the gastric body may give a slender cytoplasmic process to a parietal cell, suggesting the synaptocrine control of the former upon the latter (Fig. 13-9). It has been proposed that this morphological relation may correspond to the inhibitory action of somatostatin (the D cell's secretion) upon the hydrochloric acid secretion in the stomach (47, 48).

3. Gut endocrine cells as paraneurons

The structural and functional features of gut endocrine cells depicted above all support the view that they are closely related to neurons and deserve the name of paraneurons (39, 40). These cells are, so to speak, half receptive and half secretory in nature, and therefore are located at the middle of the category of sensory and endocrine paraneurons (40).

In this context, the secretions of the gut endocrine cells are especially important. Typical features of paraneurons include not only the occurrence of numerous neuropeptides and neuramines (hence the "brain-gut hormone"), but also their coexistence with different categories of substances (see above) within the secretory granules. In other words, these are features common to neurons.

Recent immunocytochemical studies have revealed that various kinds of neuron-specific peptides may be contained in paraneurons, including gut endocrine cells. Neuron-specific enolase (49, 50, 51), neurofilament protein (51), neurotubulin protein (52), and Purkinje cell-specific protein (spot 35 protein) (53, 54) may be detected in some populations of gut endocrine cells in humans and in other animals. This fact strengthens the view that the gut endocrine cells share different metabolic mechanisms with neurons.

The open-type gut endocrine cells are bipolar cells, recognizing the chemical information in the lumen, which is the external milieu, and, in response, releasing their secretions into the connective tissue, which is the internal milieu. Thus cells of this type are the simplest and self-completed paraneurons directly mediating the external and internal milieus and transducing stimuli into biological signals. The counterparts of the gut endocrine cells are found in the bronchopulmonary epithelium and in the urethra of mammals.

4. Ontogeny

4.1. Origin

It has long been a generally accepted view that the gut endocrine cells are of endodermal origin. Some authors, however, proposed mesenchymal (55) and neurogenic (56) derivation of the cells

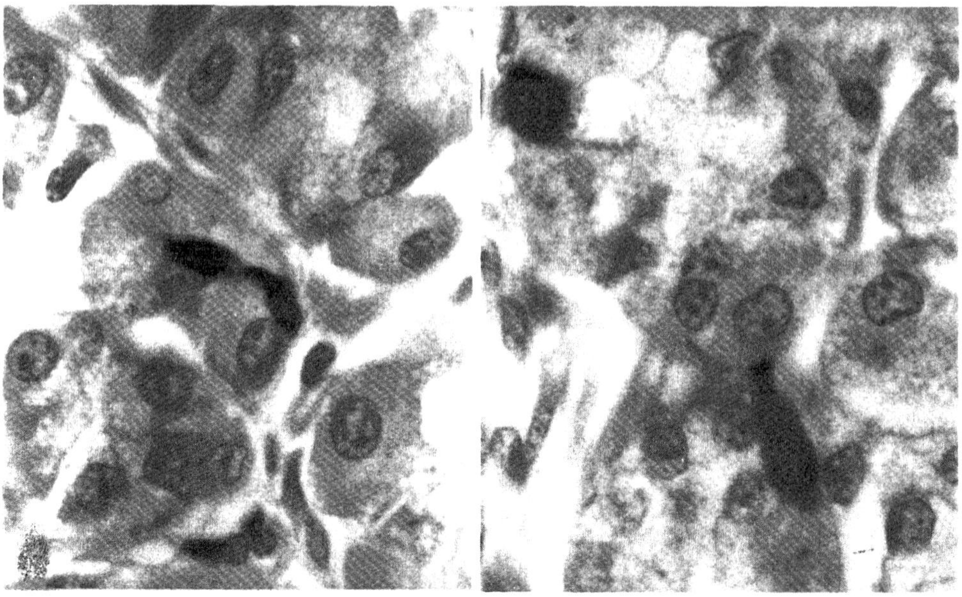

Figure 13-9. Images suggesting a synaptocrine action of D cells, immunostained with an antisomatostatin antiserum, upon vicinal parietal cells. Note the slender cell process embracing or terminating on the target cell. Dog gastric body. Micrography by Dr. Y. Kusumoto.

on the basis of their observations of fetal tissues.

Paying attention to the APUD ability of the cells in question, Pearse and Polak (38) declared that they are derived from the neural crest and migrate into the gut wall. This view attracted the attention of many research workers but later came to be neglected by experimental-embryological studies (57) using the quail and chicken chimaera (58) and other techniques (59).

In fetal intestine, endocrine cells may be located in the subepithelial tissue and are often associated with neurons and nerve fibers (60). It seems possible to presume that the endocrine cells may migrate out of the epithelium to make contact with the nervous elements, so that a nervous induction may take place here. The ontogeny of the gut endocrine cells may be summarized, therefore, as follows: the cells originate from the endoderm or the gut epithelium, but it may be possible that their differentiation is completed after receiving a nervous induction.

4.2. *Cell kinetics*

The differentiation and proliferation of the gut endocrine cells take place in conspicuously different ways according to the animal species and parts of the digestive tract. In rat fetuses, glucagon-immunoreactive cells are the first recognizable endocrine elements which appear on day 11 of gestation in the foregut epithelium (61). In fetal and neonatal rats, the GEP (gastro-entero-pancreatic) endocrine cells are distributed less site-specifically than in adult animals. Thus, gastrin cells constantly occur in the pancreas, whereas cells immunoreactive for insulin and PP (pancreatic polypeptide) are dispersed in the upper digestive tract (62).

The gut endocrine cells are believed to differentiate from "immature precursor cells" in the generative zone of the epithelium (63). Mitotic proliferation has been reported to occur after the cells have possessed their specific peptide or amine (63, 64, 65).

The endocrine cells, after differentiation and proliferation, migrate towards the site of epithelial cell degeneration, i.e., upward to the villous tip in the intestine and, in the stomach, downward to the bottom of the fundic and pyloric glands. In the fetal intestine of humans and some mammals, a large, saucer-shaped aggregation of endocrine cells is found on the top of every villus and is called the Segi's cap (66). Its functional significance and the mechanism of formation are unknown. It has been proposed that the endocrine cells live longer than ordinary enterocytes and therefore survive them after migrating to the villous top (14).

The turnover rate of the gut endocrine cells varies, largely according to the researchers rather than among species. It thus ranges from a few months (64, 65, 67) to two to four days (63, 68–70).

5. Phylogeny

5.1. *Bipolar gut paraneurons in animal evolution*

5.1.1. *Coelenterata*

The prototype of the open-type (bipolar) paraneurons can be found in the gastrodermis (endoderm) of *Hydra* (Coelenterata) (71). The cells contain aggregations of secretory granules and those of synaptic vesicles in the basal portion of the cell and extend an apical process equipped with a single cilium and a specialized cytoplasmic fold. Cells with essentially identical structures have also been demonstrated in the ectoderm of *Hydra* (72). It is unclear as to which of the hitherto known neuropeptides in *Hydra* may be contained in the gastrodermal paraneurons, but the reports by Grimmelikhuijzen and associates (73, 74) seem to suggest that at least neurotensin-like and oxytocin/vasopressinlike immunoreactivities might be contained in those cells.

5.1.2. *Protostomes*

The flatworm *Microstomum lineare* possesses paraneuronlike cells of both open and closed type dispersed in the intestinal epithelium. They seem to contain vasotocinlike and PP-like immunoreactivities (75). In molluscs, it is not clear whether the intestinal paraneurons are present or not.

Among arthropods, insects (cockroaches, butterflies, crickets, etc.) have been demonstrated to possess typical bipolar paraneurons in the epithelium of the midgut. They contain PP-, glicentin, somatostatin-, PHI- and urotensin I-like immunoreactivities (76–80). The electron microscope studies by Endo and Nishiitsutsuji-Kwo (81) have revealed that the cells extend a very long process covered by a tuft of microvilli into the intestinal lumen and store secretory granules which are, when the cells are stimulated, released by a clear exocytotic mechanism.

5.1.3. Deuterostomes

The gut epithelium of the seasquirt (Ascidacea, Protochordata) possesses numerous paraneurons, mostly of the typical open type (82, 83). In *Ciona intestinalis* these cells have been shown to contain immunoreactivities for somatostatin, gastrin, bombesin/GRP (gastrin releasing peptide), secretin, PP (pancreatic polypeptide), and neurotensin; as a monoamine, serotonin was also detected (84).

In the amphioxus *Branchiostoma japonicum* (Cephalochordata), open-type endocrine cells have been shown in the gut epithelium by electron microscopy (85). Recent immunocytochemical studies in *Branchiostoma lanceolatum* suggest the occurrence of eight brain-gut peptides; insulin, glucagon, somatostatin, PP, CCK/gastrin, VIP (vasoactive intestinal polypeptide), secretin, and neurotensin.

It is noteworthy that insulin-containing cells are present in the gut epithelium as open-type paraneurons in the amphioxus (86). Thus, they are believed to detect chemical stimuli in food, probably sugars and possibly amino acids, and react to them by releasing messenger substances. In higher animals, insulin cells leave the intestinal epithelium to gather, in connective tissue, into an islet organ. Thus they come to react to the chemical information, not directly of the food, but of the blood, which is richly supplied to the islet organ. The insulin-producing B cells are now confined to the pancreatic islets and are no longer recognizable in the gut. Only in the fetal period in mammals are occasional B cells found in the intestinal epithelium, suggesting an atavistic state.

Holocephalian, selachian, and teleosteal fishes, amphibians, and reptiles have numerous endocrine cells, mainly of the open type, containing various gut hormones with the exception of insulin. Birds possess, besides the open-type cells in the intestine, closed-type cells extending dendritic processes in the proventricle. These are presumed to be mechanoreceptive endocrine cells.

6. Concluding remarks

The knowledge of the gut endocrine cells' regularly opening into the lumen with a microvillous apex has brought about the idea that these cells are chemoreceptive, besides being secretory in nature. This view is the first to enable us to account for the mechanism of the recognition of luminal stimuli and the reactions of the digestive organs to them. This view also has visualized the cells in question as half-nervous and half-secretory cells, deserving the name of paraneurons.

Discoveries of numerous bioactive peptides and amines shared by neurons and endocrine cells (paraneurons), including the gut endocrine cells, have supported the concept of paraneurons. More recently, other secretory and nonsecretory substances have been demonstrated to occur commonly in neurons and paraneurons.

Phylogenetic investigations indicate that the bipolar paraneurons have originated in the ectoderm and endoderm of the simplest metazoa, *Hydra*, and have been preserved in the gut epithelium, both through protostomian and deuterostomian evolution.

References

1. Heidenhain R: Untersuchungen über den Bau der Labdrüsen. *Arch Mikr Anat* 6: 368–406, 1870.
2. Patzelt V: Der Darm. In: Möllendorff's Handbuch der Mikroskopischen Anatomie des Menschen, Vol 3. p 1–448, 1936.
3. Schmidt JE: Beiträge zur normalen und pathologischen Histologie einiger zellarten der Schleimhaut des memschlichen Darmkanals. *Arch Mikr Anat* 66: 12–40, 1905.
4. Ciaccio C: Sur une nouvelle espèce cellulaire dans les glands de Lieberkühn. *C R Soc Biol Ses Fil* 60; 76–77, 1906.
5. Erspamer V, Asero B: Identification of enteramine, the specific hormone of the enterochromaffin cell system as 5-hydroxy-tryptamine. *Nature* (Lond) 169: 800–801, 1952.
6. Ito S, Winchester RJ; The fine structure of the gastric mucosa in the bat. *J Cell Biol* 16: 541–577, 1963.
7. Toner PG: Fine structure of argyrophil and argentaffin

cells in the gastrointestinal tract of the fowl. *Z Zellforsch* 63: 830–839, 1964.
8. Solcia E, Vassallo G, Sampietro R: Endocrine cells in the antro-pyloric mucosa of the stomach. *Z Zellforsch* 81: 474–486, 1967.
9. Orci L, Pictet R, Forssmann WG, Renold AE, Rouiller C: The endocrine cells in the epithelium of the gastrointestinal mucosa of the rat. *Diabetologia* 4: 56–67, 1968.
10. Forssmann WG, Orci L, Pictet R, Renold AE, Rouiller C: The endocrine cells in the epithelium of the gastrointestinal mucosa of the rat. An electron microscope study. *J Cell Biol* 40: 692–715, 1969.
11. Sasagawa T, Kobayashi S, Fujita T: Electron microscope studies on the endocrine cells of the human gut and pancreas. In: Gastro-entero-pancreatic Endocrine System. A Cell-biological Approach. T Fujita (ed), Tokyo Igaku-Shoin, p 17–38, 1973.
12. Grube D, Forssmann WG: Morphology and function of the enteroendocrine cells. *Horm Metab Res* 11: 589–606, 1979.
13. Solcia A, Creutzfeldt W, Falkmer S, Fujita T, Greider MH, Grossman MI, Grube D, Håkanson R, Larsson LI, Lechago J, Lewin K, Polak JM, Rubin W: Human gastroenteropancreatic endocrineparacrine cells. Santa Monica 1980 classification. In: Cellular Basis of Chemical Messengers in the Digestive System. MI Grossman, MAB Brazier, J Lechago (eds), New York: Academic Press, p 159–165, 1981.
14. Kobayashi S, Segi M: Gut paraneurons and Segi's cap. In: Ultrastructure of Endocrine Cells and Tissues. PM Motta (ed), Boston: Martinus Nijhoff Publishers, p 127–135, 1984.
15. Fujita T, Kobayashi S: The cells and hormones of the GEP endocrine system — the current of studies. In: Gastro-entero-pancreatic Endocrine System. A Cell-biological Approach. T Fujita (ed), Tokyo: Igaku-shoin, p 1–16, 1973.
16. Fujita T, Kobayashi S: Structure and function of gut endocrine cells. *Int Rev Cytol Suppl* 6: 187–233, 1977.
17. Iwanaga T, Yamada J: Endocrinelike cells of peculiar shape in the proventriculus of the chicken — possible mechanoreceptors? *Biomed Res* 1 (suppl): 28–32, 1980.
18. Kobayashi S, Fujita T, Sasagawa T: Electron microscope studies on the endocrine cells of the human gastric fundus. *Arch Histol Jap* 32: 429–444, 1971.
19. Böck P, Gorgas K: Enterochromaffin cells and enterochromaffinlike cells in the cat pancreas. In: Endocrine Gut and Pancreas. T Fujita (ed), Amsterdam: Elsevier Scientific Publishing Company, p 13–24, 1976.
20. Nihei K, Iwanaga T, Yanaihara N, Mochizuki T, Fujita T: Preproenkephalin A occurs in the enterochromaffin (EC) cells of the procine intestine: An immunocytochemical study using antisera to met-enkephalin-Arg^6-Gly^7-Leu^8 and to serotonin. *Biomed Res* 4: 393–398, 1983.
21. Håkanson R, Owman CH, Sjöberg NO, Sporrong B: Amine mechanisms in enterochromaffin and enterochromaffinlike cells of gastric mucosa in various mammals. *Histochemie* 21: 189–220, 1970.
22. Håkanson R, Lindstrand K, Nordgren L, Owman CH: Histamine-containing epithelial cells in rat stomach: A possible storage site for the intrinsic factor. *Eur J Pharmacol* 8: 315–325, 1969.
23. Polak JM, Bloom SE, Kuzio M, Brown JC, Pearse AGE: Cellular localization of gastric inhibitory polypeptide in the duodenum and jejunum. *Gut* 14: 284, 1973.
24. Polak JM, Buchan AMJ: Heterogeneity of the D_1 cell. In: Cellular Basis of Chemical Messengers in the Digestive System. MI Grossman, MAB Brazier, J Lechago (eds), New York: Academic Press, p 121–131, 1981.
25. Fujita T, Kobayashi S: Experimentally induced granule release in the endocrine cells of dog pyloric antrum. *Z Zellforsch* 116: 52–60, 1971.
26. Kobayashi S, Fujita T: Emiocytotic granule release in the basal-granulated cells of the dog induced by intraluminal application of adequate stimuli. In: Gastro-entero-pancreatic Endocrine System. A Cell-biological Approach. T Fujita (ed), Tokyo, Igaku-Shoin, p 49–58, 1973.
27. Osaka M, Sasagawa T, Fujita T: Emiocytotic granule release in the human antral endocrine cells. In: Gastro-entero-pancreatic Endocrine System. A Cell-biological Approach. T Fujita (ed), Tokyo: Igaku-Shoin, p 59–63, 1973.
28. Osaka M, Sasagawa T, Fujita T: Granule release from endocrine cells in acidified human duodenal bulb. An electron microscope study of biopsy materials. *Arch Histol Jpn* 37: 73–94, 1974.
29. Miyagami H, Watanabe Y, Sawada Y, Kato K, Shiono K, Kondo K, Kidokoro T: Ultrastructures of G cells and the mechanism of gastrin release before and after selective vagotomy with pyloroplasty. *Arch Histol Jpn* 40: 51–62, 1977.
30. Kobayashi S, Sasagawa T: Morphological aspects of the secretion of gastro-enteric hormones. In: Endocrine, Gut and Pancreas. T Fujita (ed), Amsterdam: Elsevier Scientific Publishing Company, p 255–271, 1976.
31. Fujita T, Osaka M, Yanatori Y: Granule release of enterochromaffin (EC) cells by cholera enterotoxin in the rabbit. *Arch Histol Jpn* 36: 367–378, 1974.
32. Osaka M, Fujita T, Yanatori Y: On the possible role of intestinal hormones as the diarrhoeagenic messenger in cholera. *Virchows Arch B (Cell Pathol)* 18: 287–296, 1975.
33. Grossman MI: Integration of neural and hormonal control of gastric secretion. *Physiologist* 6: 349–357, 1963.
34. Fujita T, Muraki S, Sato K, Noguchi R, Shimoji K: Effects of atropine and tetrodotoxin upon pancreozymin release from canine duodenum in response to luminal stimuli. *Biomed Res* 1: 59–65, 1980.
35. Fujita T, Matsunai Y, Muraki S, Sato K, Shimoji K: Effect of intraluminally administered lidocaine upon the pancreozymin-producing endocrine cells of the canine duodenum. In: Integrative Control Functions of the Brain, vol 1. M Ito (ed), Tokyo: Kodansha Scientific, p 296–297, 1978.
36. Fujita T, Kobayashi S, Muraki S, Sato K, Shimoji K: Gut endocrine cells as chemoreceptors. In: Gut Peptides. Secretion, Function and Clinical Aspects. A Miyoshi (ed), Tokyo: Kodansha, p 47–52, 1979.
37. Fujita T: Messenger substances of neurons and paraneurons: Their chemical nature and the routes and ranges

of their transport to targets. *Biomed Res* 4: 239–256, 1983.
38. Pearse AGE, Polak JM: Neural crest origin of the endocrine polypeptide (APUD) cells of the gastrointestinal tract and pancreas. *Gut* 12: 783–788, 1971.
39. Fujita T: The gastro-enteric endocrine cell and its paraneuronic nature. In: Chromaffin, Enterochromaffin and Related Cells. RE Coupland, T Fujita (eds), Amsterdam: Elsevier, p 191–208, 1976.
40. Fujita T, Kobayashi S: Current views on the paraneurone concept. *TINS* 2: 27–30, 1979.
41. O'Conner DT, Burton D, Deftos LJ: Chromogranin A: Immunohistology reveals its universal occurrence in normal polypeptide hormone producing endocrine glands. *Life Sci* 33: 1657–1663, 1983.
42. Nolan JA, Trojanowski JQ, Hogue-Angeletti R: Neurons and neuroendocrine cells contain chromogranin: Detection of the molecule in normal bovine tissues by immunochemical and immunohistochemical methods. *J Histochem Cytochem* 33: 791–798, 1985.
43. Fischer-Colbrie R, Lassmann H, Hagn C, Winkler H: Immunological studies on the distribution of chromogranin A and B in endocrine and nervous tissues. *Neurosci* 16: 547–555, 1985.
44. Kobayashi S: Cellular background in gut hormone secretion. In: Gut Peptides. Secretion, Function and Clinical Aspects. A Miyoshi (ed), Tokyo: Kodansha, p 53–58, 1979.
45. Kanno T, Saito A, Yonezawa H: Unidirectional cellular processes of stimulus-secretion coupling in a cell secreting cholecystokininpancreozymin. In: Histochemistry and Cell Biology of Autonomic Neurons, SIF Cells and Paraneurons. O Eränko et al. (eds), New York: Raven Press, p 327–332, 1980.
46. Osaka M, Kobayashi S: Duodenal basal-granulated cells in the human fetus with special reference to their relationship to nervous elements. In: Endocrine Gut and Pancreas. T Fujita (ed), Amsterdam: Elsevier Scientific Publishing Company, p 145–158, 1976.
47. Kusumoto Y, Iwanaga T, Ito S, Fujita T: Juxtaposition of somatostatin cell and parietal cell in the dog stomach. *Arch Histol Jpn* 42: 459–465, 1979.
48. Larsson LI, Goltermann N, de Magistris L, Rehfeld JF, Schwartz TW: Somatostatin cell processes as pathways for paracrine secretion. *Science* 205: 1393–1395, 1979.
49. Schmechel D, Marangos PJ, Brightman M: Neurone-specific enolase is a molecular marker for peripheral and central neuroendocrine cells. *Nature* 276: 834–836, 1978.
50. Wharton J, Polak JM, Cole GA, Marangos PJ, Pearse AGE: Neuron-specific enolase as an immunocytochemical marker for the diffuse neuroendocrine system in human foetal lung. *J Histochem Cytochem* 29: 1359–1364, 1981.
51. Fujita T, Iwanaga T, Nakajima T: Immunohistochemical detection of nervous system-specific proteins in normal and neoplastic paraneurons in the gut and pancreas. In: Gut Peptides and Ulcer. A Miyoshi (ed), Tokyo: Biomedical Research Foundation, p 81–88, 1983.
52. Iwanaga T, Fujita T, Ito S: Immunohistochemical staining of enteroendocrine paraneurons with anti-brain tubulin antiserum. *Biomed Res* 3: 99–101, 1982.
53. Iwanaga T, Hozumi I, Yamakuni T, Takahashi Y, Fujita T: A cerebellar Purkinje cell-specific protein (spot 35 protein) showing wide distribution in the endocrine system of some mammals: An immunohistochemical study. *Cell Tissue Res* (in press).
54. Iwanaga T, Takahashi-Iwanaga H, Fujita T, Yamakuni T, Takahashi Y: Immunohistochemical demonstration of a cerebellar protein (spot 35 protein) in some sensory cells of guinea pig. *Biomed Res* 6: 329–334, 1985.
55. Dias-Amads L: Sur la signification des cellules de Nicolas. *C R Soc Biol* (Paris) 93: 1550–1551, 1925.
56. Danisch F: Zur Histogenese der sogenannten Appendixkarzinoide. *Beitr Pathol Anat Allgem Pathol* 72: 687–709, 1924.
57. Andrew A: APUD cells and paraneurons: Embryonic origin. *Adv Cell Neurobiol* 2: 3–32, 1981.
58. Le Douarin NM, Teillet MA: The migration of neural crest cells to the wall of the digestive tract in avian embryo. *J Embryol Morphol* 30: 31–48, 1973.
59. Andrew A: APUD cells in the endocrine pancreas and intestine of chick embryos. *Gen Comp Endocrinol* 26: 485–495, 1975.
60. Osaka M, Kobayashi S: Duodenal basal-granulated cells in the human fetus with special reference to their relationship to nervous elements. In: Endocrine Gut and Pancreas. T Fujita (ed), Amsterdam: Elsevier Scientific Publishing Company, p 145–158, 1976.
61. Fujii S: Development of pancreatic endocrine cells in the rat fetus. *Arch Histol Jpn* 42: 467–479, 1979.
62. Larsson LI: Ontogeny of peptide-producing nerves and endocrine cells of the gastro-duodeno-pancreatic region. *Histochemistry* 54: 133–142, 1977.
63. Inokuchi H, Fujimoto S, Kawai K: Cellular kinetics of gastrointestinal mucosa, with special reference to gut endocrine cells. *Arch Histol Jpn* 46: 137–157, 1983.
64. Deschner EE, Lipkin M: An autoradiographic study of the renewal of argentaffin cells in human rectal mucosa. *Exp Cell Res* 43: 661–665, 1966.
65. Lehy T, Willems G: Population kinetics of antral gastrin cells in the mouse. *Gastroenterology* 71: 614–619, 1976.
66. Kobayashi S, Iwanaga T, Fujita T: Segi's cap: Huge aggregation of basal-granulated cells discovered by Segi (1935) on the intestinal villi of the human fetus. *Arch Histol Jpn* 43: 79–83, 1980.
67. Tsubouchi S, Leblond CP: Migration and turnover of enteroendocrine and caveolated cells in the epithelium of the descending colon, as shown by radioautography after continuous infusion of ^{3}H-thymidine into mice. *Am J Anat* 156: 431–452, 1979.
68. Odartchenko N, Hedinger C, Ruzicka J, Weber E: Cytokinetics of argentaffin cells in mouse intestinal mucosa. *Virchows Arch Abt B Zellpathol* 6: 132–136, 1970.
69. Ferreira MN, Leblond CP: Argentaffin and other "endocrine" cells of the small intestine in the adult mouse. 2. Renewal. *Am J Anat* 131: 331–352, 1971.
70. Cheng H, Leblond CP: Origin and differentiation and renewal of the four main epithelial cell types in the mouse small intestine. 3. Entero-endocrine cells. *Am J Anat* 141: 503–520, 1974.

71. Fujita T, Kobayashi S, Yui R, Iwanaga T: Evolution of neurons and paraneurons. In: Hormones, Adaptation and Evolution. S Ishii et al. (eds), Tokyo: Japan Scientific Society Press/Berlin: Springer-Verlag, p 35–43, 1980.
72. Westfall JA, Kinnamon JC: A second sensory-motor-interneuron with neurosecretory granules in *Hydra. J Neurocytol* 7: 365–379, 1978.
73. Grimmelikhuijzen CJP, Carraway RE, Rökaeus Å, Sundler F: Neurotensinlike immunoreactivity in the nervous system of hydra. *Histochemistry* 72: 199–209, 1981.
74. Grimmelikhuijzen CJP, Dierickx K, Boer GJ: Oxytocin/vasopressinlike immunoreactivity is present in the nervous system of hydra. *Neuroscience* 7: 3191–3199, 1982.
75. Reuter M, Karhi T, Schot LPC: Immunocytochemical demonstration of peptidergic neurons in the central and peripheral nervous systems of the flatworm *Microstomum lineare* with antiserum to FMRF-amide. *Cell Tiss Res* 238: 431–436, 1984.
76. Nishiitsutsuji-Uwo J, Endo Y: Gut endocrine cells in insects: The ultrastructure of the endocrine cells in the cockroach midgut. *Biomed Res* 2: 30–44, 1981.
77. Iwanaga T, Fujita T, Nishiitsutsuji-Uwo J, Endo Y: Immunohistochemical demonstration of PP-, somatostatin-, enteroglucagon- and VIP-like immunoreactivities in the cockroach midgut. *Biomed Res* 2: 202–207, 1981.
78. Fujita T, Yui R, Iwanaga T, Nishiitsutsuji-Uwo J, Endo Y, Yanaihara N: Evolutionary aspects of "brain-gut peptides": An immunohistochemical study. *Peptides* 2(suppl 2): 123–131, 1981.
79. Fujita T, Yui R, Iwanaga T: Immunohistochemical studies of neuropeptides in invertebrates. *Adv Neurolog Sci* 27: 468–478, 1983.
80. Yui R, Iwanaga T, Kuramoto H, Fujita T: Neuropeptide immunocytochemistry in protostomian invertebrates, with special reference to insects and molluscs. *Peptides* 6(suppl 3): 411–415, 1985.
81. Endo Y, Nishiitsutsuji-Uwo J: Exocytotic release of secretory granules from endocrine cells in the midgut of insects. *Cell Tiss Res* 222: 515–522, 1982.
82. Bevis PJR, Thorndyke MC: A cytochemical and immunofluorescence study of endocrine cells in the gut of the ascidian *Styela clava. Cell Tiss Res* 199: 139–144, 1979.
83. Pestarino M: Occurrence of different secretinlike cells in the digestive tract of the ascidian *Styela plicata* (Urochordata, Ascidiacea). *Cell Tiss Res* 226: 231–235, 1982.
84. Fritsch HAR, Van Noorden S, Pearse AGE: Gastrointestinal and neurohormonal peptides in the alimentary tract and cerebral complex of *Ciona intestinalis* (Ascidiaceae). Their relevance to the evolution of the diffuse neuroendocrine system. *Cell Tiss Res* 223: 369–402, 1982.
85. Kataoka K, Fujita H: The occurrence of endocrine cells in the intestine of the lancelet, *Branchiostoma japonicum*. An electron microscope study. *Arch Histol Jpn* 36: 401–406, 1974.
86. Van Noorden S, Pearse AGE: The localization of immunoreactivity to insulin, glucagon and gastrin in the gut of *Amphioxus (Branchiostoma) lanceolatus*. In: The Evolution of the Pancreatic Islets. TAI Grillo, L Liebson, A Epple (eds), Oxford: Pergamon Press, p 163–178, 1976.

Author's address:
Department of Anatomy
Niigate University Meohicol School
Asahi-mechi 51 Niigata, Japan

CHAPTER 14

Morphology and porosity of the alimentary epithelial basal lamina

SAMUEL G. MCCLUGAGE & FRANK N. LOW

1. Introduction

Basal laminas occupy a key position in the morphology and physiology of the animal organism. They are so located as to separate the cells of epithelium, muscle, nerve, and fat from the cellular and extracellular components of the connective tissues (1). The formed elements of the connective tissues are true residents of the interstitial space (tissue space, connective tissue space, milieu interne). They exist within it, not being separated from it by basal laminas (2).

Certain events in the history of discovery and recognition of basal laminas merit brief notice. Todd and Bowman in 1845 (3) were the first to discover and describe what they then called basement membranes. Their holistic thinking recognized a master plan of the animal body comprised of an epithelium and its underlying basement membrane. Their description of this body plan is essentially valid today:

> This external integument is a part only of a great physiological system, which comprehends also the mucous membranes and the true or secreting glands; all of which, taken together, and reduced to their most simple expression, are a continuous membrane, more or less involuted, more or less modified in the elementary tissues which compose it or are in connexion with it, and within which all the rest of the animal is contained. This expanse consists of two elements: a basement tissue composed of simple membrane, uninterrupted, homogeneous and transparent, covered by an epithelium, or pavement of nucleated particles. Underneath the basement membrane vessels, nerves and areolar tissue are placed.

This recognition of basement membranes as subepithelial structures became an essential part of histology and was both accepted and taught until the middle of the twentieth century. Then in 1949, Gersh and Catchpole (4), using the recently developed periodic acid-Schiff technique (5), demonstrated PAS-positive basement membranes. These existed not only under epithelium but also surrounded the cells of muscle, nerve, and fat. With the development of electron microscopy of sectioned biological tissue in the 1950s, a system of basal laminas occupying the same positions as the PAS-positive membranes of Gersch and Catchpole was discovered. They were much thinner than the PAS-positive membranes but occupied exactly the same positions. These are the basal laminas that we recognize today.

The embryology of basal laminas testifies to their ancient phylogeny. They can be recognized in the chick embryo at the time incubation begins, on the basal surface of the epiblast which will form the epithelium of the dorsal surface of the embryo. During the first 12 hours of incubation they develop a recognizable lamina lucida and lamina densa and become indistinguishable from adult basal laminas toward the end of the first day of incubation (6). During this same period their substructure of fine primary fibrils (4 nm in diameter) becomes visible (7). These fibrils become decked with the minute particulate or "fuzzy" material that is essentially the only structural component demonstrable in the ordinary preparations of transmission electron microscopy. Entodermal basal laminas are formed around the foregut (pharynx) by 36 hours. A basal lamina is also present at this stage around the notochord and the medial surfaces of the somites. Mesothelial and endothelial basal laminas follow later but appear quite early in embryonic development. The first microfibrils in the chick embryo appear on the interstitial surface of the epithelial basal lamina at 24 hours (7). They are later found free in the tissue space, particularly around the notochord. Unit collagen fibrils with characteristic periodicity appear to develop from microfibrillar growth at about four days incubation age. Elastic fibers do not appear until later, at about one week incubation age in the chick aorta. In resumé one may recall Ernst Haeckel who formulated the now well-known basic principle of embryonic development, "Ontogeny recapitulates phylogeny." Thinking along Haeckel's lines, one may recognize basal laminas as the true phyletic forerunners of the extracellular connective tissues.

The epithelial basal lamina of the alimentary canal has received special attention from a number of investigators (8–12). In general its contours are determined by the basal surface of the epithelium lining the gut. Its notable structural modification is the regional presence of pores which penetrate the entire thickness of the membrane, occurring mainly in absorptive areas (12) and where interposed between lymph nodules and epithelium (12, 13). In fetal tissues the presence of comparable pores is somewhat more extensively documented (14–17).

2. Regional variations of basal laminas

2.1. *Esophagus*

The epithelial basal lamina of the esophagus can be revealed by epithelial removal (although stratified squamous epithelium resists removal more strongly than the simple columnar type found elsewhere in the gut). Folds such as characterize

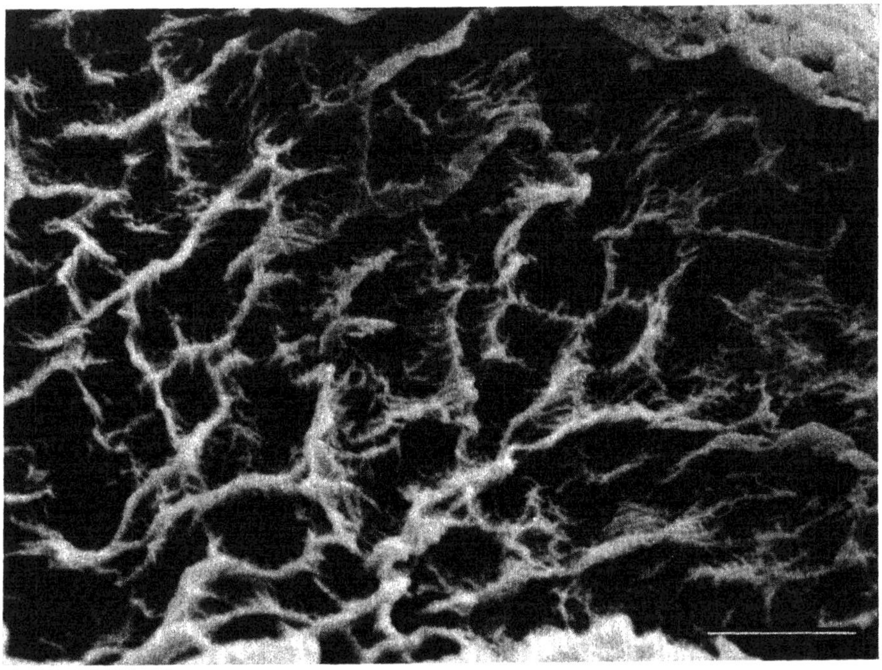

Figure 14-1. Esophagus, rat. The basal lamina is thrown into coarse, irregular folds which represent the connective tissue papillae which underly the epithelium. The smaller folds may mark the bases of individual cells. Remnants of the epithelium may be seen above right and below. The bar is 13 μm long.

the basal surface of the epithelium are present (Fig. 14-1). These may be longitudinal in the empty esophagus but are for the most part irregular. They represent the connective tissue papillae that underlie the epithelial lining of the tube (11). Still smaller elevations of the basal lamina such as might surround the bases of single epithelial cells are present throughout the contours of the epithelial basal lamina of the esophagus.

2.2. Stomach

2.2.1. Fundic part

During life, the fundic part of the stomach is lined with simple columnar epithelium, both on the surface areas and in the gastric glands. This epithelium is removed by our technique, revealing the luminal surface of the underlying epithelial basal lamina (Fig. 14-2). When viewed in SEM this luminal surface is generally smooth and without pores. Its contours reveal the distinction between gastric pits and the openings of individual gastric glands (Fig. 14-3). The connective tissue partitions between gastric glands are much thinner and more deeply located than the interstices between gastric pits. Occassionally there are fairly broad areas of basal lamina facing the lumen between individual gastric pits.

2.2.2. Pyloroduodenal junction

The transition between pyloric stomach and duodenum is abrupt (Fig. 14-4). After epithelial removal the ostium of each pyloric gland is empty, leading into a cul-de-sac formed by the basal lamina of the original gland. The mosaic pattern formed by the pyloric ostia ends abruptly in a position underlying the pyloric sphincter. It is replaced by the connective tissue cores of villi covered by basal lamina (herein called villous cores). Each villous core is essentially clavate in shape and one may run into another in the upper portion of the small intestine. A continuous circular villous core marks the sharp transition between pyloric stomach and duodenum (Fig. 14-4).

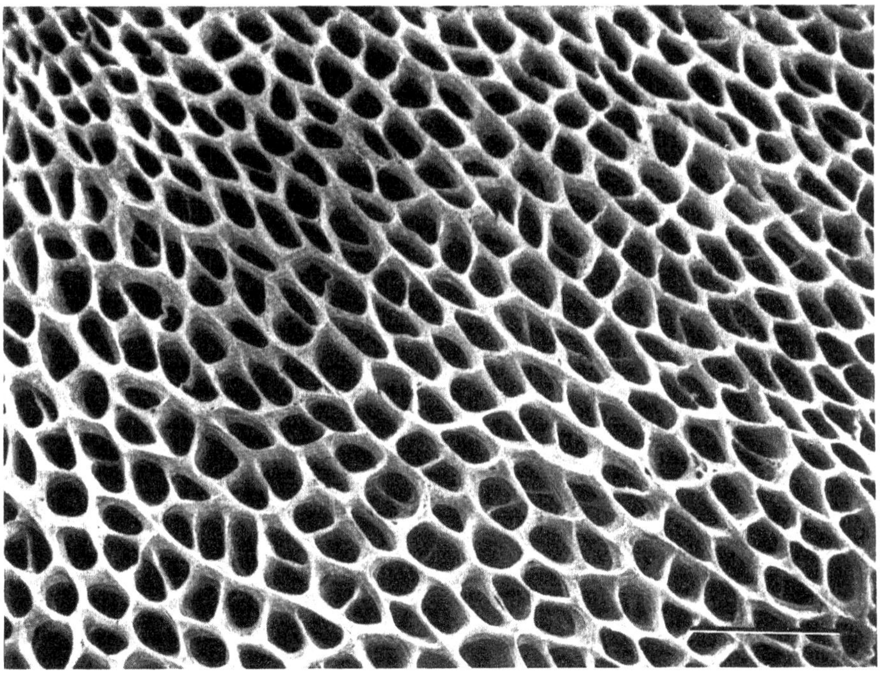

Figure 14-2. Gastric stomach, rat. Each dark rounded area represents the ostium of a gastric gland from which the epithelium has been removed. The basal lamina forms a cul-de-sac around each gland and shows white in the interstices between them. The bar is 140 μm long.

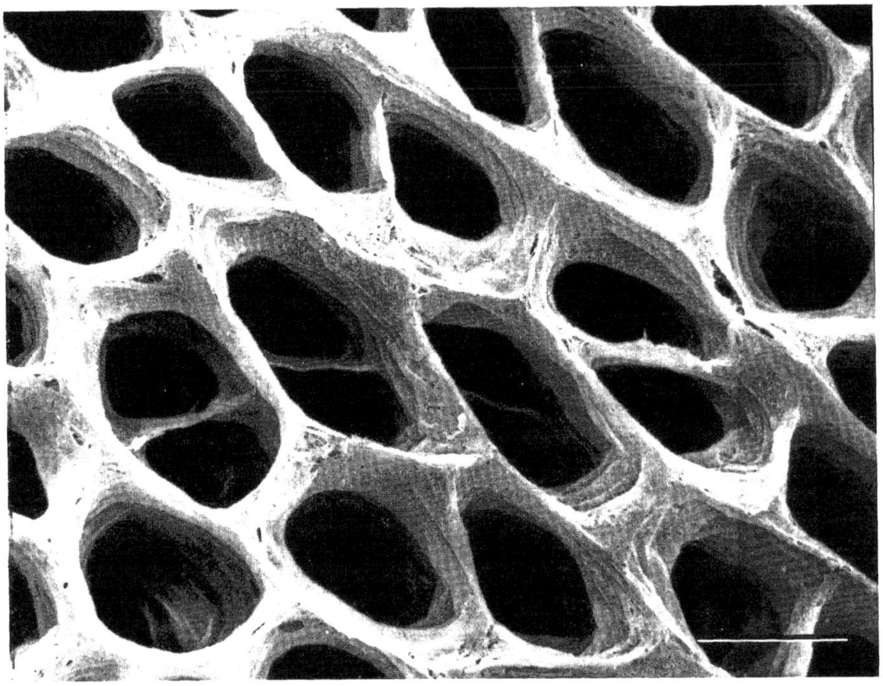

Figure 14-3. Gastric stomach, rat. Most gastric glands open individually into the stomach lumen. Where two glands are separated by a narrow partition between them, they open into a gastric pit. The bar is 37 μm long.

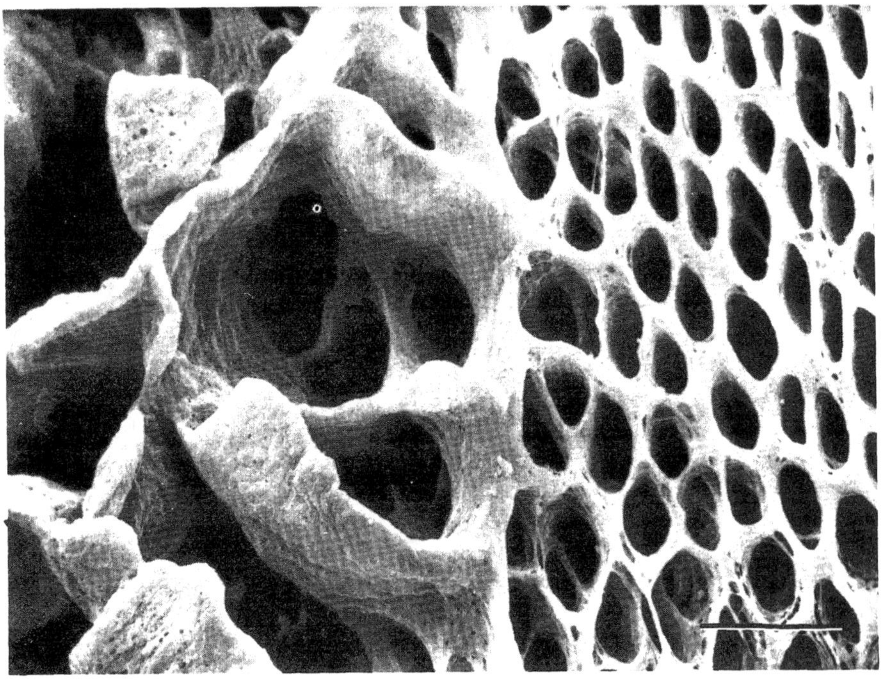

Figure 14-4. Gastro-duodenal junction, rat. Ostia of gastric glands (right) end abruptly where the connective tissue cores of duodenal villi first appear. In this portion of the duodenum the villi are confluent, often forming a ringlike structure around the ostia of Brunner's glands. The bar is 130 μm long.

2.3. Duodenum

Between each of the villous cores of the duodenum the ostia of Brunner's glands are visible (Fig. 14-5). These are of much greater diameter than the ostia of pyloric glands of the stomach. Villous cores vary considerably in size and shape, but their overall form in the rat is clavate rather than cylindrical. Between their bases the ostia of Brunner's glands are arranged, usually haphazardly rather than with regularity. There may be a single row of ostia between two villous cores or the line may be doubled with irregular variants of placement in between.

2.4. Jejunum and ileum

As one progresses from the duodenum to the jejunum and finally the ileum there is no essential change in the morphological arrangement described above. The villous cores remain essentially clavate in shape (Fig. 14-6); however, there appears to be species differences. Villous cores in the monkey are cylindrical (Fig. 14-7) and in the human many are bifurcate (Fig. 14-8). There is perhaps less tendency for them to run one into the other as in the upper duodenum. As one changes from Brunner's glands to crypts of Lieberkühn, the only visible change is a somewhat leser number of glandular ostia in the jejunum and ileum as compared to the duodenum.

The most striking feature of the epithelial basal lamina of the villous cores is the presence of pores (Fig. 14-6) which extend from the subepithelial area, through the membrane, into the interstitium containing the capillary network and the lacteal of the villus (12). These pores are conspicuous along the entire length of the small intestine. They occupy each villous core, except at the very tip and extend to its base into close proximity with the glandular ostia. They are a conspicuous feature of this particular part of the basal lamina in comparison to other, nonporous areas.

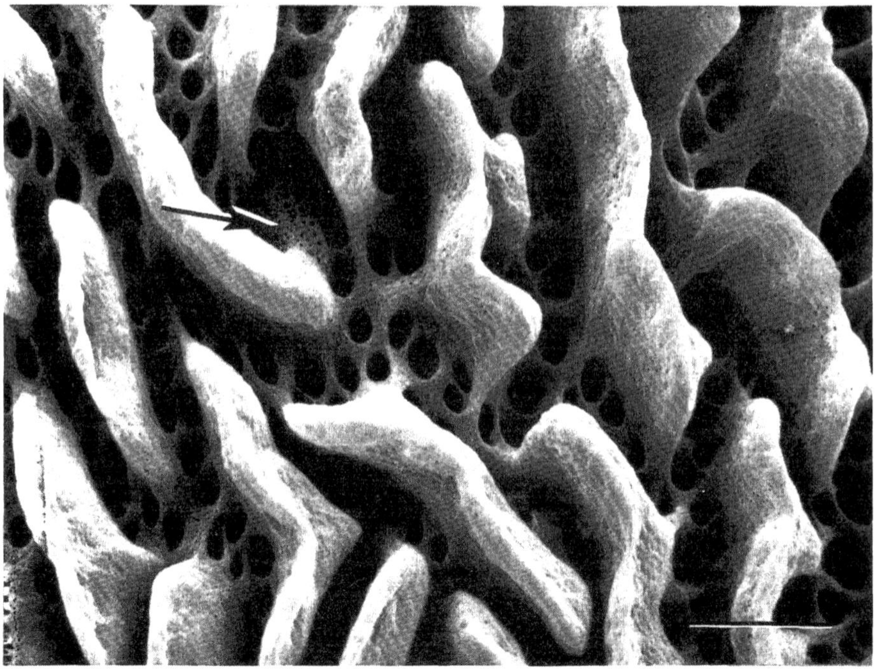

Figure 14-5. Duodenum, rat. This view, taken from the duodenal lumen, looks directly down on the tips of villous cores which are completely denuded of their epithelium. The ostia of Brunner's glands, also without their epithelium, appear as dark, rounded areas between the bases of the villous cores. The small area showing pores in the basal lamina (arrow) may mark the position of an underlying lymph nodule. The bar is 200 μm long.

Figure 14-6. Ileum, rat. This "side-on" view of villous cores emphasizes their clavate or tongue-shaped form. Note the presence of pores in their lateral walls. There is also some indication of the capillary network underlying the basal lamina (arrow) in the interstitial space. The bar is 130 μm long.

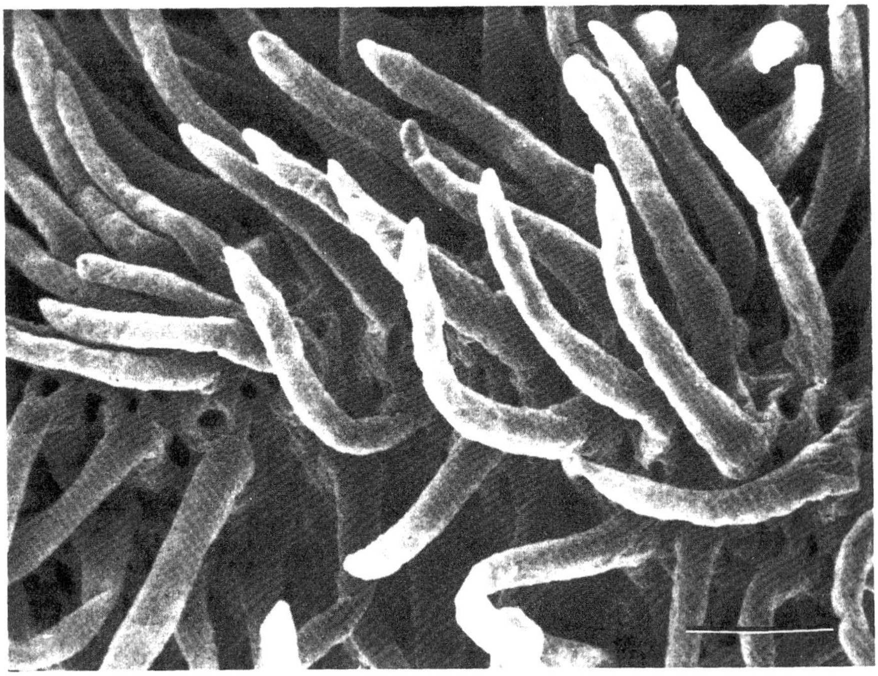

Figure 14-7. Small intestine, monkey. The villous cores in this form are slender and cylindrical. The ostia of crypts of Lieberkühn are clearly visible. The bar is 240 μm long.

2.4.1. Peyer's patches

The uniform pattern of villous cores separated by glandular ostia is occasionally interrupted by lymph nodules either solitary or, more frequently, forming Peyer's patches (Fig. 14-9). In these areas the lymph nodule, here superficial under a specialized epithelium containing M cells, lies directly under the epithelial basal lamina. The basal lamina is highly specialized, being characterized by large pores. These are more numerous and prominent on the periphery of the nodule. The ostia of the crypts of Lieberkühn on the periphery of the nodule appear to have been forcibly displaced through the aggressive action of the nodule.

2.4.2. Fat absorption

The pathway taken by chylomicra of absorbed fat is particularly related to the pores found in the epithelial basal lamina of small intestinal villi (8, 18). When the minute particles of absorbed fat have passed through the columnar epithelial cells of the small intestine, they are extruded into an intercellular compartment that abuts directly on the porous basal lamina of the villous cores. Here they may be seen passing through these pores and entering the connective tissue space of the core, in close proximity to the lacteal. Passage into the lacteal can also be observed. There is no doubt that the pores in the villous cores accommodate the process of fat absorption. It must not be assumed, however, that this is the only function of these pores. They are known to be occupied by processes of epithelial cells which can extend through them, directly into the underlying tissues space (12). It is entirely possible that these pores thereby accommodate nonlipid forms of digestive absorption.

2.4.3. Ileocecal valve

The transition from terminal ileum to the colon is gradual rather than abrupt and takes place along the luminal surface of the ileocecal valve (Fig.

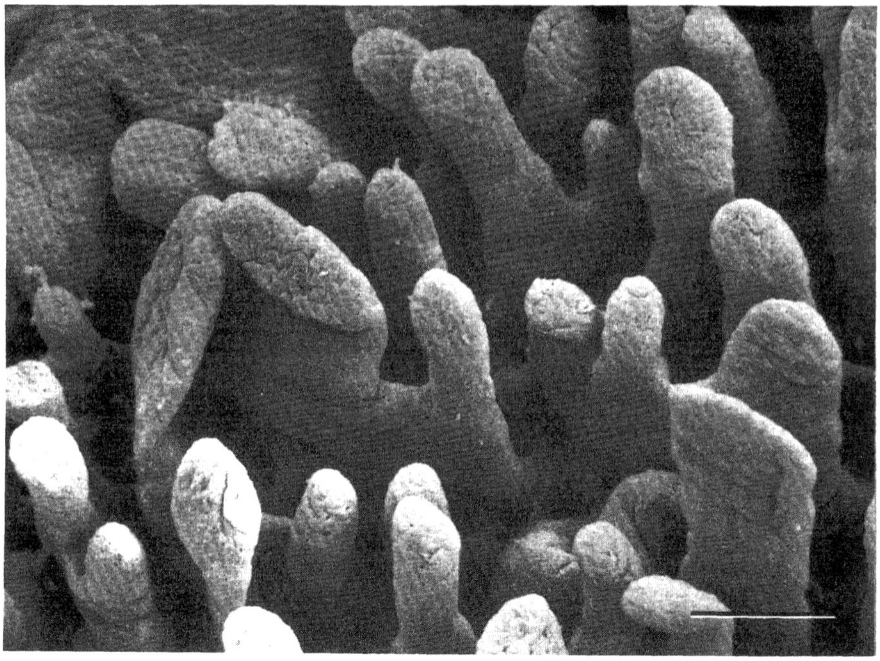

Figure 14-8. Small intestine, human. The villous cores illustrated here are both clavate and cylindrical. They often bifurcate, a situation seldom seen in laboratory animals. Other features of the human basal lamina do not differ substantially from those of other vertebrates. The bar is 190 μm long.

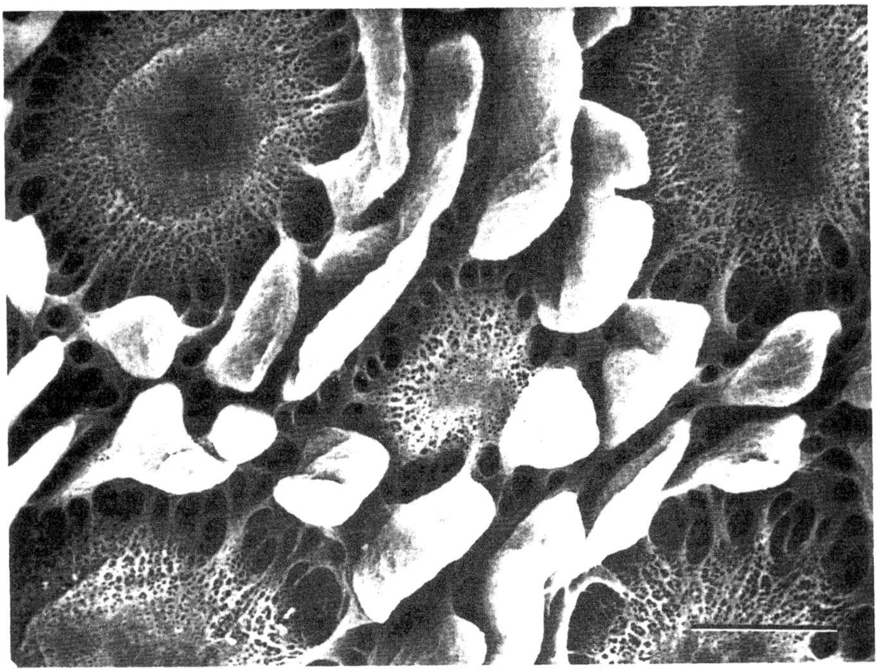

Figure 14-9. Peyer's patch in ileum, rat. Lymph nodules underlying the epithelium are not covered by villous cores. The basal lamina separating them from the epithelium is extremely porous, especially at their margins close to the ostia of crypts of Lieberkühn. The bar is 370 μm long.

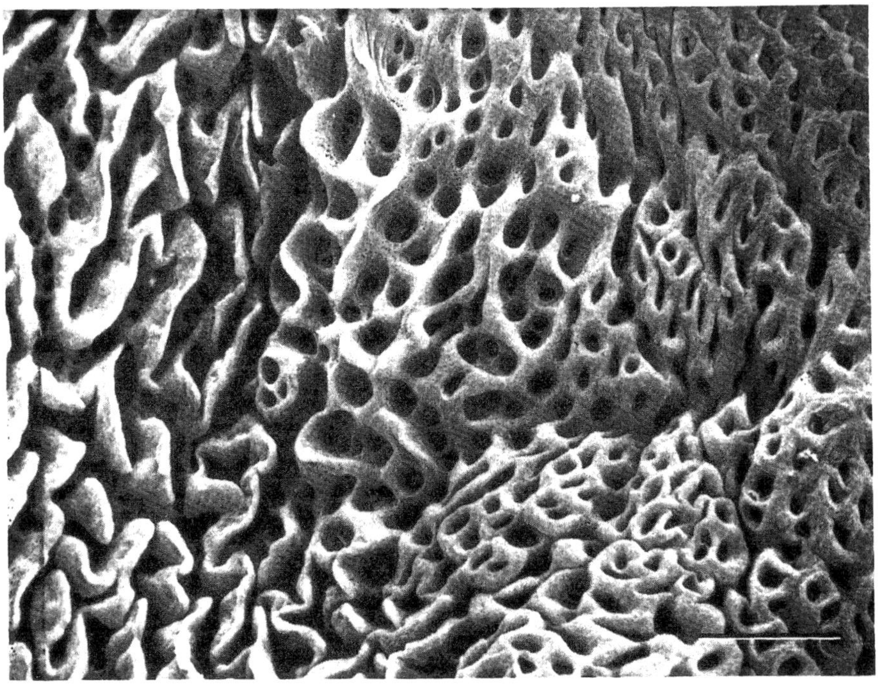

Figure 14-10. Ileo-cecal junction, rat. The transition from the villous ileum (left) to the nonvillous cecum (right) is gradual rather than abrupt. Proceeding to the right the villi become less and less elevated until the basal lamina does not project above the ostia of the glands of Lieberkühn. The bar is 390 μm long.

14-10). As this valve is approached from the ileum, the villous cores become shorter and shorter until, at the crest of the mucosal fold that forms the valve, they have disappeared completely. On the colonic side of the fold little more change takes place since the basic features of the colonic basal lamina have already been achieved. Ostia of glands of Lieberkühn are separated from each other by flat areas of epithelial basal lamina. The general appearance is quite similar to that presented by the gastric and pyloric stomach.

2.5. *Cecum*

The cecum presents very little that distinguishes it from the colon. Partitions between individual glands of Lieberkühn are generous so that the population of glands is not as high as in the colon (Fig. 14-11). A few pores may be present. They are located in the interstices between glands of Lieberkühn and extend for some distance into the glands (Fig. 14-12). Their arrangement tends to be haphazard.

2.6. *Colon*

There is no essential difference between the morphology of the cecal basal lamina and that of the colon. In the colon the interglandular spaces tend to be broader. The pattern of pores is essentially the same without any very orderly arrangement.

3. Functional considerations

Basal laminas are found outside the plasmalemma of all histologic cell types except connective tissue cells and their derivatives. They are composed of type IV collagen, laminin, and heparan sulfate proteoglycan. The percentage of these components that each basal lamina contains varies from one part of an organism to another. Furthermore, other glycoproteins such as fibronectin and entactin have been demonstrated in some basal laminas. Basal laminas are known to contain, by transmission electron microscopy, a lamina rara adjacent to the plasmalemma and a lamina densa adjacent to the connective tissue stroma. Transmission electron microscopy and electron immunohistochemistry have both demonstrated much structural and biochemical heterogeneity between basal laminas.

Our observations by SEM demonstrate that the porosity of the epithelial basal lamina lining the gut varies from one part to another. No pores are evident in the basal lamina of esophagus or stomach. However, they can be easily observed in the large and small intestine, being more numerous in the latter (12). The porosity appears to be greatest within the basal lamina overlying lymph nodules (13). Thus, there appears to be a morphologic heterogeneity of epithelial basal laminas within the gastrointestinal tract. Furthermore, the functional significance of these pores and how they are formed becomes an important consideration in the physiologic response of an organism.

One of the more interesting observations is the marked porosity of the basal lamina overlying lymph nodules (12, 13). Nonepithelial cells, mostly lymphocytes, comprise 10% to 20% of the cell population of the intestinal mucosa (19, 20). The percentage is even higher within the mucosa overlying lymph nodules. The intestinal mucosa is known to play a major role in the immune response of an organism, since many foreign substances and organisms have access to the body through the gut wall (21, 22). Since lymph nodules within Peyer's patches are known to be sites of antigen sampling (23), the pores are probably formed by migrating lymphocytes. Similar pores have been observed during passage of lymphocytes through cutaneous basal laminas in normal skin (24) and of rat intestinal villi (25).

The presence of pores may help to explain other biologic events which occur either within isolated lymph nodules or those within Peyer's patches. It is known that the origin of the columnar epithelial cells of intestinal villi is from undifferentiated cells located in the crypts of Lieberkühn (26). Similarly, the origin of the M cells, which are only found on the surfaces of lymph nodules, appears to be from stem cells within the crypts of Lieberkühn (27). There may be some types of cell sorting within the crypts which determine whether a cell will migrate onto a villus or onto a lymph nodule, with the potential in the latter to become an M cell. In this regard, there is a subepithelial capillary network directly under the surface of the nodule which intercon-

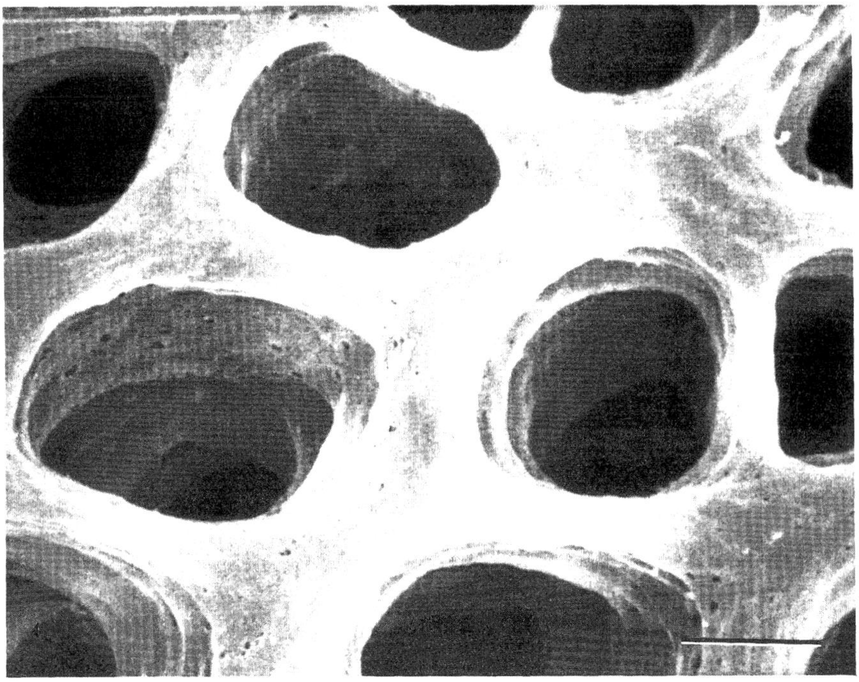

Figure 14-11. Cecum, rat. The ostia of glands of Lieberkühn are separated by areas of basal lamina. The cluster of pores may represent the location of a lymph nodule in the underlying interstitium. The bar is 72 μm long.

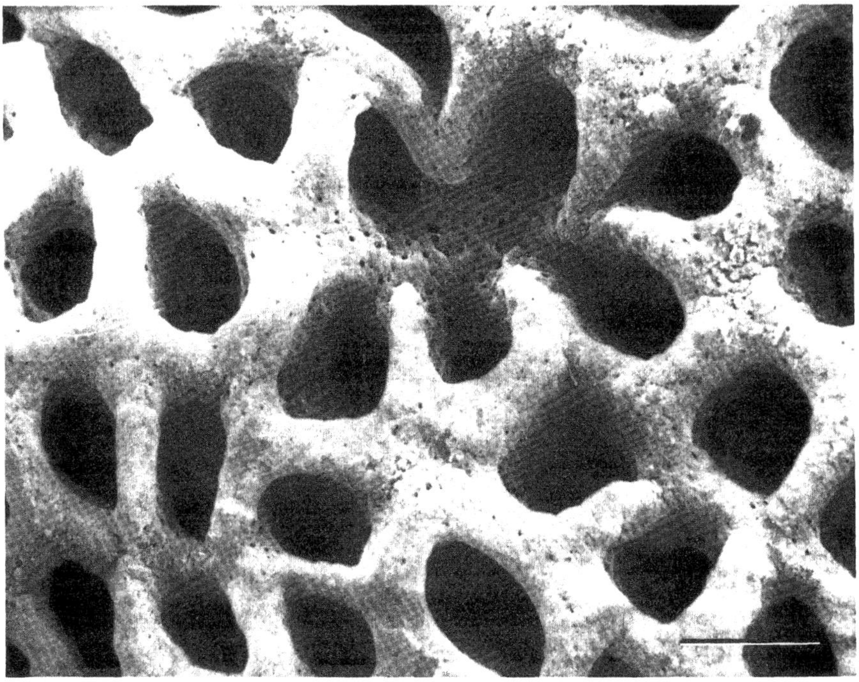

Figure 14-12. Colon, rat. The cul-de-sacs formed by the basal lamina around glands of Lieberkühn possess a few small pores. In general, the glands are spaced less farther apart in the colon than in the cecum. The bar is 37 μm long.

nects with baskets of capillaries surrounding adjacent crypts (28).

The highly porous structure of the basal lamina of the nodules and their associated crypts may provide a means by which various cellular products from the surface of the nodule have easy access to the underlying capillary network. This, then, may partially explain the unique cell sorting that takes place within the crypts of Lieberkühn surrounding lymph nodules. The possibility of a feedback control between the surface of the nodule and the crypt may explain how the antigen-trapping epithelium of the surface of the nodule replenishes itself (28). It may further explain why the surface population of cells of villi and lymph nodules is so different (28).

The highly porous nature of the epithelial basal lamina of the lymph nodules may also permit more bidirectional interaction between constituents on both sides of this tissue boundary. For example, migration of lymphocytes out of post-capillary venules within a nodule may be affected by the ability of lymphocytes or their secretory products to pass into the subepithelial capillary network (28). The lymphocytes leaving the post-capillary venules, in turn, cross the basal lamina to pass into the overlying mucosal epithelium (29). Furthermore, some mucosal intraepithelial lymphocytes appear to be derived directly from the lymph nodule below (29).

The porosity of the basal lamina overlying a lymph nodule increases centrifugally (12, 13). M cells appear to be more frequent toward the center or apex of the surface of the nodule (30), the area where the underlying basal lamina is least porous. On the periphery of the surface, lymphocytes occur in association with M cells and other epithelial cells (30). Here the porosity of the basal lamina is greater. The significance of these differences in porosity is puzzling. If lymphocyte migration into the mucosal epithelium is greater on the periphery of the nodule, then this could account for the greater porosity peripherally. Also, if the interaction of lymphocytes with immature M cells leads to the transformation of the latter into mature M cells, the presence of many migrating lymphocytes might be necessary (30).

Basal processes of epithelial cells are often observed passing through the intestinal epithelial basal lamina (12, 19, 31–35). These processes are in close proximity to the underlying capillary network of the interstitium (12, 34) and may play a role in the absorption of substances from the intestinal lumen (8). In this regard, pores may play a role in the passage of fat particles from the intercellular clefts of the absorptive epithelium to the interstitium where they pass into lacteals or capillaries (8, 18, 31–33). It is interesting to speculate that the basal processes of the absorptive epithelium withdraw from the pores in the basal lamina in order to facilitate absorption. This may be due to a pressure-dependent deformation of the overlying epithelium due to the increased presence of intercellular lipid. This remains to be proven, however.

The basal lamina appears to be a self-repairing, penetrable tissue boundary that permits passage of various cells and their products. The morphologic heterogeneity of the pore population within epithelial basal laminas suggests that at any one time, pores are in different stages of repair or formation (36). Basal laminas must maintain their own integrity by repairing any defect such as a pore within it. Warfel and Hull (24) described sequential morphologic deformations within cutaneous basal laminas during passage of lymphocytes. After passage, these defects (pores) disappeared. This was interpreted as a repair of the defect.

One question which arises in this regard is how does a pore form in the first place. Simpson (35, 37) believes that basal laminas are thixotropic gels, capable of undergoing pressure-dependent changes in viscosity in order to facilitate passage of a cell or molecule. The application of pressure to the basal lamina such as would occur with passage of a cell may cause a localized change in viscosity due to deformation of a collagenous molecular lattice within the basal lamina. Other investigators (38, 39) believe that type IV collagen forms a three-dimensional network within basal laminas which is flexible enough to permit passage of cells or various molecular aggregates without disassembly of the collagenous network. It may be that thixotropy is simply an expression of this flexible nature of the collagenous network as described by Timpl et al. (38).

The regional differences in porosity of the epithelial basal lamina of the gut support the current belief that basal laminas are important structural components of the extracellular matrix. Basal laminas display a morphologic adaptation to a specific tissue or organ which may, in turn, play an integral role in the physiologic response of an organism (12, 13, 40). Even though the process by which pores are formed and subsequently repaired still eludes us, we must acknowledge that such a dynamic process does exist. Further studies on pore formation and repair will enable us to have a better understanding of how much the basal lamina acts as a tissue barrier under normal conditions and how this may change in various pathologic states.

4. Concluding remarks

The epithelial basal lamina lining the alimentary canal demonstrates much morphologic heterogeneity throughout its course. In the esophagus, it appears as folds which closely follow the contours of the epithelium. This gradually changes to a perforated appearance in the stomach due to evaginations of the basal lamina forming the cul-de-sacs of the gastric glands. In the small intestine, the basal lamina surrounds villous cores separated at their base by the ostia of the intestinal crypts. The surface of the large intestine appears perforated, similar to that of stomach. Ostia of intestinal crypts are separated from each other by partitions of basal lamina. The most conspicuous feature of the basal lamina of the gut is the presence of pores which penetrate its entire thickness. They are absent in esophagus and stomach, more numerous in small than in large intestine, and are greatest in number within the basal lamina overlying lymph nodules. Migratory cells, processes of cells, or molecular aggregates such as fat appear to induce pore formation in epithelial basal laminas. The process by which these defects (pores) repair themselves awaits further study.

References

1. Low FN, Burkel WE: A boundary membrane concept of ultrastructural morphology. *Anat Rec* 151: 489–490, 1965.
2. Low FN: A boundary membrane concept of ultrastructure applicable to the total organism. *Proc Third European Regional Conference on Electron Microscopy, Prague* p 115–116, 1964.
3. Todd RB, Bowman W: The Physiological Anatomy and Physiology of Man. London: John W. Parker, p 404, 1845.
4. Gersh I, Catchpole HR: The organization of ground substance and basement membrane and its significance in tissue injury, disease and growth. *Am J Anat* 85: 457–521, 1949.
5. Hotchkiss RD: A microchemical reaction resulting in the staining of polysaccharide structures in fixed tissue preparations. *Arch Bioch* 16: 131–141, 1948.
6. Low FN: Developing boundary (basement) membranes in the chick embryo. *Anat Rec* 159: 231–238, 1967.
7. Low FN: Extracellular connective tissue fibrils in the chick embryo. *Anat Rec* 160: 93–108, 1968.
8. Palay SL, Karlin LJ: An electron microscopic study of the intestinal villus. II. The pathway of fat absorption. *J Biophys Biochem Cytol* 5: 373–384, 1959.
9. Allen L, Weatherford T: Role of fenestrated basement membrane in lymphatic absorption from peritoneal cavity. *Am J Physiol* 197: 551–554, 1959.
10. Highison GJ, Low FN: Microdissection by ultrasonication after prolonged OsO_4 fixation: A technique for scanning electron microscopy. *J Submicrosc Cytol* 14: 161–170, 1982.
11. Low FN, McClugage SG: Microdissection by ultrasonication: Scanning electron microscopy of the epithelial basal lamina of the alimentary canal in the rat. *Am J Anat* 169: 137–147, 1984.
12. McClugage SG, Low FN: Microdissection by ultrasonication: Porosity of the intestinal epithelial basal lamina. *Am J Anat* 171: 207–216, 1984.
13. McClugage SG, Low FN, Zimny ML: Porosity of the basement membrane overlying Peyer's patches in rats and monkeys. *Gastroenterology* 91: 1128–1133, 1986.
14. Cutler LS, Chaudry AP: Intercellular contacts at the epithelial-mesenchymal interface during the prenatal development of the rat submandibular gland. *Dev Biol* 33: 229–240, 1973.
15. Coughlin MD: Early development of parasympathetic nerves in mouse submandibular gland. *Dev Biol* 43: 123–139, 1975.
16. Mathan M, Hermos JA, Trier JS: Structural features of the epithelial-mesenchymal interface of rat duodenal mucosa during development. *J Cell Biol* 52: 577–588, 1972.
17. Brody JS, Vasscaro CA, Gill PJ, Silbert JE: Alterations in alveolar basement membranes during postnatal lung growth. *J Cell Biol* 95: 394–402, 1982.
18. Dearing BD, McClugage SG, Low FN: A SEM study of the epithelial basal lamina of the small intestine during fat absorption. *Anat Rec* 208: 43A, 1984.
19. Toner PG, Ferguson A: Intraepithelial cells in the human intestinal mucosa. *J Ultrastruct Res* 34: 329–344, 1971.
20. Darlington D, Rogers AW: Epithelial lymphocytes in the small intestine of the mouse. *J Anat* 100: 813–830, 1966.

21. Fichtelius KE: The gut epithelium — a first level lymphoid organ? *Exp Cell Res* 49: 87–104, 1968.
22. Bienenstock J, Befus D, Braciale T, McDermott M, Rosenthal K, Tagliabue A: Some aspects of mucosal effector cells. In: Regulation of the Immune Response. PLO Buffalo, DMJ Buffalo (eds), Basel: Karger, p 99–106, 1983.
23. Owen RL, Allen CL, Stevens DP: Phagocytosis of *Giardia muris* by macrophages in Peyer's patch epithelium in mice. *Infect Immunol* 33: 591–602, 1981.
24. Warfel KA, Hull MT: Migration of lymphocytes through the cutaneous basal lamina in normal skin: An ultrastructural study. *Anat Rec* 208: 349–355, 1984.
25. Komuro T: Fenestrations of the basal lamina of intestinal villi of the rat. Scanning and transmission electron microscopy. *Cell Tissue Res* 239: 183–188, 1985.
26. Messier B, Leblond CP: Cell proliferation and migration as revealed by radioautography after injection of thymidine-^{3}H into male rats and mice. *Am J Anat* 106: 247–285, 1960.
27. Bye WA, Allen CH, Trier JS: Structure, distribution and origin of M cells in Peyer's patches in mouse ileum. *Gastroenterology* 86: 789–801, 1984.
28. Bhalla DK, Murakami T, Owen RL: Microcirculation of intestinal lymphoid follicles in rat Peyer's patches. *Gastroenterology* 81: 481–491, 1981.
29. Bhalla DK, Owen RL: Migration of B and T lymphocytes to M cells in Peyer's patch follicle epithelium: An autoradiographic and immunocytochemical study in mice. *Cell Immunol* 81: 105–117, 1983.
30. Smith MW, Peacock MA: "M" cell distribution in follicle-associated epithelium of mouse Peyer's patches. *Am J Anat* 159: 157–166, 1980.
31. Trier JS, Rubin CE: Electron microscopy of the small intestine: A review. *Gastroenterology* 49: 574–603, 1965.
32. Rubin CE: Electron microscopic studies of triglyceride absorption in man. *Gastroenterology* 50: 65–77, 1966.
33. Trier JS: Morphology of the epithelium of the small intestine. In: Handbook of Physiology, Section 6: Alimentary Canal, Vol. III, Intestinal Absorption. Baltimore: Williams & Wilkins, p 1125–1175, 1968.
34. Partridge BT, Simpson LO: Epithelial cell basal processes in the duodenum of NZB mice. In: Proc 9th Int Congr Electron Microsc, Toronto, Canada, p 534–535, 1978.
35. Simpson LO: Basement membranes and biological thixotropy: A new hypothesis. *Pathology* 12: 377–389, 1980.
36. Grenier CP, McClugage SG, Low FN: A SEM study of the fibrous microskeleton of pores within epithelial basement membranes. *Anat Rec* 211: 73A, 1985.
37. Simpson LO: Glomerular permeability: An alternative to the pore theory. *Lancet* 2: 251–252, 1981.
38. Timpl R, Wiedemann H, Van Delden V, Furthmayr H, Kuhn K: A network model for the organization of type IV collagen molecules in basement membranes. *Eur J Biochem* 120: 203–211, 1981.
39. Inoué S, Leblond CP, Laurie GW: Ultrastructure of Reichert's membrane, a multilayered basement membrane in the parietal wall of the rat yolk sac. *J Cell Biol* 97: 1524–1537, 1983.
40. Carlson EC, Kenney MC: An ultrastructural analysis of isolated basement membrane in the acellular renal cortex: A comparative sutdy of human and laboratory animals. *J Morphol* 171: 195–211, 1982.

Author's address:
Dr. Samuel G. McClugage
Department of Anatomy
Louisiana State University
Medical Center
1901 Perdido Street
New Orleans, LA 70112-1393, USA
Phone: (504) 568-4011

CHAPTER 15

The peritoneum

F. BARBERINI, S. CORRER & P.M. MOTTA

1. Introduction

The walls of the body cavities of coelomic origin and the splanchnic organs are covered with serous membranes. These membranes consist of mesothelial cells arranged in a single layer, resting on a thin basal lamina to which is attached a small amount of loose connective tissue (subserosal connective tissue). This allows a variable degree of slipping over the underlying layers. In physiological conditions, the free surface of the membrane is provided with a thin coat of serous fluid containing some floating cells (1, 2), which likely serves to protect the thin subjacent mesothelium from surface friction damage as one organ moves over another (3, 4). In fact, if these structures are injured, adhesions between adjacent tissues may follow (5–7). The first cytological studies by light microscopy within the nineteenth century revealed that the serous cavities were lined by a characteristic type of flattened cells (8–10).

Early microscopists also noted a close similarity between serous cells and the endothelial cells of blood and lymph vessels and, therefore, considered the serous cells to be endothelial in nature (9, 11–13). Furthermore, since the serous membranes also resembled flat epithelia, their cells were regarded as simple squamous epithelium (8, 14, 15).

The term "mesothelium" was introduced first by Minot (16) and later by Mall (17) to indicate the cell layer lining the serous cavities, while the term "endothelium" was reserved only for those cells lining walls of both blood and lymph vessels. Von Recklinghausen (9), using a fixative solution containing silver nitrate, was the first to show that the peritoneal surface of the diaphragm is covered by a single layer of mesothelial cells, arranged to form a kind of "carpet" of flattened elements with characteristic boundaries.

The first electron microscopical observations on a serosa were reported in a study by Odor (18) on the mesothelium covering the rat oviduct, in which special attention was paid to the form, distribution, and structure of the mesothelial microvilli. Odor's study, as well as others (19–21), revealed that the serosal microvilli project from the cell surface and are long, slender, and less packed than those seen on the luminal side of the intestine and proximal convoluted kidney tubules. Many controversial points raised by previous light microscopic reports (22–28) concerning the structure and function of mesothelium have not yet been clarified satisfactorily. These included: (a) the nature of intercellular boundaries, (b) the presence of cell surface projections, such as cilia or brush border, and (c) the mechanism of absorption and transport of soluble substances or particulate matter. Therefore, the advent of electron microscopy permitted a reinvestigation of these controversial aspects at a more detectable ultrastructural level. For example, electron microscopical investigations have shed light on the structure and significance of mesothelial vesicles and surface cell invaginations, interpreted as pinocytotic in nature (18–20, 29–33), and also contributed toward the clarification of the conformation of intercellular spaces

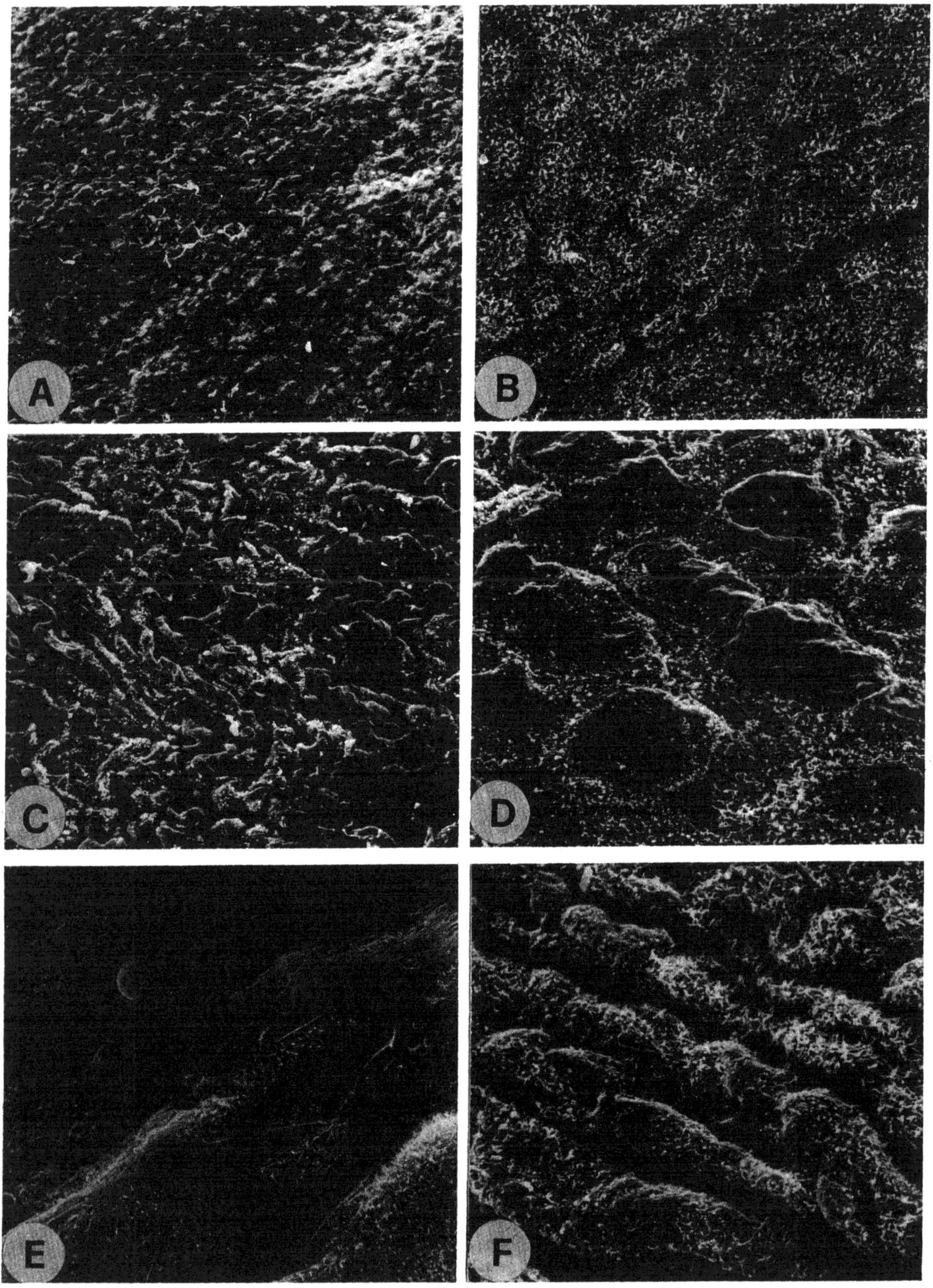
A
B
C
D
E
F

(21, 33) and areas of cell-to-cell apposition (20, 21, 34–41). Moreover, mesothelial cells seemed to be involved very little in the synthesis of serous exudate, or in any other protein synthesis, since their rough endoplasmic reticulum is scanty and poorly developed (20).

In addition, given that the serosa consists of a flattened mesothelial layer with complex intercellular relationships, a question arose in the literature in regard to mesothelial permeability, an intriguing phenomenon studied in its morphological and physiological involvements and therapeutically exploited.

Ultrastructural features of the peritoneal membrane indeed have been investigated recently in experimental models of peritoneal dialysis in rodents (39, 40, 42, 43) and in cases of therapeutic dialysis in humans in order to evaluate the long-term effects of this treatment on the structure of the serosa in patients with advanced chronic renal failure (39). Peritoneal permeability is obviously related to fluid exchanges between serosal cells and underlying vessels. Thus, special attention has been paid, for example, to the parietal peritoneum covering the abdominal surface of the diaphragm of small rodents, with special reference to the lymphatic lacunae (38, 44–49), as well as to the rat omentum (50), rabbit mesentery (41), pelvic peritoneum (51), and mesentery in man (52). These morphological aspects are noteworthy if one takes into consideration the active traffic of lymphocytes entering or leaving the peritoneal cavity, where they may perform an immunological function (50).

In this chapter, the fine morphology of various parietal and visceral serosa seen by correlated scanning (SEM) and transmission electron microscopy (TEM) will be reported. In addition, the relationships between mesothelial cells, with their complex intercellular clefts, and subjacent structures will also be reviewed in relation to peritoneal permeability.

2. Ultrastructure of serosa

The peritoneal mesothelium in different areas of the body, as viewed topographically by SEM, appears to be composed of variably arranged cells forming either a relatively uniform carpet covering the parenchymal organs (liver, Fig. 15-1A and spleen, Fig. 15-1B) or functionally oriented layers on the outer aspect of hollow organs (digestive tract, Fig. 15-1C). Whereas light microscopic reports were inevitably limited to the generic descriptions of major morphological aspects of serous linings, only electron microscopic techniques have helped to mitigate controversial points of classical light microscopic literature. In accordance with these results, it has been commonly accepted that the mesothelial layers are composed of polyhedral cells tending to assume a flattened shape (epithelioid layers).

Similar ultrastructural features also have been reported in human pelvic peritoneum (51), mesentery (52), and parietal and visceral peritoneal layers (39). A number of TEM and/or SEM findings have nevertheless demonstrated that such a characterization cannot be excessively generalized, since differences in mesothelial cell shape do occur (38, 46, 53). Thus, a certain variability in cell shape in different and even in contiguous peritoneal areas was noted (33, 53), while other ultrastructural reports (38, 45, 46) on mouse diaphragmatic mesothelium described the presence of two discrete cell populations, cuboidal and flattened cells, provided with a variable amount of microvillous projections. It has been affirmed, for example, that in areas covered with tall cells, the number of microvilli increases; but, in zones where serosal cells are flattened, microvilli are considerably less numerous (33, 53). The serosal cells show irregularly polygonal contours.

Intercellular boundaries can be more easily recognized in areas where they are flattened (Fig. 15-1D). Under SEM, the surfaces of the mesothe-

←

Figure 15-1. Different areas of the peritoneal mesothelium consist of variously arranged cells, either (A, B) forming a relatively uniform "carpet" or (C) functionally oriented layers on the walls of the digestive tract. (D) The mesothelial cells, especially when flattened, show generally well delineated boundaries (→). (E) The serosal surface often is concealed in part by a small amount of serous fluid; in fluid-deprived zones, a dense microvillous lining may protrude (Mv→). (F) Since the cell shape is probably related to the dynamics of the underlying organs, the cells may assume a prismatic appearance. SEM: (A). rabbit liver, X 200; (B). rabbit spleen, X 1,000; (C). rabbit intestine, X 500; (D). rabbit stomach, X 2,000; (E). rat abdominal wall, X 1,750; (F). rabbit intestine, X 2,000.

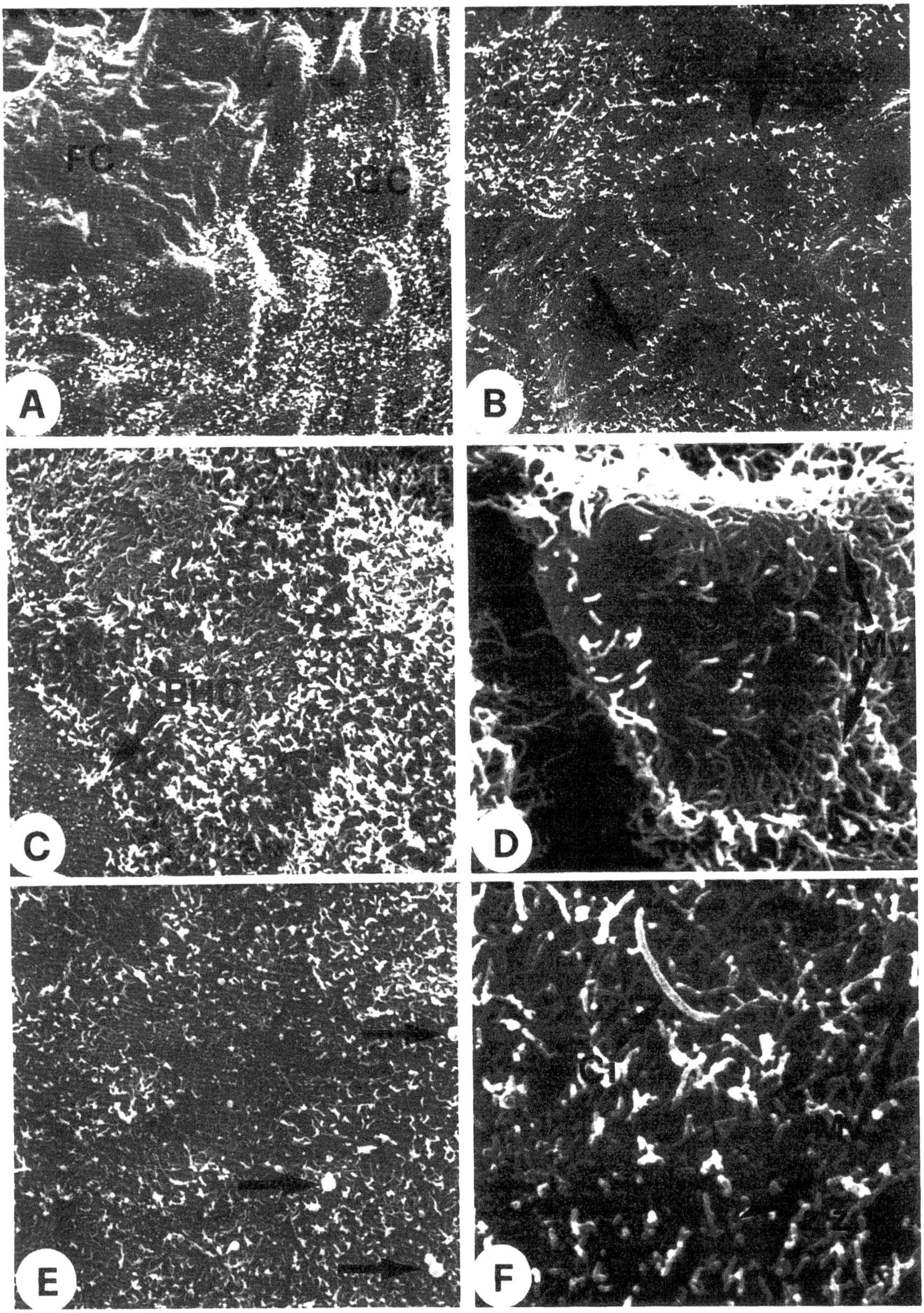
FC
GC
A
B
BMC
C
Mv
D
E
F

lial cells are frequently obscured by a discontinuous, small amount of serous fluid. In areas free of fluid, a dense population of microvilli can be detected (Fig. 15-1E). It has been revealed that number, density, and distribution of serosal microvilli may vary in several species (4, 20, 33, 39, 53–55) including humans (52). The cell-shape variability and number of microvilli likely reflect functional dynamics related to intraperitoneal organs and topographical connections with subjacent tissues, such as muscle and vascular components (Fig. 15-1F).

In some areas, isoprismatic cells rich in microvilli are contiguous with flattened cells having smooth surfaces, provided with few microvilli and overlying blood vessels (Fig. 15-2A). Higher SEM magnifications have demonstrated that microvilli in the flat cells may be short and scanty in relation to the cell body, whereas, along the intercellular borders, they may exhibit a more densely packed arrangement. Consequently, cell outlines may be better defined by elongated or irregularly circular microvillous rows (Fig. 15-2B). The uneven distribution of the microvilli particularly stands out when, in a population of mesothelial cells provided with numerous microprojections, the central zone of some cells appears relatively bare ("bald-headed cells"), showing a few stubby microvillous evaginations (33, 53). In contrast, the mesothelial cell contact areas contain a great number of microvilli, perpendicularly oriented to the cell surfaces. These peripheral microprojections are long and interdigitate with those of neighboring cells, nearly concealing the intercellular spaces (Fig. 15-2C).

Frequently, as seen by SEM, the central area of serosal cells is covered by a coat of coagulated serum which occasionally may be perforated by a few microvillous tips, whereas along the cell borders a dense interlacement of microvilli, arranged in coronal fashion, seems to entrap the serum in the middle (Fig. 15-2D). Similar findings were also reported by Baradi and Rao (55) who described, in various areas of parietal and visceral mouse peritoneum, bare mesothelial surfaces lined by microvilli, sometimes interlocking to constitute coronal (crown-like) formations.

Such features, in addition to the occurrence of serosal cells possessing a dense marginal microvillous interlacement, holding a layer of coagulated serous fluid, corroborate the assumption of Andrews and Porter (4) that the naked areas may exist in vivo and that adjacent mesothelial microvilli may create small compartments in which serous fluid becomes entrapped. It may be that the strong water-binding capabilities of the negatively-charged acid glycosaminoglycans of the glycocalyx coating the microvilli and intervillar surface retain the fluid within these spaces (56). Such entanglement might serve to provide a slippery, liquid cushion layer, which could defend the underlying thin mesothelium against surface friction damage and so prevent adhesion formation (4). In light of this hypothesis, these authors (4), in a systematic study on the mesothelial layers covering the major serosal body cavities, reported that more motile organs, such as the heart, lungs, and diaphragm, are provided with a greater number of these microprojections. Indeed, it is probable that actively moving organs are subjected to relatively more surface friction. It also has been demonstrated that the visceral mesothelium suffers severe damage even during brief periods of drying. In this case, en face silver staining and electron microscopy exhibited loss of mesothelium and exposure of underlying connective tissue. At the same time, the normal serosal fibrinolytic activity was also lost (7).

←

Figure 15-2. (A). Two distinct populations of mesothelial cells may occupy even adjacent areas of the peritoneal surface: right, cubic microvillous elements (CC); left, flat cells (FC) with very few such microevaginations. These two cell types appear to be in direct continuity with each other. (B). The flattened elements are scarcely provided with microvilli, which sometimes are arranged in peripheric rows outlining the cell borders (→). (C). Among serosal cells showing many microvillous projections, some centrally denuded elements ("bald-headed cells") (BHC→) are present. (D). A marginal network of microvilli (Mv→) appears to "crown" the cell and seems to retain an amount of coagulated serum (S) on the cell surface. (E). The mesothelial microvilli possess either smooth or relatively rough surfaces, some of them even being provided with club-tipped features (→). (F). A single isolated cilium (Ci→) surrounded by a dense microvillous population (Mv→), juts over the surface of a mesothelial cell. SEM: (A). rat intestine, X 1,0000 (B). rat intestine, X 2,100; (C). rat spleen, X 4,860; (D). rabbit stomach, X 6,800; (E). rat intestine, X 2,400; (F). rabbit spleen, X 10,000.

The average measurements of serosal microvilli are: 0.05–0.08 μm in diameter and 1.0–2.5 μm in length. Microvilli have either smooth or relatively rough surfaces, and sometimes their tips are swollen to form a small terminal protuberance ("club-tipped" microvilli, Fig. 15-2E) or are fused in groups by a very thin, interconnecting amorphous material (Figs 15-2C and 15-2E) (4, 33, 39, 53).

Occasionally, here and there on the serosal surface, such as in spleen and liver, single isolated cilia of about 5–10 μm in length were observed to arise from the cell surface and project towards the peritoneal cavity beyond the surrounding microvilli (Fig. 15-2F) (4, 57). These cell projections are considered to be probable vestigial remnants of those occurring on the serosa of lower vertebrates (14, 58, 59) and, as such, nonfunctional. It is possible, however, that they produce movement and participate in serous liquid renewal (4). Interest in the mobility of isolated cilia has been recently renewed by a study by Odor and Blandau

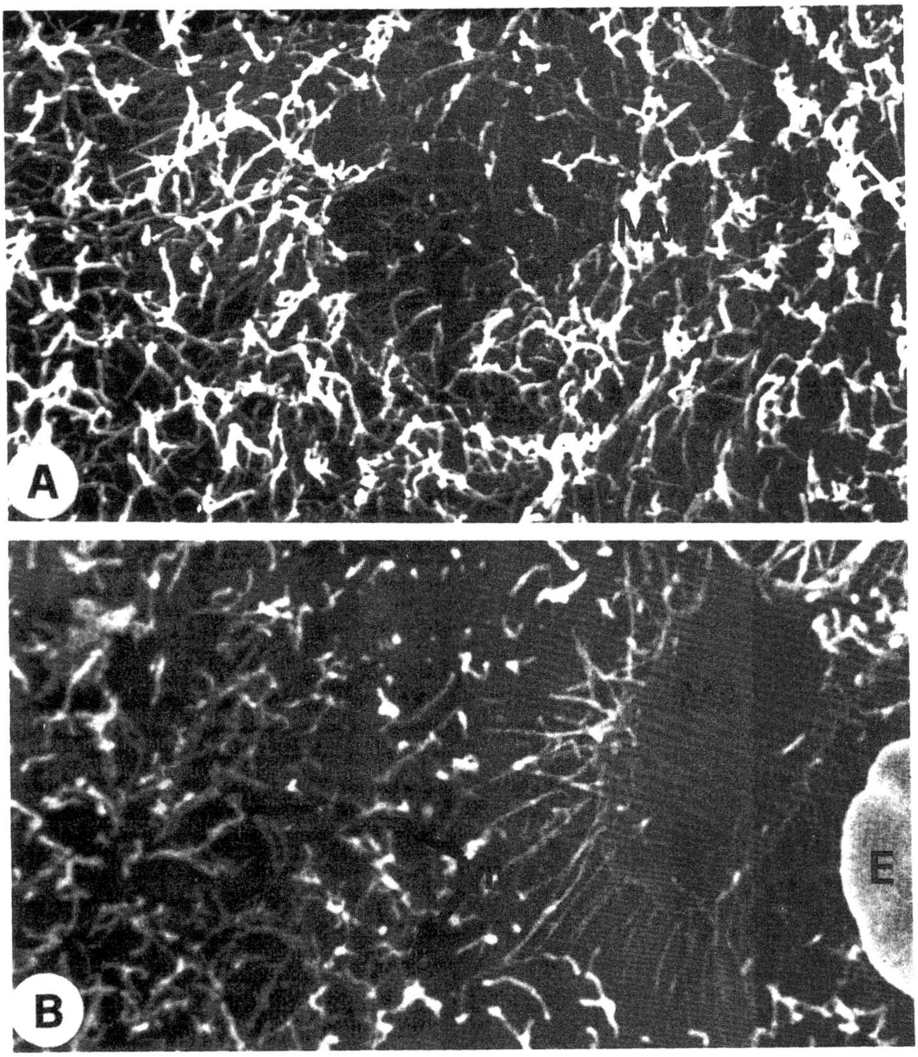

Figure 15-3. (A). An open pit in the splenic serosa, seen among numerous microvilli, forms a deep, saccular invagination, appearing circular in section. Many of these microprojections line its wall, extending into the lumen (Mv→). (B). In various areas of the mesothelial surface, free elements of a diverse nature, including erythrocytes (E) and macrophages (Ma), are visible resting on the microvillous carpet (Mv→). SEM: (A). rabbit spleen, X 7,000; (B). rat abdominal wall, X 5,000.

(60), who recorded cinematographically movements of single cilia on fimbrial epithelial cells in tissue cultures.

Sometimes, in various areas of the mesothelial serosa, at the sites of intersection of three or four intercellular spaces, rimmed circular holes were found (33, 38, 46–50, 53). These are variable in width, and, by SEM, appear as roundish entrance portals of deep saccular invaginations, in direct continuity with the intercellular clefts and going down below the serosal surface (Fig. 15-3A). A number of microvilli line the walls of these invaginations and project inward, toward the lumen, so that their course cannot be followed as far as the basal lamina (33, 53).

On the peritoneal surface, isolated free cells, such as erythrocytes, lymphocytes, and, sometimes, large irregular ameboid elements of a macrophagic nature scattered among the microvilli, have been encountered (Figs. 15-1E and 15-3B) (33, 53).

In parallel TEM observations, the shape variability of mesothelial cells in different areas of the peritoneum has been confirmed; the cells appear isoprismatic, cuboidal (Fig. 15-4A), or flattened (Figs. 15-4B and 15-4C). Similarly, taller cells possess a large number of microvilli (Fig. 15-4A), whereas flat mesothelial cells often exhibit only short and sparse microvilli (Fig. 15-4C) along with endocytotic vesicles. The latter fill almost the entire cytoplasm and have been observed to fuse with the cell membrane, especially with the luminal and abluminal sides, forming clusters of vesicles (Fig. 15-4C).

As seen by TEM (33), the spaces between individual cells may appear as relatively straight channels, where adjacent cell membranes run coarsely parallel (Fig. 15-4A). In other cases, they take the form of long convoluted tubules, in which adjacent cell membranes form very intricate interdigitations (Fig. 15-4D) (20, 33, 36, 39). The junctional complexes closing and/or restricting such spaces may consist of tight junctions ("zonulae occludentes"), desmosomes ("maculae adherentes"), and "zonulae adherentes," similar to those occurring in other epithelia.

The junctional complexes vary in location in relation to the cell shape. If cells are isoprismatic (Fig. 15-4A), the junctional complexes are scattered throughout various levels along the cell interfaces. In flat cells or flattened extensions of taller cells, on the contrary, the junctional complexes are located closer to the serosal cavity (Figs 15-4D and 15-4E). When all three types of junctional complexes are present in flattened cells, the tight junctions have been reported (39) to be located nearest the free mesothelial surface, while the zonulae adherentes lie deeper. Finally, the deepest elements are the desmosomes. TEM observations (20, 33, 39) also have revealed that the irregular, meandering tubular channels (intermesothelial clefts) beneath the junctional complex may dilate to form wide intercellular lacunae and side branches (Fig. 15-4B), apparently opening on the upper aspect of the subjacent basal lamina (Fig. 15-4E). The main ultrastructural characteristics of the peritoneal serosa are summarized in Fig. 15-5.

3. Morphophysiological implications of mesothelial ultrastructure

An important aspect of the mesothelium that has often been discussed in the literature, especially concerning the structure of the peritoneal mesothelium, is its permeability. In fact, the transport mechanism for solutes and/or particles has recently assumed great significance in clinical practice, particularly in regard to peritoneal dialysis, a maintenance treatment for chronically uremic patients with end-stage renal disease (39, 41, 43, 61, 62).

Since the early ultrastructural observations of the mesothelium of the peritoneum of rat oviduct (18), intracytoplasmic vesicles were seen even in the apical microvilli (20). Furthermore, it was suggested that microvillous projections greatly increase the surface area for the exchange of soluble substances between cells and serous cavity and that vesicles represent the morphological evidence of such transfers. In addition, it has been reported that omental microvilli and subjacent cytoplasmic regions have an internal cytoskeleton, similar to that occurring in resorptive epithelia (37).

The peritoneum undoubtedly acts as a selective membrane, consisting of mesothelial cells, interstitium, and blood and/or lymph capillary vessels (40, 41, 43). Although the capillaries, together

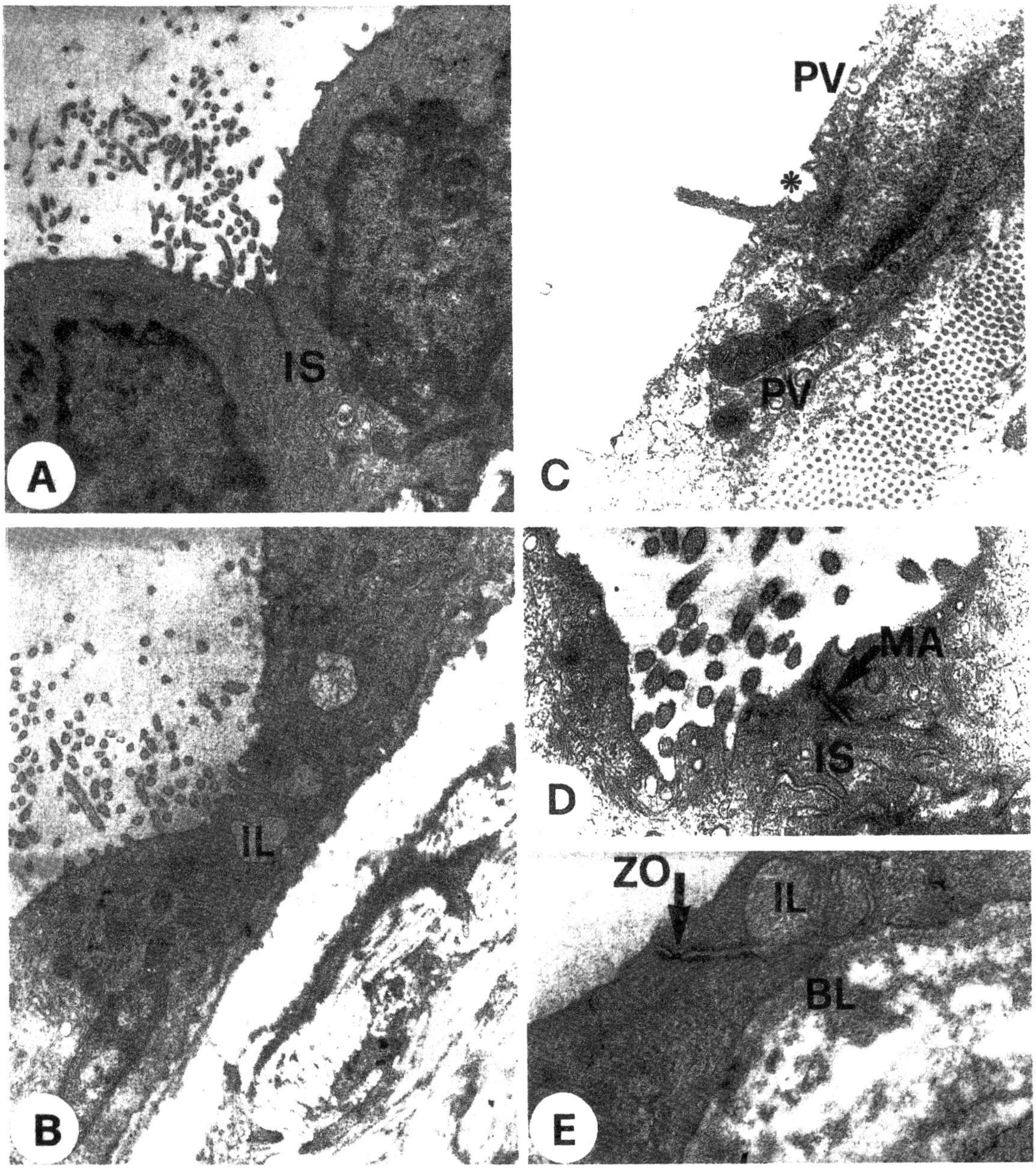

Figure 15-4. TEM observations have confirmed the variable morphology of the serosal cells: (A) the cubic, and (B, C) the flattened shapes are the ones most frequently observed; in the latter case (C), a great number of pinocytotic vesicles (PV) scattered throughout the cytoplasm and/or fused with the cell membrane, either singularly or en masse (*), are seen. The mesothelial cells are connected by a variety of junctional complexes, such as (D) desmosomes ("maculae adherentes") (MA→) and (E) tight junctions ("zonulae occludentes") (ZO→), and are separated by spaces (IS) either (A) relatively regular in outline, or (D) describing extremely tortuous courses, and (B, E) often dilating in large intercellular lacunae (IL), sometimes (E) apparently opening towards the basal lamina (BL). TEM: (A). rat spleen, X 6,000; (B). rat spleen, X 6,000; (C). rat abdominal wall, X 18,000; (D). rat spleen, X 18,000; (E). rat abdominal wall, X 18,000).

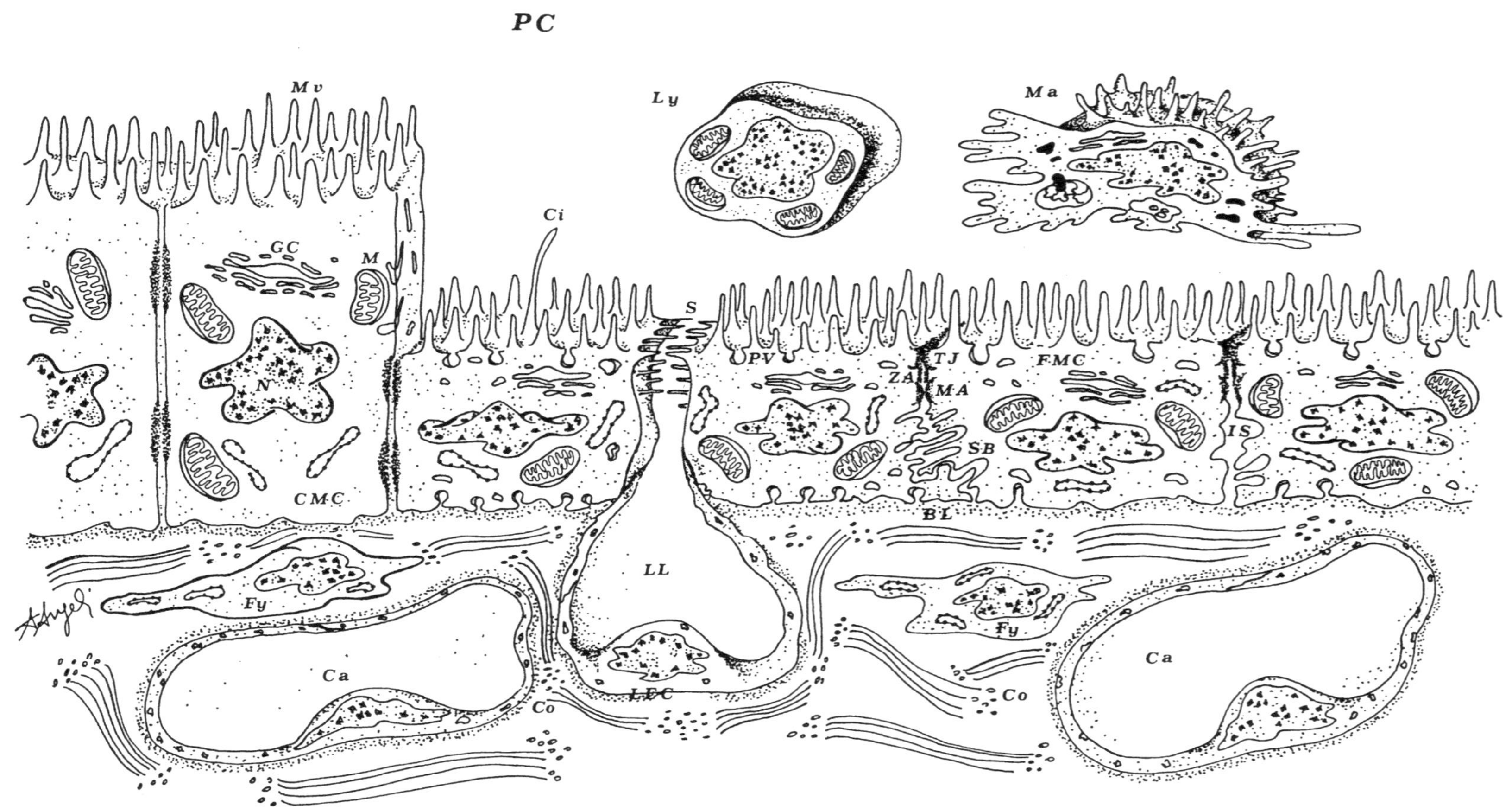

Figure 15-5. This diagram summarizes the main ultrastructural characteristics of the peritoneal serosa. PC, peritoneal cavity; Ly, lymphocyte; Ma, macrophage; CMC, cuboidal mesothelial cell; FMC, flattened mesothelial cell; Mv, microvilli; Ci, cilium; N, nucleus; GC, Golgi complex; M, mitochondrion; PV, pinocytotic vesicles; TJ, tight junction; ZA, zonula adherens; MA, macula adherens; IS, intercellular space; SB, side branch; BL, basal lamina; S, stoma, LL, lymphatic lacuna; LEC, lymphatic endothelial cell; Ca, blood capillary; Fy, fibrocyte; Co, collagenous fibers.

with the associated basal lamina and interstitial spaces, may restrict to some extent the passage of large molecules, such as proteins, the mesothelial layer presumably represents the major rate-limiting diffusion barrier (63) because of its junctional belt, which separates the peritoneal cavity from the interstitium (40).

3.1. Solute transport across the peritoneal membrane

Studies of peritoneal kinetics have addressed the question of how solutes come to cross the mesothelium. Two types of passageway have been considered: (a) intracellular (i.e., vesicular transport) (19, 29, 42) and (b) intercellular (paracellular), through discontinuous intercellular junctions (21, 35, 64, 65).

According to many A.A., solute transport across the peritoneal membrane is an active process which takes place via intracytoplasmic vesicles (pinocytosis) (19, 20, 29–32, 39, 66, 67). Mesothelial vesicles are usually abundant, and have been detected on both apical and basal surfaces of cells, where they may open through the plasma membrane, assuming a characteristic flask-shaped configuration (41). Sometimes, they even appear to open into the intercellular clefts (20, 41, 42, 52). Recently, peritoneal permeability has been regarded as a property related to the mesothelial cytoskeleton, therefore being a phenomenon modifiable by microfilament-active agents (43). These investigators, in a model of rat peritoneal dialysis, noted extensive development of membrane protuberances on the cell surface ("zeiotic knobs"), presumably due to loss of cytoskeletal anchorage of the plasmalemma.

Moreover, the permeability rates of urea, inulin, and plasma proteins were increased by the addition to the dialysate of substances such as cytochalasins, which, by inhibiting elongation of actin filaments, produce changes in microfilament assembly with consequent alteration in cell motility, including locomotion and cytokynesis (68), and mesothelial permeability and intercellular junctions.

Although there is a general consensus regarding the transfer of solutes across the cytoplasm by means of pinocytotic vesicles, intercellular passage remains controversial, with much of the question revolving around intermesothelial junctions. Some A.A. have detected both closed and open apparatuses (18, 20, 21, 35, 54, 64, 65, 69), while others have suggested that the mesothelial cells are provided with adhering and occluding junctions, sealing the intercellular spaces (41, 70). Hama (34), in a study of frog mesothelium, reported the presence of desmosomes.

In order to explain the high degree of permeability of the mesothelial membrane, many A.A. suggested that water-soluble molecules might take a primarily paracellular route, crossing the mesothelium by passive diffusion, probably through intercellular spaces and junctional complexes (21, 35, 64, 65, 71–73). Furthermore, during the past century it has been observed that water and salt solutions containing vital dyes were absorbed from the abdominal cavity more rapidly than particulate material (74).

Recently, evidence has been provided that show that mesothelial paracellular regions may be the main pathway for the transport of ions and small molecules. In fact, by using microelectrode surface mapping, it has been possible to record variations in electric potential from point to point along frog mesentery; these were attributed to low resistance sites, possibly located over junctional regions (75).

Simionescu and Simionescu (37), on the basis of freeze-fracture replicas, reported the frequent occurrence of free-ending junctional strands in the peritoneal mesothelium of rats and mice and suggested that these are simple attachment devices rather than continuous occluding junctions. Their extensive polymorphism may be an expression of some lability, perhaps dependent upon successive stages of junctional assembly and/or disassembly. These results seem to be in agreement with those of Leak and Rahil (38), who described numerous discontinuous junctional ridges in the diaphragmatic peritoneum of the mouse and postulated that many of these are labile and easily separable. In addition, numerous peripheral cell portions, simply placed over one another and lacking specialized structures for cell-to-cell adhesion, provided for the free diffusion of large molecules within the same peritoneal area (44, 46).

Cotran and Karnovsky (21) demonstrated the passage of the tracer horseradish peroxidase

(HRP) (molecular weight: 40.000 daltons; radius 30 Å) through junctions of the mesothelium lining the cremasteric sac of the rat. Due to the tortuosity of the junctions, these investigators encountered some difficulty in assessing the ultrastructure of the junctional complexes. Many of these did not display fusion of external leaflets and obliteration of intercellular spaces, and a great number of tight junctions were not found along the entire cell circumference. Similar observations were reported in vascular endothelium, where it was suggested that junctions consisted of "maculae" rather than "zonulae occludentes" (76).

On the basis of these results, Cotran and Karnovsky (21) attributed the movements of the tracer from the peritoneal cavity to the interstitium to the presence of open junctions (windows) of about 2–4 nm in width. However, their attempt to reproduce passage of HRP across the mesentery was not successful. It was also proposed in other studies that mesothelial junctions could be similar to those of venular endothelium, not only for their general geometry but also for the appearance of the cleavage plane at their level (77).

The junctions of venular endothelium show permeability and lability under normal (78) and pathological conditions (79, 80). Mesothelium, as well as the venular endothelium, reacts to 5-hydroxytryptamine by showing increased permeability to large molecules and particles and, in the case of the endothelium, this enhancement is due to focal separations along the intercellular junctions (79).

Baradi and Rayns (36), in a freeze-etching study of junctional complexes in mouse parietal peritoneum, reported the occurrence of tortuous tubular profiles, often associated with vesicular surfaces. These tubules may course into the cytoplasm or form part of a complex interdigitating cell-to-cell apposition. In some views, the ridges and furrows characteristic of zonulae occludentes or tight junctions have been seen. However, their macular or zonular nature was not fully clarified, and the A.A. proposed that the junctions could be of intermediate permeability; if they are proven to be of a macular nature, however, they might indeed be extremely permeable.

Although occluding junctions are usually considered to be strongly adhesive (81, 82), the amount of literature suggesting their possible detachment continues to increase. These studies have proposed that such detachment may be a consequence of various processes including normal differentiation (83). Tight junctions can also separate to accommodate the structural changes that normally accompany changes in cell shape (84). Moreover, there is firm electrophysiological evidence supporting variable permeability in zonular types of tight junctions in different epithelia, and there appears to be a relationship between permeability and number of tight junctional ridges (85). Kaye and Pappas (86), while studying corneal endothelium, suggested that the bypass of constriction foci through focal vesicular transport might be the mechanism that accounts for intercellular cleft permeability. Yet, even if tight junctions of mesothelial cells may indeed be relatively "leaky," they may still be the rate-limiting barrier for medium-sized molecules such as inulin (37). In addition, it is more likely that the peritoneal regions overlying the diaphragm, where discontinuities were found (46), represent the main site of protein flux into the peritoneal cavity (43).

3.2. Intercellular clefts

Another important aspect associated with peritoneal absorption via the paracellular route concerns the nature and pattern of the intercellular clefts.

These frequently form long, irregularly tortuous courses, often labyrinthic in appearance (20, 36, 39, 52), being somewhat similar to those present in the visceral yolk-sac epithelium of the guinea pig (87), where intercellular spaces are continuous down to the basal lamina, permitting the passage of nondiffusable particles. On the contrary, in an ultrastructural study of rat visceral mesothelium (18), these spaces, corresponding to those detected by light microscopy and termed "zwischenrawum" by Kolossow (23), were neither continuous with the serous cavity nor with the subjacent connective tissue spaces. Similarly, in rabbit parietal serosa, Baradi and Hope (20) were unable to demonstrate the entire course of the intercellular boundaries. However, they recognized a variety of highly complicated cell boundaries and reported the presence of side channels. Some of these, which originated from main intercellular channels, opened separately in the peritoneal cavity or reached the connective tissue boundary. Moreover, the basal lamina appeared discontinuous in most cases and was connected to a thick lamina of an amorphous, intercellular substance, showing uneven density and interruptions by communicating spaces and vacuoles.

In a subsequent study on mouse parietal peritoneum, Baradi and Rayns (36) described long, meandering tubular structures which extended considerably into the cytoplasm and suggested that they may represent either intercellular clefts or intracellular channels. Spindle-shaped dilatations, side branches of intercellular clefts, and cellular interdigitations beneath the junctional complexes, were also reported by Dobbie et al. (39) in the peritoneal mesothelium in rats, mice, and humans, including uremic patients.

König et al. (52), in a comparative study of the fine morphology of the rat and human mesenteries, confirmed the presence of intercellular spaces with a labyrinthine and tortuous course but concluded that the channels cannot open into the subjacent connective tissue due to the presence of a continuous basal lamina throughout the whole extension of the mesentery.

Many physiological studies regard the peritoneal membrane as a mosaic consisting of populations of small and large "pores." Since dextran molecules (molecular weight 10,000 to 90,000 daltons) undergo significant clearance across rabbit peritoneum, the existence of at least two different types of pores has been suggested. Most of the peritoneal surface is made up of pores having an effective radius of 15 Å to 20 Å, while another population of pores of 70 Å (effective radius) constitutes the remaining 0.01% of the total pore surface area (88).

In a series of equilibrium studies on isolated peritoneum and mesentery using diffusional transfer of radiorubidium and ^{32}P orthophosphate as markers, it has been concluded that mesothelial permeability to lipid-insoluble substances diffusing paracellularly could be accounted for by pores or channels occupying about 0.2% to 0.6% of the macroscopic area of the peritoneal membrane (71, 72, 89).

Cotran and Karnovsky (21) suggested that the intermesothelial clefts might represent the small and large "pores" described by physiologists. Although these studies conceded a route of intracellular transport, they postulated that this process must be slow and likely plays a minor role if compared to the intercellular route. It should be noted, however, that according to Baradi and Rayns (36), whether such pores follow an intracellular or paracellular course is functionally immaterial to the validity of the pore theory, so long as they span the cell between the apical and basal surface as potentially open thoroughfares. Thus, these researchers applied to mesothelial pores the idea put forth by Landis and Pappenheimer (90) (who introduced the pore theory for capillary endothelium), which states that there are no reasons to assume that the channels extending through capillary walls are either cylindrical or perfectly uniform.

3.3. Stomata

Another morphological question, related mainly to the passage of large particles and migration of cells to and from the peritoneal cavity, concerns the occurrence and significance of openings (fenestrae) along the peritoneal mesothelium. In the past, movement of particles through the spaces between adjacent mesothelial cells, particularly in the diaphragmatic peritoneum, has been reported (24, 91, 92), but it was von Recklinghausen (9) who first described "stomata" interposed between mesothelial cells. The original concept of stomata, based upon the appearance of circular deposits of silver nitrate within the intercellular junctions, was correlated to the observation that various tracers such as cinnabar, cobalt blue, etc. entered large terminal lymphatics of the diaphragm via these openings. On the other hand, it was well demonstrated that particles and cells of various sizes can be rapidly removed from the peritoneal cavity by these wide lymphatic capillaries (9, 26, 93–97). Von Recklinghausen (9) termed these terminal lymphatics "lacunae," and believed that stomata opening into them were stable structures. Other A.A. nevertheless regarded the peritoneal cavity as an entirely closed structure. They explained these circular deposits, or stomata, as artifacts produced by the interaction of silver salts with serous fluid which adhered to the cell boundaries, giving the impression of pores (22–24, 93, 98–100).

Some investigators (26, 93) disputed the assumption of stomata stability and raised the possibility that these were instead transient. Allen (95) reported that openings in the supporting basal lamina, through which the peritoneal cavity was in direct contact with the intercellular junctions of the underlying lymphatic endothelium, corresponded to stomata. Recently, ultrastructural findings (47–49, 101) have confirmed the absence of a basal lamina at the sites of such mesothelial-endothelial juxtaposition ("lacunar roof" by MacCallum) (93), thereby lending support to the suggestion that these sites constitute communicating channels between serosal cavity and lymph vessels. Moreover, since the layer of submesothelial connective tissue in these areas is interrupted, it has been suggested that where a channel forms, adjacent fibroblasts secrete enzymes necessary for the local removal of tissue (47). In addition, collagenous fibers outside the lacunae are densely packed, but in the lacunar roof they run in periodically arranged bundles (49).

Leak and Rahil (38) reported the features of diaphragmatic stomata in mice. They observed two populations of pores: (a) small, irregularly-shaped pores (i.e., intercellular gaps) appearing as shallow pits circumscribed by filamentous processes arising from the cell boundaries, and (b) large circular pores, ranging from 4 to 10 μm in diameter, occurring at the confluence of several intercellular spaces, and enfolded to form true, deep channels which appeared to open into the underlying lymphatic lacunae. The latter pores, probably corresponding to the stomata of von Recklinghausen (9), were characterized by a collar-like edge, often projected above the mesothelial surface in a funnel-shaped structure.

Both mesothelial and endothelial lacunar cells, interconnected by junctional complexes, may take part in formation of the channel walls (38, 47, 48, 102). These channels, originating from stomata, may exhibit a tortuous course and an irregular lumen, due to the presence of invaginated cells of submesothelial stroma (46) or flattened cytoplasmic projections of bordering lymphatic endothelia (48). The occurrence of numerous microvilli protruding within the lumen of the stomata partially or entirely obscures their SEM visualization (33, 53), and, even by TEM, it is often difficult to demonstrate the entire course of the channels due to the tortuosity of their longitudinal axis. Other studies have confirmed, however, that the channels are indeed continuous, since tracers such as ferritin or India ink pass from the peritoneal cavity to the lacunae via the spaces between the interdigitating processes crisscrossing the channels (47). Some A.A. (49, 95) have proposed that the mesothelial stomata, passively adapting themselves to the respiratory movements of the diaphragm, act as valves. Casley-Smith (103) hypothesized that the endothelial processes extending within the lumen of the channels may play such a role.

On the basis of these data, it has been suggested that the regurgitation of fluid from the lacunae towards the peritoneal cavity would be averted if the endothelial flaps of the lymphatic capillary walls were involved in such a mechanism (48). In fact, contractions of both mesothelial cells of the peritoneum and lymphatic endothelial cells have been demonstrated (104–106) and, using the technique of actin-decoration with the S1 subfragment of myosin, numerous actin filaments were identified in mesothelial and endothelial cells bordering stomata and channels (47, 48). Such findings have documented that these structures perform a dynamic role in the regulation and drainage of particles and cells via the lymphatic lacunae, actively adapting their patency to the changing states of contraction and/or relaxation of the diaphragm and attendant variations in intra-abdominal pressure (48).

Recently, the occurrence of lymphatic stomata in the ovarian bursa of the golden hamster has been reported (101). These openings seem to vary in number depending upon the phases of the sexual cycle, and they have been functionally related to the lymphatic resorption of erythrocytes discharged into the bursal cavity during ovulation. Therefore, it was suggested that the patency of these stomata might depend upon changing conditions in the bursal cavity and that the morphological basis of such a regulation likely consists of the dehiscence of the interendothelial lymphatic junctions and, according to Tsilibary and Wissig (48), in valvular mechanisms involving endothelial contractile flaps.

Intercellular stomata similar to those encountered in the diaphragm were also reported in rat omentum (50). These stomata lie above certain discrete aggregates of lymphoreticular cells (the classical "milky spots" of Ranvier) (107, 108), and have been considered open portals through which lymphocytes may travel to and/or from the peritoneal cavity. Occasionally, lymphocytes were also observed traversing these stomata, and it was suggested that they reach the milky spots from the omental postcapillary venules (109, 110). Once these lymphocytes have entered the peritoneal cavity in order to "patrol" it, they could ultimately exit the cavity principally via the diaphragmatic lymphatics which have similar stomata overlying them (38, 44–49).

Similar openings also were reported on the serosal surface of the spleen (33, 53). Therefore, since the spleen is essentially a lymphoid organ, the possibility that its stomata might perform a similar immunological defense should not be excluded. Baradi and Rao (55) described intracellular stomata in the parietal and visceral layers of the mouse peritoneum through SEM and suggested that they might correspond to the long tortuous tubular profiles seen in their freeze-fracture views.

The peritoneal serosa exhibits different morphological features from one area to another. Its main functions essentially are absorption, exudation, and plasticity (19). Thus, it is rather safe to assume that there are structurally specialized zones in which these functions may prevail over one another (41, 42). For example, Digenis et al. (42) found particles of iron dextran only in those mesothelial cells overlying vessels. This observation supports the concept that only small peritoneal portions, i.e., the parts of the peritoneal membrane in juxtaposition to the capillaries, participate in solute exchange (62). The major sites of absorption from the peritoneal cavity may be the diaphragm and omentum (38, 70, 111). Fluids and small molecules are rapidly removed from the peritoneal cavity by mesenteric and omental blood vessels, while particulate matter and cells cross the submesothelial lymphatic lacunae of the diaphragm using major passageways (45, 49). In a certain sense, then, the peritoneal cavity may be regarded as an intricate section of the lymphatic system where absorption takes place primarily through the submesothelial lymphatics (38); in fact, those of the diaphragm are now considered to be the principal route for drainage (46).

4. Concluding remarks

This chapter has limited its scope mainly to the description of ultrastructural aspects of peritoneal tissue. Although we have not outlined in depth the enzyme content of the peritoneum, which has been studied by standard histochemical techniques (112), certain points should be noted. For example, positive reactions for adenosine triphosphatase (ATPase) present in abundance at the junctional regions between adjacent cells, could be related to active transport of materials along intercellular clefts beyond foci of constriction, such as tight junctions.

Positive reactions for acid phosphatase correlate well with a phagocytic function of the mesothelium, demonstrated ultrastructurally by intracytoplasmic, membrane-closed vacuoles ("segresomes" by Tanaka) (113, 114). Positive reactions for the mitochondria-associated enzymes, such as succinate dehydrogenase and cytochrome oxidase, indicate oxidative metabolism within the cell which may be related to the provision of energy for active transport of materials across the cell. In conclusion, these and other results clearly reveal that the peritoneal mesothelium does not play a merely passive role as a simple, inert membrane, but must be regarded as a complex and selective barrier whose sophisticated functions can be approached through the integrated application of ultrastructural methods (39, 66).

Acknowledgments

The authors would like to thank Mr. Gianfranco Franchitto, Mr. Ezio Battaglione, and Mr. Ettore Pizzoli for preparation of the plates, Ms. Alessandra Angeli, M.D. for drawing the diagram, and Mrs. Eleonora Vergnano for typing the manuscript.

References

1. Vogel A: Zur struktur des peritoneal-mesothelium. *Experientia* 13: 54, 1957.
2. Policard A, Baud CA: Les Structures Inframicroscopiques Normales et Pathologiques des Cellules et des Tissus. Paris: Masson, 1958.
3. Draper MH, Chalmers J: The Supporting Tissues. In: A Companion to Medical Studies, Vol 1. R Passmore and JS Robson (eds), Philadelphia: FA Davis Co, p 16.1–16.16, 1968.
4. Andrews PM, Porter KR: The ultrastructure morphology and possible functional significance of mesothelial microvilli. *Anat Rec* 177: 409–426, 1973.
5. Bengt F: Polyphloretin phosphate — a hyaluronidast inhibitor — and hyaluronidase in prevention of intraperitoneal adhesions. *Acta Chir Scand* (suppl) 217: 1–97, 1956.
6. Schade DS, Williamson JR: The pathogenesis of peritoneal adhesions: An ultrastructural study. *Ann Surg* 167: 500–510, 1968.
7. Ryan GB, Grobety J, Majno G: Mesothelial injury and recovery. *Am J Path* 71: 93–102, 1973.
8. Bichat MFX: Traité des Membranes en General et des Diverses Membranes en Particular. Paris: Gabon, 1827.
9. Von Recklinghausen F: Zur fettresorption. *Virchows Arch Path Anat* 26: 172, 1863.
10. Henle J: Handbuch der Systematischen Anatomie des Menschen. Braunschweig: Viemeg, 1866.
11. Schweigger-Seidel F, Dogiel J: Ueber die peritonealhöhle bei forschen und ihren zusammenhang mit dem lymphgefassysteme. *Arb ad Phys Inst z Leipz*: 68, 1867.
12. Paladino G: Sur l'endothelium vibratile chez les mammifères et sur quelques faits physiologiques relatifs aux formation endotheliale. *Arch Ital de Biol* 3: 43, 1883.
13. His W: Die häute und höhlen des körpers. *Arch f Anat u Physiol Anat*, Abt 368, 1903.
14. Neumann E: Die beziehungen des flimmerepithels der bauchhöhle zum eileiterepithel beim frosche. *Arch f Mikr Anat* 11: 354–378, 1875.
15. Waldeyer W: Archiblast und parablast. *Arch f Mikr Anat* 22: 1, 1883.
16. Minot CS: The mesoderm and the coelom of vertebrates. *Amer Natur* 24: 877, 1890.
17. Mall FP: Development of the lesser peritoneal cavity in birds and mammals. *J Morph* 5: 165, 1891.
18. Odor L: Observations of the rat mesothelium with the electron and phase microscopes. *Am J Anat* 95: 433, 1954.
19. Fukata H: Electron microscopic study on normal rat peritoneal mesothelium. *Acta Path Jpn* 13: 309–325, 1963.
20. Baradi AJ, Hope J: Observations on ultrastructure of rabbit mesothelium. *Exp Cell Res* 34: 33–34, 1964.
21. Cotran RS, Karnovsky MJ: Ultrastructural studies on the permeability of the mesothelium to horseradish peroxidase. *J Cell Biol* 37: 123–137, 1968.
22. Tourneux F, Herrmann G: Recherches sur quelques épithéliums plats dans la série animale (Deuxieme partie). *J Anat Physiol* 12: 386–424, 1876.
23. Kolossow A: Ueber die struktur des pleuroperitoneal und gefässepithels (endothels). *Arch S Mikr Anat* 42: 318–383, 1893.
24. Muscatello G: Ueber den bau und das aufsaugungsvermögen des peritonäum. *Virchows Arch Path Anat* 142: 327, 1895.
25. Prenant MA: Sur la morphologie des cellules épithéliales ciliées qui recouvrent le péritoine des amphibiens. *Arch d'Anat Micr* 7: 473–482, 1904–1905.
26. Cunningham RS: The physiology of the serous membranes. *Physiol Rev* 6: 242, 1926.
27. Maximov A: Bundegewebe und blutbildende gewebe. In: Handbuch der Mikroskopisches Anatomie des Menschen, pt 1. A V Möllendorff, (ed). 289–297, 1927.
28. Chlopin NG: Über regenerationsprozesse im mesothel und die bedeutung der serosadeckzellen. *Beit z Path Anat u z Allg Path* 98: 35–64, 1936–1937.
29. Odor DL: Uptake and transfer of particulate matter from the peritoneal cavity of the rat. *J Biophys Biochem Cytol* 2 (suppl): 105–108, 1956.
30. Staubesand J: Zur histophysiologie des herzbeutels. II. Mitteilung. Elektronmikroskopische untersuchungen uber die passage von mettal-solen durch mesotheliale membranen. *Z Zellforsch Mikroskop Anat* 58: 915–934, 1963.
31. Marchesi VT: The role of pinocytic vesicles in the transport of material across the walls of small blood vessels. *J Invest Ophthamol* 1: 1111, 1965.
32. Fedorko ME, Hirsch JG, Fried B: Studies on transport of macromolecules and small particles across mesothelial cell of the mouse omentum. *Exp Cell Res* 69: 313–323, 1971.
33. Barberini F, Carpino F, Renda T, Motta P: Observations by scanning electron microscopy (SEM) on the surface of mesothelial cells of different peritoneal areas of rat. *Anat Anz* 142: 486–496, 1977.
34. Hama K: The fine structure of the desmosomes in frog mesothelium. *J Biophys Biochem Cytol* 7: 575–579, 1960.
35. Cotran RS, Majno G: Studies on intercellular junctions of mesothelium and endothelium. *Protoplasma* 63: 45–51, 1967.
36. Baradi AF, Rayns DG: Mesothelial intercellular junctions and pathways. *Cell Tiss Res* 173: 133–138, 1976.
37. Simionescu M, Simionescu N: Organization of cell junctions in the peritoneal mesothelium. *J Cell Biol* 74: 98–110, 1977.
38. Leak LV, Rahil K: Permeability of the diaphragmatic mesothelium: The ultrastructural basis for "stomata". *Am J Anat* 151: 557–594, 1978.
39. Dobbie JW, Zaki M, Wilson L: Ultrastructural studies on the peritoneum with special reference to chronic ambulatory peritoneal dialysis. *Scot Med J* 26: 213–223, 1981.
40. Alavi N, Lianos E, Andres G, Bentzel C: Effect of protamine on permeability and structure of rat peritoneum. *Kidney Int* 21: 44–53, 1982.
41. Gotloib L, Digenis GE, Rabinovich S, Medline A,

Oreopoulos DG: Ultrastructure of normal rabbit mesentery. *Nephron* 34: 248–255, 1983.

42. Digenis GE, Rabinovich S, Medline A, Rodella H, Wu G, Oreopoulos DG: Electron microscopic study of the peritoneal kinetics of iron dextran during peritoneal dialysis in the rabbit. *Nephron* 37: 108–112, 1984.
43. Alavi N, Lianos E, Van Liew JB, Mookerjee BK, Bentzel CJ: Peritoneal permeability in the rat: Modulation by microfilament-active agents. *Kidney Int* 27: 411–419, 1985.
44. Leak LV: Permeability of peritoneal mesothelium: A TEM and SEM study. *J Cell Biol* 70 (2, Pt 2): 423 a (Abstr), 1976.
45. Leak LV: Topographical changes in mesothelial cells following intraperitoneal injections. *Anat Rec* 187: 635, 1977.
46. Tsilibary EC, Wissig SL: Absorption from the peritoneal cavity: SEM study of mesothelium covering the peritoneal surface of the mouse diaphragm. *Am J Anat* 149: 127–133, 1977.
47. Tsilibary EC, Wissig SL: Structural plasticity in the pathway for lymphatic drainage from the peritoneal cavity. *Microvasc Res* 17: S 144, 1979.
48. Tsilibary EC, Wissig SL: Lymphatic absorption from the peritoneal cavity: Regulation of patency of mesothelial stomata. *Microvasc Res* 25: 22–39, 1983.
49. Bettendorf U: Raster- und transmissionelektronenmikroskopische untersuchungen über die resorption korpuskulärer partikel aus der peritonealhöhle der ratte. *Verh Dtsch Ges Path* 62: 344, 1978.
50. Mironov VA, Gusev SA, Baradi AF: Mesothelial stomata overlying omental milky spots: Scanning electron microscopic study. *Cell Tiss Res* 201: 327–330, 1979.
51. Blaustein A, Lee H: Surface cells of the ovary and pelvic peritoneum: A histochemical and ultrastructure comparison. *Gynecol Oncol* 8: 34–43, 1979.
52. König B, De Camargo AM, Ferraz de Carvalho CA: Morphological aspects and observations about the permeability of the rat mesentery to the lanthanium nitrate. Some comparisons with the morphology of the human mesentery. *Anat Anz* 149: 365–374, 1981.
53. Motta P, Barberini F: Stomata in the rat peritoneum as revealed by scanning electron microscopy. In: Problèmes de Morphologie et d'Embryologie Expérimentales et Comparées. Sofia: Editions de l'Académie Bulgare des Sciences, p 315–324, 1977.
54. Kluge T, Hovig T: The ultrastructure of human and rat pericardium. 2. Intercellular spaces and junctions. *Acta Path Microbiol Scand* 71: 547–563, 1967.
55. Baradi AF, Rao SN: A scanning electron microscope study of mouse peritoneal mesothelium. *Tissue and Cell* 8: 159–162, 1976.
56. Bennett HS: Morphological aspects of extracellular polysaccharides. *J Histochem Cytochem* 11: 14–23, 1963.
57. Motta P: The three-dimensional fine structure of the liver as revealed by scanning electron microscopy. *Int Rev Cytol Suppl* 6: 347–390, 1977.
58. Prenant MA: Sur la morphologie des cellules épithéliales ciliées qui recouvrent le péritoine hépatique des amphibiens. *C R Soc de Biol* 55: 1044–1046, 1903.
59. Delaney PA: On specialized phagocytic mesothelium in the cholemia. *Anat Rec* 43: 65, 1929.
60. Odor DL, Blandau RJ: Observations on the solitary cilium of rabbit oviductal epithelium: Its motility and ultrastructure. *Am J Anat* 174: 437–453, 1985.
61. Boen ST: Kinetics of peritoneal dialysis. *Medicine* 40: 243–250, 1961.
62. Henderson LW: The problem of peritoneal membrane area and permeability. *Kidney Int* 3: 409–410, 1973.
63. Nolph KD: Peritoneal Clearances. *J Lab Clin Med* 94: 519–525, 1979.
64. Casley-Smith JR: An electron microscopical study of the passage of ions through the endothelium of lymphatics and blood capillaries and through the mesothelium. *Q J Exp Physiol Cogn Med Sci* 52: 105–113, 1967.
65. Kluge T: The permeability of mesothelium to horseradish peroxidase. *Acta Path Microb Scand* 75: 257–269, 1969.
66. Shear J, Harvey JD, Barry KG: Peritoneal sodium transport: Enhancement by pharmacologic and physical agents. *J Lab Clin Med* 67: 181, 1966.
67. Rasio E: The permeability of isolated mesentery. Effect of temperature. In: Capillary Permeability. C Crone, NA Lassen (eds). New York: Academic Press, p 643–646, 1970.
68. Brown SS, Spudich JA: Cytochalasin inhibits the rate of elongation of actin filament fragments. *J Cell Biol* 83: 657–662, 1979.
69. Baradi AF, Campbell WG: Exudative peritonitis induced in mice by bovine serum albumin: An ultramicroscopic study. *Arch Path* 97: 2–12, 1974.
70. Fedorko ME, Hirsch JG: Studies on transport of macromolecules and small particles across mesothelial cells of the mouse omentum. I. Morphological aspects. *Exp Cell Res* 69: 113–127, 1971.
71. Berndt WO, Gosselin RE: Physiological factors influencing radiorubidium fluxe across isolated rabbit mesentery. *Am J Physiol* 200: 454–458, 1961.
72. Gosselin RE, Berndt WO: Diffusional transport of solutes through mesentery and peritoneum. *J Theoret Biol* 3: 487–495, 1962.
73. Cascarano J, Rubin A, Chick W, Zweifach BW: Metabolically induced permeability changes across mesothelium and endothelium. *Am J Physiol* 206: 373–382, 1964.
74. Starling EH, Tubby AH: On the paths of absorption from the peritoneal cavity. *J Physiol* 16: 140, 1894.
75. Michel CC: Direct observation of the sites of permeability for ions and small molecules in mesothelium and endothelium. In: Capillary Permeability. C Crone, NA Lassen (eds). New York: Academic Press, p 628–642, 1970.
76. Karnovsky MJ: Ultrastructural basis of capillary permeability studied with peroxidase as a tracer. *J Cell Biol* 35: 213, 1967.
77. Simionescu M, Simionescu N, Palade GE: Segmental differentiations of cell junctions in the vascular endothelium. The Microvasculature. *J Cell Biol* 67: 863–885, 1975.
78. Simionescu N, Simionescu M, Palade GE: Structural

basis of permeability in sequential segments of the microvasculature. *J Cell Biol* 70 (2, pt 2): 186a (abstr), 1976.
79. Majno G, Palade GE: Studies on inflammation. I. The effect of histamine and serotonin on vascular permeability: An electron microscopic study. *J Biophys Biochem Cytol* 11: 571–605, 1961.
80. Majno G, Palade GE, Schoefl G: I. Studies on inflammation. The effect of histamine and serotonin on vascular permeability: An electron microscopic study. *J Biophys Biochem Cytol* 11: 607, 1961.
81. Berry MN, Friend DS: High-yield preparation of isolated rat liver parenchymal cells. A biochemical and fine structural study. *J Cell Biol* 43: 506–520, 1969.
82. Goodenough DA, Gilula NB: The splitting of hepatocyte junctions and zonulae occludentes with hypertonic disaccharides. *J Cell Biol* 61; 575–590, 1974.
83. Decker RS, Friend DS: Assembly of gap junctions during amphibian neurulation. *J Cell Biol* 62: 32–47, 1974.
84. Pitelka DR, Hamamoto ST, Duafala JG, Nemanic MK: Cell contacts in the mouse mammary gland. I. Normal gland in postnatal development and secretory cycle. *J Cell Biol* 56: 797–818, 1973.
85. Claude P, Goodenough DA: Fracture faces of zonulae occludentes from tight and leaky epithelia. *J Cell Biol* 58: 390–400, 1973.
86. Kaye GI, Pappas GD: Studies on the cornea. I. The fine structure of rabbit cornea and the uptake and transport of colloidal particles by the cornea in vivo. *J Cell Biol* 12: 457, 1962.
87. Dempsey EW: Electron microscopy of the visceral yolk-sac epithelium of the guinea pig. *Am J Anat* 93: 331–364, 1953.
88. Arturson G: Permeability of peritoneal membrane. In: Sixth Eur Conf Microcir, Aalburg. p 197–202, 1970.
89. Berndt WO, Gosselin RE: Differential changes in permeability of the mesentery to rubidium and phosphate. *Am J Physiol* 202: 761–767, 1962.
90. Landis EM, Pappenheimer JR: Circulation. In: Handbook of Physiology. Amer Physiol Soc Washington, p 961–1034, 1963.
91. Baillif RN: Reaction of the rat omentum to injections of particulate matter. *Proc Soc Exp Biol Med* 47: 409–414, 1941.
92. Simer PH: The passage of particulate matter from the peritoneal cavity into the lymph vessels of the diaphragm. *Anat Rec* 101: 333–351, 1948.
93. MacCallum WG: On the relation of the lymphatics to the peritoneal cavity in the diaphragm and the mechanism of absorption of granular material from the peritoneum. *Anat Anz* 23: 157, 1903.
94. Cunningham RS: Studies on absorption from serous cavities. III. The effect of dextrose upon the peritoneal mesothelium. *Am J Physiol* 53: 488–494, 1920.
95. Allen L: The peritoneal stomata. *Anat Rec* 67: 89–103, 1936.
96. Yoffey JM, Courtice FC: Lymphatics, Lymph and Lymphoid Tissues. E Arnold, (ed), London: Harvard University Press, 1956.
97. French JE, Florey HW, Morris B: The absorption of particles by the lymphatics of the diaphragm. *Quart J Exp Physiol* 215: 88, 1960.
98. Afonassiew N: Ueber den anfang von lymphgefassen in der serösen höhlen. *Virchow's Arch* 44: 37, 1868.
99. Foa' P: Ueber die beziehung des blut-ud lymphgefasse zum saktanal-system. *Virchow's Arch* 65: 284, 1875.
100. Bizzozzero G, Salvioli G: Sulla struttura delle membrane sierose e particolarmente del peritoneo diaframmatico. *Giorn R Accad Med Torino* 19: 466–470, 1876.
101. Shinohara H, Nakatani T, Matsuda T: The presence of lymphatic stomata in the ovarian bursa of the golden hamster. *Anat Rec* 213; 44–52, 1985.
102. Wang N: The preformed stomas connecting the pleural cavity and the lymphatic in the parietal pleura. *Am Rev Respir Dis* 111: 12–20, 1975.
103. Casley-Smith JR: Endothelial permeability — the passage of particles into and out of diaphragmatic lymphatics. *Q J Exp Physiol* 49: 365–383, 1964.
104. Leak LV, Kato F: Electron microscopic studies of lymphatic capillaries during early inflammation. I. Mild and severe thermal injuries. *Lab Invest* 26: 572–588, 1972.
105. Casley-Smith JR, Bolton T: Electron microscopy of the effects of histamine and thermal injury on the blood and lymphatic endothelium and the mesothelium of the mouse diaphragm, together with the influence of coumarin and rutin. *Experientia* 29; 1386–1388, 1973.
106. Casley-Smith JRL: The response of the microcirculation to inflammation. In: Handbook of Inflammation, Vol 2, The Cell Biology of Inflammation. G. Weissman (ed). Amsterdam: Elsevier/North-Holland Biomedical Press, p 53–89, 1980.
107. Ranvier L: Recherches sur la formation des mailles du grand epiploon. *Arch de Physiol* 2nd series. 1: 421–428, 1874a.
108. Ranvier L: Du développement et de l'accroissement des vaisseaux sanguins. *Arch de Physiol* 2nd series. 1: 429–450, 1874b.
109. Gowans JL, Knight EJ: The route of recirculating small lymphocytes in the rat. *Proc R Soc Series B* 159: 257–282, 1964.
110. Orenstein JM, Shelton E: Surface topography of leukocytes in situ: Cells of mouse peritoneal milky spots. *Exp Mol Pathol* 24: 415–423, 1976.
111. Felix MD: Observations on the surface cells of the mouse omentum as studied with the phase contrast and electron microscopes. *J Nat Cancer Inst* 27: 713–745, 1961.
112. Raftery AT: An enzyme histochemical study of mesothelial cells in rodents. *J Anat* 115: 365–373, 1973.
113. Tanaka H: *Proc of J Soc Ret Syst* 1: 45, 1961 (in Japanese).
114. Tanaka H: *Nihon Ketsuckigaku Zensyo*. 1: 249, 1963 (in Japanese).

Author's address:
Dr. F. Barberini
Department of Anatomy
Faculty of Medicine
University of Rome
Viale R.Elena, 289 00161
Rome, Italy

Index

MIX
Papier aus verantwortungsvollen Quellen
Paper from responsible sources
FSC® C105338

If you have any concerns about our products,
you can contact us on
ProductSafety@springernature.com

In case Publisher is established outside the EU,
the EU authorized representative is:
Springer Nature Customer Service Center GmbH
Europaplatz 3, 69115 Heidelberg, Germany

Printed by Libri Plureos GmbH
in Hamburg, Germany